FUNDAMENTALS OF BIOTECHNOLOGY

Edited by
Ian Moore

ArclerPress

Fundamentals of Biotechnology

Edited by Ian Moore

Edition 2018

ArclerPress

Published by

Arcler Press LLC

708 3rd Avenue, 6th Floor

New York

NY 10017

United States of America

www.arclerpress.com

ISBN: 978-1-68094-523-2 (Hardback)

Printed in United States of America

Library of Congress Control Number: 2017933226

About the Editor

Ian Moore

Ian received his Bachelor of Science from the University of Michigan-Dearborn in 2014. He has interests that include the role of histone modifications in the regulation of gene transcription and the identification of novel protein interactions and their roles in biological processes. He is currently at Wayne State University working as a teaching assistant and completing his graduate thesis.

Table of Contents

3 CELL: THE FUNDAMENTAL UNIT OF LIFE

4 ALLELE AND SELECTIVE BREEDING

5 DNA MOLECULE

6 RNA MOLECULE

7 FERMENTATION

8 GENE

9 PLASMID AND PROTEIN

10 BASIC TECHNIQUSES AND TOOLS IN BIOLOGY FIELD

11 BASIC TECHNIQUES AND TOOLS CHEMICAL FIELD

List of Abbreviations

AC	Alternating current
ATCC	American Type Culture Collection
ALS	Amyotrophic lateral Sclerosis
ADC	Analog-to-Digital Conversion
AUC	Analytical Ultracentrifugation
AES	Atomic Emission Spectroscopy
BOD	Biochemical Oxygen Demand
BIO	Biotechnology Industry Organization
BTPs	Biotechnology Training Programs
CEC	Capillary Electrochromatography
CD	Cluster of Differentiation
CSTR	Continuous Flow Stirred-tank Reactor
CBA	Cytometric Bead Array
DNA	Deoxyribonucleic Acid
DC	Direct Current
DAC	Director's Advisory Committee
dsDNA	Double Stranded Structure
dsRNA	Double-stranded RNA
ER	Endoplasmic Reticulum
ECM	Extracellular Matrix
FBS	Fetal bovine Serum
FACS	Fluorescence-Activated Cell Sorting
FDA	Food and Drug Administration ,
FC	Forced Circulation

FSC	Forward-scattered Light
FTIR	Fourier Transform Infrared Spectroscopy
GMO	Genetically Modified Organisms
hPL	Human Platelet Lysate
HIC	Hydrophobic Interaction Chromatography
IPC	Inducement Prize Contest
IETS	Inelastic Electron Tunneling Spectroscopy
IHHNV	Infectious Hypodermal and Haematopoietic Necrosis Virus
IPNV	Infectious Pancreatic Necrosis Virus
Kbp	Kilobase Pairs
LIBS	Laser-Induced Breakdown Spectroscopy
MLM	Magnetic Levitation Method
MBBR	Moving Bed Biofilm Reactor
MDR	Multidrug Resistance
NIGMS	National Institute of General Medical Sciences National Institutes of Health
NIH	National Institutes of Health
NMR	Nuclear Magnetic Resonance
PETA	People for Ethical Treatment of Animals
PBR	Photobioreactor
PSK	Postsegregational Killing System
rDNA	Recombinant DNA
RPM	Revolutions per Minute
RNA	Ribonucleic Acid ,
rRNA	Ribosomal RNA ,
STR	Short Tandem Repeat
SSC	Side-scattered Light
ssDNA	Single-stranded DNA Molecules
SFC	Supercritical-fluid Chromatography,
TSV	Taura Syndrome Virus
TPa	Tissue plasminogen Activator
tRNA	Transfer RNA tRNA ,
TtC	Trillion tons of Carbon

UPS	Ultraviolet Photoelectron Spectroscopy
VFAs	Various Volatile Organic Acids
XRF	X-ray Fluorescence
XPS	X-ray Photoelectron Spectroscopy

Preface

Biotechnology is technology that utilizes biological systems, living organisms or parts of this to develop or create different products. Biotechnology is a broad discipline in which biological processes, organisms, cells or cellular components are exploited to develop new technologies. New tools and products developed by biotechnologists are useful in research, agriculture, industry and the clinic. Modern biotechnology provides breakthrough products and technologies to combat debilitating and rare diseases, reduce our environmental footprint, feed the hungry, and use less and cleaner energy, and have safer, cleaner and more efficient industrial manufacturing processes.

Brewing and baking bread are examples of processes that fall within the concept of biotechnology (use of yeast (= living organism) to produce the desired product). Such traditional processes usually utilize the living organisms in their natural form (or further developed by breeding), while the more modern form of biotechnology will generally involve a more advanced modification of the biological system or organism. With the development of genetic engineering in the 1970s, research in biotechnology (and other related areas such as medicine, biology etc.) developed rapidly because of the new possibility to make changes in the organisms' genetic material (DNA). Today, biotechnology covers many different disciplines (e.g. genetics, biochemistry, molecular biology, etc.). New technologies and products are developed every year within the areas of e.g. medicine, agriculture or industrial biotechnology (production of chemicals, paper, textiles and food).

Content Coverage

Chapter One presents the basic concept of biotechnology as the application of scientific and engineering principles to the processing of materials by biological agents to provide goods and services.

Chapter Two is the history of biotechnology begins with zymotechnology, which commenced with a focus on brewing techniques for beer. Biotechnology has maintained a close relationship with society. Although now most often associated with the development of drugs, historically biotechnology has been principally associated with food, addressing such issues as malnutrition and famine.

Chapters Three and *Four* focus on the understanding of the tissues and organisms that cells compose, and the process of allele and selective breeding, respectively

Chapter Five and *Six* present the study of DNA and RNA molecule, respectively. RNA and DNA are nucleic acids, and, along with proteins and carbohydrates, constitute the four major macromolecules essential for all known forms of life.

Chapters Seven and *Eight* discuss about fermentation process and concept of a gene, respectively. Using a POS system makes a business much more efficient, lowering the costs of doing business while improving customer service and making the environment more pleasant to work in.

Chapters Nine provides the theory of plasmids and proteins. However, plasmids, like viruses, are not generally classified as life. Proteins perform a vast array of functions within organisms, including catalysing metabolic reactions, DNA replication, responding to stimuli, and transporting molecules from one location to another.

Chapters Ten and *Eleven* provide the basic techniques and tools in biology and chemical field.

CHAPTER 1

Biotechnology

Chapter Contents

- **Introduction**
- **Definitions**
- **History**
- **Examples**
- **Learning**

At its simplest, biotechnology is technology based on biology - biotechnology harnesses cellular and biomolecular processes to develop technologies and products that help improve our lives and the health of our planet. We have used the biological processes of microorganisms for more than 6,000 years to make useful food products, such as bread and cheese, and to preserve dairy products.

INTRODUCTION

Biotechnology is the use of living systems and organisms to develop or make products, or "any technological application that uses biological systems, living organisms, or derivatives thereof, to make or modify products or processes for specific use" (UN Convention on Biological Diversity, Art. 2). Depending on the tools and applications, it often overlaps with the (related) fields of bioengineering, biomedical engineering, biomanufacturing, molecular engineering, etc.

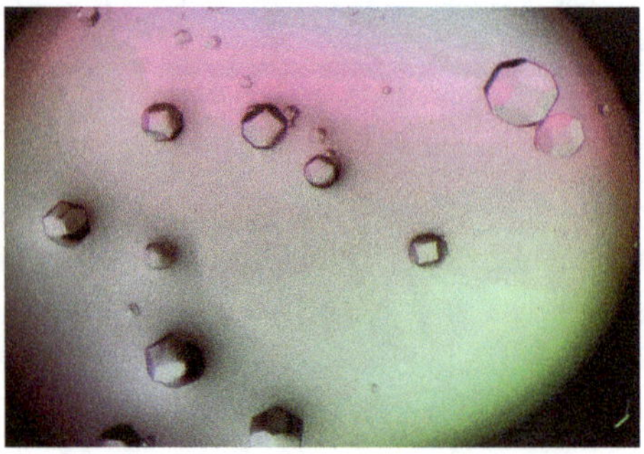

Insulin crystals

For thousands of years, humankind has used biotechnology in agriculture, food production, and medicine. The term is largely believed to have been coined in 1919 by Hungarian engineer Károly Ereky. In the late 20th and early 21st centuries, biotechnology has expanded to include new and diverse sciences such as genomics, recombinant gene techniques, applied immunology, and development of pharmaceutical therapies and diagnostic tests.

DEFINITIONS

The wide concept of "biotech" or "biotechnology" encompasses a wide range of procedures for modifying living organisms according to human purposes, going back to domestication of animals, cultivation of the plants, and "improvements" to these through breeding programs that employ artificial selection and hybridization. Modern usage also includes genetic engineering as well as cell and tissue culture technologies. The American Chemical Society defines biotechnology as the application of biological organisms, systems, or processes by various industries to learning about the science of life and the improvement of the value of materials and organisms such as pharmaceuticals, crops, and livestock. As per European Federation of Biotechnology, biotechnology is the integration of natural science and organisms, cells, parts thereof, and molecular analogues for products and services. Biotechnology also writes on the pure biological sciences (animal cell culture, biochemistry, cell biology, embryology, genetics, microbiology, and molecular biology). In many instances, it is also dependent on knowledge and methods from outside the sphere of biology including:

- bioinformatics, a new brand of computer science
- bioprocess engineering
- biorobotics
- chemical engineering

Conversely, modern biological sciences (including even concepts such as molecular ecology) are intimately entwined and heavily dependent on the methods developed through biotechnology and what is commonly thought of as the life sciences industry. Biotechnology is the research and development in the laboratory using bioinformatics for exploration, extraction, exploitation and production from any living organisms and any source of biomass by means of biochemical engineering where high value-added products could be planned (reproduced by biosynthesis, for example), forecasted, formulated, developed, manufactured, and marketed for the purpose of sustainable operations (for the return from bottomless initial investment on R & D) and gaining durable patents rights (for exclusives rights for sales, and prior to this to receive national and international approval from the results on animal experiment and human experiment, especially on the pharmaceutical branch of biotechnology to prevent any undetected side-effects or safety concerns by using the products).

By contrast, bioengineering is generally thought of as a related field that more heavily emphasizes higher systems approaches (not necessarily the altering or using of biological materials *directly*) for interfacing with and utilizing living things. Bioengineering is the application of the principles of engineering and natural sciences to tissues, cells and molecules. This can be considered as the use of knowledge from working with and manipulating biology to achieve a result that can improve functions in plants and animals. Relatedly, biomedical engineering is an overlapping field that often draws upon and applies *biotechnology* (by various definitions), especially in certain sub-fields of biomedical and/or chemical engineering such as tissue engineering, biopharmaceutical engineering, and genetic engineering.

HISTORY

Although not normally what first comes to mind, many forms of human-derived agriculture clearly fit the broad definition of "'utilizing a biotechnological system to make products". Indeed, the cultivation of plants may be viewed as the earliest biotechnological enterprise.

Brewing was an early application of biotechnology

Agriculture has been theorized to have become the dominant way of producing food since the Neolithic Revolution. Through early biotechnology, the earliest farmers selected and bred the best suited crops, having the highest yields, to produce enough food to support a growing population. As crops and fields became increasingly large and difficult to maintain, it was discovered that specific organisms and their by-products could effectively fertilize, restore nitrogen, and control pests. Throughout the history of agriculture, farmers have inadvertently altered the genetics of their crops through introducing them to new environments and breeding them with other plants — one of the first forms of biotechnology.

These processes also were included in early fermentation of beer. These processes were introduced in early Mesopotamia, Egypt, China and India, and still use the same basic biological methods. In brewing, malted grains (containing enzymes) convert starch from grains into sugar and then adding specific yeasts to produce beer. In this process, carbohydrates in the grains were broken down into alcohols such as ethanol. Later other cultures produced the process of lactic acid fermentation which allowed the fermentation and preservation of other forms of food, such as soy sauce. Fermentation was also used in this time period to produce leavened bread. Although the process of fermentation was not fully understood until Louis Pasteur's work in 1857, it is still the first use of biotechnology to convert a food source into another form.

Before the time of Charles Darwin's work and life, animal and plant scientists had already used selective breeding. Darwin added to that body of work with his scientific observations about the ability of science to change species. These accounts contributed to Darwin's theory of natural selection.

For thousands of years, humans have used selective breeding to improve production of crops and livestock to use them for food. In selective breeding, organisms with desirable characteristics are mated to produce offspring with the same characteristics. For example, this technique was used with corn to produce the largest and sweetest crops.

In the early twentieth century scientists gained a greater understanding of microbiology and explored ways of manufacturing specific products. In 1917, Chaim Weizmann first used a pure microbiological culture in an industrial process, that of manufacturing corn starch using *Clostridium acetobutylicum,* to produce acetone, which the United Kingdom desperately needed to manufacture explosives during World War I.

Biotechnology has also led to the development of antibiotics. In 1928, Alexander Fleming discovered the mold *Penicillium*. His work led to the purification of the antibiotic compound formed by the mold by Howard Florey, Ernst Boris Chain and Norman Heatley – to form what we today know as penicillin. In 1940, penicillin became available for medicinal use to treat bacterial infections in humans.

The field of modern biotechnology is generally thought of as having been born in 1971 when Paul Berg's (Stanford) experiments in gene splicing had early success. Herbert W. Boyer (Univ. Calif. at San Francisco) and Stanley N. Cohen (Stanford) significantly advanced the new technology in 1972 by transferring genetic material into a bacterium, such that the imported material would be reproduced. The commercial viability of a biotechnology industry was significantly expanded on June 16, 1980, when the United States Supreme Court ruled that a genetically modified microorganism could be patented in the case of *Diamond v. Chakrabarty*. Indian-born Ananda Chakrabarty, working for General Electric, had modified a bacterium (of the *Pseudomonas* genus) capable of breaking down crude oil, which he proposed to use in treating oil spills. (Chakrabarty's work did not involve gene manipulation but rather the transfer of entire organelles between strains of the *Pseudomonas* bacterium.

Revenue in the industry is expected to grow by 12.9% in 2008. Another factor influencing the biotechnology sector's success is improved intellectual property rights legislation—and enforcement—worldwide, as well as strengthened demand for medical and pharmaceutical products to cope with an ageing, and ailing, U.S. population.

Rising demand for biofuels is expected to be good news for the biotechnology sector, with the Department of Energy estimating ethanol usage could reduce U.S. petroleum-derived fuel consumption by up to 30% by 2030. The biotechnology sector has allowed the U.S. farming industry to rapidly increase its supply of corn and soybeans—the main inputs into biofuels—by developing genetically modified seeds which are resistant to pests and drought. By boosting farm productivity, biotechnology plays a crucial role in ensuring that biofuel production targets are met.

EXAMPLES

Biotechnology has applications in four major industrial areas, including health care (medical), crop production and agriculture, non food (industrial) uses of crops and other products (e.g. biodegradable plastics, vegetable oil, biofuels), and environmental uses.

A rose plant that began as cells grown in a tissue culture

For example, one application of biotechnology is the directed use of organisms for the manufacture of organic products (examples include beer and milk products). Another example is using naturally present bacteria by the mining industry in bioleaching. Biotechnology is also used to recycle, treat waste, clean up sites contaminated by industrial activities (bioremediation), and also to produce biological weapons.

A series of derived terms have been coined to identify several branches of biotechnology; for example:

- Bioinformatics is an interdisciplinary field which addresses biological problems using computational techniques, and makes the rapid organization as well as analysis of biological data possible. The field may also be referred to as *computational biology*, and can be defined as, "conceptualizing biology in terms of molecules and then applying informatics techniques to understand and organize the information associated with these molecules, on a large scale." Bioinformatics plays a key role in various areas, such as functional genomics, structural genomics, and proteomics, and forms a key component in the biotechnology and pharmaceutical sector.
- Blue biotechnology is a term that has been used to describe the marine and aquatic applications of biotechnology, but its use is relatively rare.
- Green biotechnology is biotechnology applied to agricultural processes. An example would be the selection and domestication of plants via micro propagation. Another example is the designing of transgenic plants to grow under specific environments in the presence (or absence) of chemicals. One hope is that green biotechnology might produce more environmentally friendly solutions than traditional industrial agriculture. An example of this is the engineering of a plant to express a pesticide, thereby ending the need of external application of pesticides. An example of this would be Bt corn. Whether or not green biotechnology products such as this are ultimately more environmentally friendly is a topic of considerable debate.
- Red biotechnology is applied to medical processes. Some examples are the designing of organisms to produce antibiotics, and the engineering of genetic cures through genetic manipulation.
- White biotechnology, also known as industrial biotechnology, is biotechnology applied to industrial processes. An example is the designing of an organism to produce a useful chemical. Another example is the using of enzymes as industrial catalysts to either produce valuable chemicals or destroy hazardous/polluting chemicals. White biotechnology tends to consume less in resources than traditional processes used to produce industrial goods.

The investment and economic output of all of these types of applied biotechnologies is termed as "bioeconomy".

Medicine

In medicine, modern biotechnology finds applications in areas such as pharmaceutical drug discovery and production, pharmacogenomics, and genetic testing (or genetic screening).

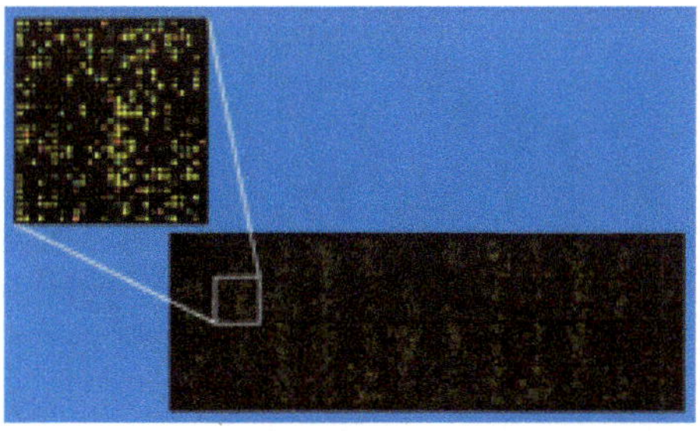

DNA microarray chip – some can do as many as a million blood tests at once

Pharmacogenomics (a combination of pharmacology and genomics) is the technology that analyses how genetic makeup affects an individual's response to drugs. It deals with the influence of genetic variation on drug response in patients by correlating gene expression or single-nucleotide polymorphisms with a drug's efficacy or toxicity. By doing so, pharmacogenomics aims to develop rational means to optimize drug therapy, with respect to the patients' genotype, to ensure maximum efficacy with minimal adverse effects. Such approaches promise the advent of "personalized medicine"; in which drugs and drug combinations are optimized for each individual's unique genetic makeup.

Biotechnology has contributed to the discovery and manufacturing of traditional small molecule pharmaceutical drugsas well as drugs that are the product of biotechnology – biopharmaceutics. Modern biotechnology can be used to manufacture existing medicines relatively easily and cheaply. The first genetically engineered products were medicines designed to treat human diseases. To cite one example, in 1978 Genentech developed synthetic humanized insulin by joining its gene with a plasmid vector inserted into the bacterium *Escherichia coli*. Insulin, widely used for the treatment of diabetes, was previously extracted from the pancreas of abattoir animals (cattle and/or pigs). The resulting genetically engineered bacterium enabled the production of vast quantities of synthetic human insulin at relatively low cost. Biotechnology has also enabled emerging therapeutics like gene therapy. The application of biotechnology to basic science (for example through the Human Genome Project) has also dramatically improved our understanding of biology and as our scientific knowledge of normal and disease biology has increased, our ability to develop new medicines to treat previously untreatable diseases has increased as well.

Genetic testing allows the genetic diagnosis of vulnerabilities to inherited diseases, and can also be used to determine a child's parentage (genetic mother and father) or in general a person's ancestry. In addition to studying chromosomesto the level of individual genes, genetic testing in a broader sense includes biochemical tests for the possible presence of genetic diseases, or mutant forms of genes associated with increased risk of developing genetic disorders. Genetic testing identifies changes in chromosomes, genes, or proteins. Most of the time, testing is used to find changes that are associated with inherited disorders. The results of a genetic test can confirm or rule out a suspected genetic condition or help determine a person's chance of developing or passing on a genetic disorder. As of 2011 several hundred genetic tests were in use. Since genetic testing may open up ethical or psychological problems, genetic testing is often accompanied by genetic counseling.

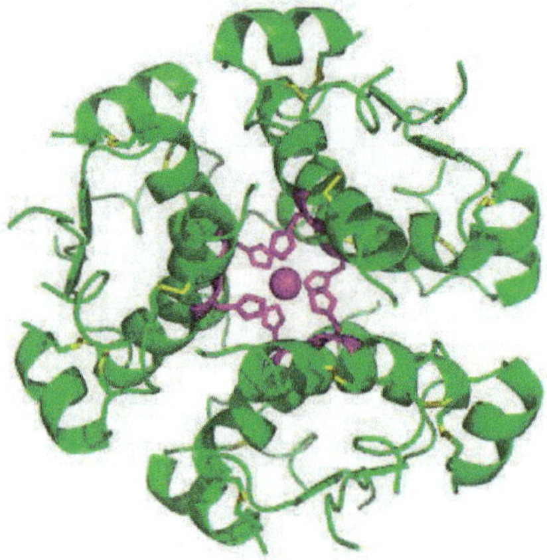

Computer-generated image of insulin hexamers highlighting the threefold symmetry, the zinc ions holding it together, and the histidine residues involved in zinc binding.

Agriculture

Genetically modified crops ("GM crops", or "biotech crops") are plants used in agriculture, the DNA of which has been modified with genetic engineeringtechniques. In most cases the aim is to introduce a new trait to the plant which does not occur naturally in the species.

Examples in food crops include resistance to certain pests, diseases, stressful environmental conditions, resistance to chemical treatments (e.g. resistance to a herbicide), reduction of spoilage, or improving the nutrient profile of the crop. Examples in non-food crops include production of pharmaceutical agents, biofuels, and other industrially useful goods, as well as for bioremediation.

Farmers have widely adopted GM technology. Between 1996 and 2011, the total surface area of land cultivated with GM crops had increased by a factor of 94, from 17,000 square kilometers (4,200,000 acres) to 1,600,000 km^2 (395 million acres). 10% of the world's crop lands were planted with GM crops in 2010. As of 2011, 11 different transgenic crops were grown commercially on 395 million acres (160 million hectares) in 29 countries such as the USA, Brazil, Argentina, India, Canada, China, Paraguay, Pakistan, South Africa, Uruguay, Bolivia, Australia, Philippines, Myanmar, Burkina Faso, Mexico and Spain.

Genetically modified foods are foods produced from organisms that have had specific changes introduced into their DNA with the methods of genetic engineering. These techniques have allowed for the introduction of new crop traits as well as a far greater control over a food's genetic structure than previously afforded by methods such as selective breeding and mutation breeding. Commercial sale of genetically modified foods began in 1994, when Calgene first marketed its Flavr Savr delayed ripening tomato. To date most genetic modification of foods have primarily focused on cash crops in high demand by farmers such as soybean, corn, canola, and cotton seed oil. These have been engineered for resistance to pathogens and herbicides and better nutrient profiles. GM livestock have also been experimentally developed, although as of November 2013 none are currently on the market.

There is a scientific consensus that currently available food derived from GM crops poses no greater risk to human health than conventional food, but that each GM food needs to be tested on a case-by-case basis before introduction. Nonetheless, members of the public are much less likely than scientists to perceive GM foods as safe. The legal and regulatory status of GM foods varies by country, with some nations banning or restricting them, and others permitting them with widely differing degrees of regulation.

GM crops also provide a number of ecological benefits, if not used in excess. However, opponents have objected to GM crops per se on several grounds, including environmental concerns, whether food produced from GM crops is safe, whether GM crops are needed to address the world's food needs, and economic concerns raised by the fact these organisms are subject to intellectual property law.

Industrial

Industrial biotechnology (known mainly in Europe as white biotechnology) is the application of biotechnology for industrial purposes, including industrial fermentation. It includes the practice of using cells such as micro-organisms, or components of cells like enzymes, to generate industrially useful products in sectors such as chemicals, food and feed, detergents, paper and pulp, textiles and biofuels. In doing so, biotechnology uses renewable raw materials and may contribute to lowering greenhouse gas emissions and moving away from a petrochemical-based economy.

Environmental

The environment can be affected by biotechnologies, both positively and adversely. Vallero and others have argued that the difference between beneficial biotechnology (e.g. bioremediation to clean up an oil spill or hazard chemical leak) versus the adverse effects stemming from biotechnological enterprises (e.g. flow of genetic material from transgenic organisms into wild strains) can be seen as applications and implications, respectively. Cleaning up environmental wastes is an example of an application of environmental biotechnology; whereas loss of biodiversity or loss of containment of a harmful microbe are examples of environmental implications of biotechnology.

Regulation

The regulation of genetic engineering concerns approaches taken by governments to assess and manage the risks associated with the use of genetic engineering technology, and the development and release of genetically modified organisms (GMO), including genetically modified crops and genetically modified fish. There are differences in the regulation of GMOs between countries, with some of the most marked differences occurring between the USA and Europe. Regulation varies in a given country depending on the intended use of the products of the genetic engineering. For example, a crop not intended for food use is generally not reviewed by authorities responsible for food safety. The European Union differentiates between approval for cultivation within the EU and approval for import and processing. While only a few GMOs have been approved for cultivation in the EU a number of GMOs have been approved for import and processing. The cultivation of GMOs has triggered a debate about coexistence of GM and non GM crops. Depending on the coexistence regulations incentives for cultivation of GM crops differ.

LEARNING

In 1988, after prompting from the United States Congress, the National Institute of General Medical Sciences (National Institutes of Health) (NIGMS) instituted a funding mechanism for biotechnology training. Universities nationwide compete for these funds to establish Biotechnology Training Programs (BTPs). Each successful application is generally funded for five years then must be competitively renewed. Graduate students in turn compete for acceptance into a BTP; if accepted, then stipend, tuition and health insurance support is provided for two or three years during the course of their Ph.D. thesis work. Nineteen institutions offer NIGMS supported BTPs. Biotechnology training is also offered at the undergraduate level and in community colleges.

REFERENCES AND NOTES

1. Text of the CBD. CBD.int. Retrieved on 2013-03-20.
2. "Incorporating Biotechnology into the Classroom What is Biotechnology?", from the curricula of the ‹Incorporating Biotechnology into the High School Classroom through Arizona State University's BioREACH program', accessed on October 16, 2012). Public.asu.edu. Retrieved on 2013-03-20.
3. Biotechnology. Portal.acs.org. Retrieved on 2013-03-20.
4. What is biotechnology?. Europabio. Retrieved on 2013-03-20.
5. KEY BIOTECHNOLOGY INDICATORS (December 2011). oecd.org
6. Biotechnology policies – Organization for Economic Co-operation and Development. Oecd.org. Retrieved on 2013-03-20.
7. What Is Bioengineering?. Bionewsonline.com. Retrieved on 2013-03-20.
8. See Arnold, John P. (2005) [1911]. Origin and History of Beer and Brewing: From Prehistoric Times to the Beginning of Brewing Science and Technology. Cleveland, Ohio: BeerBooks. p. 34. ISBN 978-0-9662084-1-2. OCLC 71834130.
9. *Cole-Turner, Ronald (2003). "Biotechnology". Encyclopedia of Science and Religion. Retrieved 7 December 2014.*
10. *Thieman, W.J.; Palladino, M.A. (2008). Introduction to Biotechnology. Pearson/ Benjamin Cummings. ISBN 0-321-49145-9.*
11. *Springham, D.; Springham, G.; Moses, V.; Cape, R.E. (24 August 1999). Biotechnology: The Science and the Business. CRC Press. p. 1. ISBN 978-90-5702-407-8.*

12. "Diamond v. Chakrabarty, 447 U.S. 303 (1980). No. 79-139." *United States Supreme Court.* June 16, 1980. Retrieved on May 4, 2007.
13. VoIP Providers And Corn Farmers Can Expect To Have Bumper Years In 2008 And Beyond, According To The Latest Research Released By Business Information Analysts At IBISWorld. Los Angeles (March 19, 2008)
14. The Recession List — Top 10 Industries to Fly and Fl... (ith anincreasing share accounted for by ...), bio-medicine.org
15. Gerstein, M. "Bioinformatics Introduction." *Yale University.* Retrieved on May 8, 2007.
16. Ermak G., Modern Science & Future Medicine (second edition), 164 p., 2013
17. *Wang L (2010). "Pharmacogenomics: a systems approach". Wiley Interdiscip Rev Syst Biol Med.* ***2*** *(1): 3–22. doi:10.1002/wsbm.42. PMC 3894835. PMID 20836007.*
18. *Becquemont L (June 2009). "Pharmacogenomics of adverse drug reactions: practical applications and perspectives". Pharmacogenomics.* ***10*** *(6): 961–9. doi:10.2217/pgs.09.37. PMID 19530963.*
19. *"Guidance for Industry Pharmacogenomic Data Submissions" (PDF). U.S. Food and Drug Administration. March 2005. Retrieved 2008-08-27.*
20. *Squassina A, Manchia M, Manolopoulos VG, Artac M, Lappa-Manakou C, Karkabouna S, Mitropoulos K, Del Zompo M, Patrinos GP (August 2010). "Realities and expectations of pharmacogenomics and personalized medicine: impact of translating genetic knowledge into clinical practice". Pharmacogenomics.* ***11*** *(8): 1149–67. doi:10.2217/pgs.10.97. PMID 20712531.*
21. *Bains, W. (1987). Genetic Engineering For Almost Everybody: What Does It Do? What Will It Do?. Penguin. p. 99. ISBN 0-14-013501-4.*
22. U.S. Department of State International Information Programs, "Frequently Asked Questions About Biotechnology", USIS Online; available from USinfo.state.gov Archived September 12, 2007, at the Wayback Machine., accessed 13 September 2007. Cf. *Feldbaum, C. (February 2002). "Some History Should Be Repeated". Science.* ***295*** *(5557): 975. doi:10.1126/science.1069614. PMID 11834802.*
23. *"What is genetic testing? – Genetics Home Reference". Ghr.nlm.nih.gov. 2011-05-30. Retrieved 2011-06-07.*
24. *"Genetic Testing: MedlinePlus". Nlm.nih.gov. Retrieved 2011-06-07.*
25. *"Definitions of Genetic Testing". Definitions of Genetic Testing (Jorge Sequeiros and Bárbara Guimarães). EuroGentest Network of Excellence Project. 2008-09-11. Archived from the original on 4 February 2009. Retrieved 2008-08-10.*
26. Genetically Altered Potato Ok'd For Crops Lawrence Journal-World – 6 May 1995
27. *National Academy of Sciences (2001). Transgenic Plants and World Agriculture. Washington: National Academy Press.*
28. Paarlburg, Robert Drought Tolerant GMO Maize in Africa, Anticipating

Regulatory Hurdles Archived December 22, 2014, at the Wayback Machine. International Life Sciences Institute, January 2011. Retrieved 25 April 2011

29. Carpenter J. & Gianessi L. (1999). Herbicide tolerant soybeans: Why growers are adopting Roundup Ready varieties. AgBioForum, 2(2), 65–72.

30. *Haroldsen, Victor M.; Paulino, Gabriel; Chi-ham, Cecilia; Bennett, Alan B. (2012). "Research and adoption of biotechnology strategies could improve California fruit and nut crops" (PDF). California Agriculture. **66** (2): 62–69. doi:10.3733/ca.v066n02p62.*

31. About Golden Rice Archived November 2, 2012, at the Wayback Machine.. Irri.org. Retrieved on 2013-03-20.

32. Gali Weinreb and Koby Yeshayahou for Globes May 2, 2012. FDA approves Protalix Gaucher treatment

33. Carrington, Damien (19 January 2012) GM microbe breakthrough paves way for large-scale seaweed farming for biofuels The Guardian. Retrieved 12 March 2012

34. *van Beilen, Jan B.; Yves Poirier (May 2008). "Harnessing plant biomass for biofuels and biomaterials:Production of renewable polymers from crop plants". The Plant Journal. **54** (4): 684–701. doi:10.1111/j.1365-313X.2008.03431.x. PMID 18476872.*

35. Strange, Amy (20 September 2011) Scientists engineer plants to eat toxic pollution The Irish Times. Retrieved 20 September 2011

36. *Diaz E (editor). (2008). Microbial Biodegradation: Genomics and Molecular Biology (1st ed.). Caister Academic Press. ISBN 1-904455-17-4.*

37. ^ Jump up to:[a] [b] [c] *James, C (2011). "ISAAA Brief 43, Global Status of Commercialized Biotech/GM Crops: 2011". ISAAA Briefs. Ithaca, New York: International Service for the Acquisition of Agri-biotech Applications (ISAAA). Retrieved 2012-06-02.*

38. GM Science Review First Report Archived October 16, 2013, at the Wayback Machine., Prepared by the UK GM Science Review panel (July 2003). Chairman Professor Sir David King, Chief Scientific Advisor to the UK Government, P 9

39. *James, Clive (1996). "Global Review of the Field Testing and Commercialization of Transgenic Plants: 1986 to 1995" (PDF). The International Service for the Acquisition of Agri-biotech Applications. Retrieved 17 July 2010.*

40. *"Consumer Q&A". Fda.gov. 2009-03-06. Retrieved 2012-12-29.*

41. *Nicolia, Alessandro; Manzo, Alberto; Veronesi, Fabio; Rosellini, Daniele (2013). "An overview of the last 10 years of genetically engineered crop safety research" (PDF). Critical Reviews in Biotechnology: 1–12. doi:10.3109/07388551.2013.823595. We have reviewed the scientific literature on GE crop safety for the last 10 years that catches the scientific consensus matured since GE plants became widely cultivated worldwide, and we can conclude that the scientific research conducted so far has not detected any significant hazard directly connected with the use of GM crops.*

42. The literature about Biodiversity and the GE food/feed consumption has sometimes

resulted in animated debate regarding the suitability of the experimental designs, the choice of the statistical methods or the public accessibility of data. Such debate, even if positive and part of the natural process of review by the scientific community, has frequently been distorted by the media and often used politically and inappropriately in anti-GE crops campaigns.

43. *"State of Food and Agriculture 2003–2004. Agricultural Biotechnology: Meeting the Needs of the Poor. Health and environmental impacts of transgenic crops". Food and Agriculture Organization of the United Nations. Retrieved February 8, 2016. Currently available transgenic crops and foods derived from them have been judged safe to eat and the methods used to test their safety have been deemed appropriate. These conclusions represent the consensus of the scientific evidence surveyed by the ICSU (2003) and they are consistent with the views of the World Health Organization (WHO, 2002). These foods have been assessed for increased risks to human health by several national regulatory authorities (inter alia, Argentina, Brazil, Canada, China, the United Kingdom and the United States) using their national food safety procedures (ICSU). To date no verifiable untoward toxic or nutritionally deleterious effects resulting from the consumption of foods derived from genetically modified crops have been discovered anywhere in the world (GM Science Review Panel). Many millions of people have consumed foods derived from GM plants - mainly maize, soybean and oilseed rape - without any observed adverse effects (ICSU).*

44. *Ronald, Pamela (May 5, 2011). "Plant Genetics, Sustainable Agriculture and Global Food Security". Genetics.* ***188****: 11–20. doi:10.1534/genetics.111.128553. There is broad scientific consensus that genetically engineered crops currently on the market are safe to eat. After 14 years of cultivation and a cumulative total of 2 billion acres planted, no adverse health or environmental effects have resulted from commercialization of genetically engineered crops (Board on Agriculture and Natural Resources, Committee on Environmental Impacts Associated with Commercialization of Transgenic Plants, National Research Council and Division on Earth and Life Studies 2002). Both the U.S. National Research Council and the Joint Research Centre (the European Union's scientific and technical research laboratory and an integral part of the European Commission) have concluded that there is a comprehensive body of knowledge that adequately addresses the food safety issue of genetically engineered crops (Committee on Identifying and Assessing Unintended Effects of Genetically Engineered Foods on Human Health and National Research Council 2004; European Commission Joint Research Centre 2008). These and other recent reports conclude that the processes of genetic engineering and conventional breeding are no different in terms of unintended consequences to human health and the environment (European Commission Directorate-General for Research and Innovation 2010).*

45. *Domingo, José L.; Bordonaba, Jordi Giné (2011). "A literature review on the safety assessment of genetically modified plants" (PDF). Environment International.* ***37****: 734–742. doi:10.1016/j.envint.2011.01.003. In spite of this, the number of studies specifically focused on safety assessment of GM plants is still limited. However, it is important to remark that for the first time, a*

certain equilibrium in the number of research groups suggesting, on the basis of their studies, that a number of varieties of GM products (mainly maize and soybeans) are as safe and nutritious as the respective conventional non-GM plant, and those raising still serious concerns, was observed. Moreover, it is worth mentioning that most of the studies demonstrating that GM foods are as nutritional and safe as those obtained by conventional breeding, have been performed by biotechnology companies or associates, which are also responsible of commercializing these GM plants. Anyhow, this represents a notable advance in comparison with the lack of studies published in recent years in scientific journals by those companies.

46. *Krimsky, Sheldon (2015). "An Illusory Consensus behind GMO Health Assessment" (PDF). Science, Technology, & Human Values: 1–32. doi:10.1177/0162243915598381. I began this article with the testimonials from respected scientists that there is literally no scientific controversy over the health effects of GMOs. My investigation into the scientific literature tells another story.*
47. *Panchin, Alexander Y.; Tuzhikov, Alexander I. (January 14, 2016). "Published GMO studies find no evidence of harm when corrected for multiple comparisons". Critical Reviews in Biotechnology. doi:10.3109/07388551.2015.1130684. ISSN 0738-8551. Here, we show that a number of articles some of which have strongly and negatively influenced the public opinion on GM crops and even provoked political actions, such as GMO embargo, share common flaws in the statistical evaluation of the data. Having accounted for these flaws, we conclude that the data presented in these articles does not provide any substantial evidence of GMO harm.*
48. *The presented articles suggesting possible harm of GMOs received high public attention. However, despite their claims, they actually weaken the evidence for the harm and lack of substantial equivalency of studied GMOs. We emphasize that with over 1783 published articles on GMOs over the last 10 years it is expected that some of them should have reported undesired differences between GMOs and conventional crops even if no such differences exist in reality.*
49. *Yang, Y.T.; Chen, B. (2016). "Governing GMOs in the USA: science, law and public health". Journal of the Science of Food and Agriculture.* ***96****: 1851–1855. doi:10.1002/jsfa.7523. It is therefore not surprising that efforts to require labeling and to ban GMOs have been a growing political issue in the USA (citing Domingo and Bordonaba, 2011).*
50. *Overall, a broad scientific consensus holds that currently marketed GM food poses no greater risk than conventional food... Major national and international science and medical associations have stated that no adverse human health effects related to GMO food have been reported or substantiated in peer-reviewed literature to date.*
51. *Despite various concerns, today, the American Association for the Advancement of Science, the World Health Organization, and many independent international science organizations agree that GMOs are just as safe as other foods. Compared*

with conventional breeding techniques, genetic engineering is far more precise and, in most cases, less likely to create an unexpected outcome.

52. "*Statement by the AAAS Board of Directors On Labeling of Genetically Modified Foods" (PDF). American Association for the Advancement of Science. October 20, 2012. Retrieved February 8, 2016. The EU, for example, has invested more than 300 million in research on the biosafety of GMOs. Its recent report states: "The main conclusion to be drawn from the efforts of more than 130 research projects, covering a period of more than 25 years of research and involving more than 500 independent research groups, is that biotechnology, and in particular GMOs, are not per se more risky than e.g. conventional plant breeding technologies." The World Health Organization, the American Medical Association, the U.S. National Academy of Sciences, the British Royal Society, and every other respected organization that has examined the evidence has come to the same conclusion: consuming foods containing ingredients derived from GM crops is no riskier than consuming the same foods containing ingredients from crop plants modified by conventional plant improvement techniques.*
53. Pinholster, Ginger (October 25, 2012). "AAAS Board of Directors: Legally Mandating GM Food Labels Could "Mislead and Falsely Alarm Consumers"". American Association for the Advancement of Science. Retrieved February 8, 2016.
54. "*A decade of EU-funded GMO research (2001–2010)" (PDF). Directorate-General for Research and Innovation. Biotechnologies, Agriculture, Food. European Commission, European Union. 2010. doi:10.2777/97784. ISBN 978-92-79-16344-9. Retrieved February 8, 2016.*
55. "*AMA Report on Genetically Modified Crops and Foods (online summary)". American Medical Association. January 2001. Retrieved March 19, 2016. A report issued by the scientific council of the American Medical Association (AMA) says that no long-term health effects have been detected from the use of transgenic crops and genetically modified foods, and that these foods are substantially equivalent to their conventional counterparts. (from online summary prepared by ISAAA)" "Crops and foods produced using recombinant DNA techniques have been available for fewer than 10 years and no long-term effects have been detected to date. These foods are substantially equivalent to their conventional counterparts. (from original report by AMA:)*
56. "REPORT 2 OF THE COUNCIL ON SCIENCE AND PUBLIC HEALTH (A-12): Labeling of Bioengineered Foods" (PDF). American Medical Association. 2012. Archived from the original on September 7, 2012. Retrieved March 19,2016. Bioengineered foods have been consumed for close to 20 years, and during that time, no overt consequences on human health have been reported and/or substantiated in the peer-reviewed literature.
57. "*Restrictions on Genetically Modified Organisms: United States. Public and Scholarly Opinion". Library of Congress. June 9, 2015. Retrieved February 8, 2016. Several scientific organizations in the US have issued studies or statements regarding the safety of GMOs indicating that there is no evidence*

that GMOs present unique safety risks compared to conventionally bred products. These include the National Research Council, the American Association for the Advancement of Science, and the American Medical Association. Groups in the US opposed to GMOs include some environmental organizations, organic farming organizations, and consumer organizations. A substantial number of legal academics have criticized the US's approach to regulating GMOs.

58. *Genetically Engineered Crops: Experiences and Prospects". The National Academies of Sciences, Engineering, and Medicine (US). 2016. p. 149. Retrieved May 19, 2016. Overall finding on purported adverse effects on human health of foods derived from GE crops: On the basis of detailed examination of comparisons of currently commercialized GE with non-GE foods in compositional analysis, acute and chronic animal toxicity tests, long-term data on health of livestock fed GE foods, and human epidemiological data, the committee found no differences that implicate a higher risk to human health from GE foods than from their non-GE counterparts.*
59. *Frequently asked questions on genetically modified foods". World Health Organization. Retrieved February 8, 2016. Different GM organisms include different genes inserted in different ways. This means that individual GM foods and their safety should be assessed on a case-by-case basis and that it is not possible to make general statements on the safety of all GM foods.*
60. GM foods currently available on the international market have passed safety assessments and are not likely to present risks for human health. In addition, no effects on human health have been shown as a result of the consumption of such foods by the general population in the countries where they have been approved. Continuous application of safety assessments based on the Codex Alimentarius principles and, where appropriate, adequate post market monitoring, should form the basis for ensuring the safety of GM foods.
61. *Haslberger, Alexander G. (2003). "Codex guidelines for GM foods include the analysis of unintended effects". Nature Biotechnolgy. **21**: 739–741. doi:10.1038/nbt0703-739. These principles dictate a case-by-case premarket assessment that includes an evaluation of both direct and unintended effects.*
62. Some medical organizations, including the British Medical Association, advocate further caution based upon the precautionary principle:
63. *"Genetically modified foods and health: a second interim statement" (PDF). British Medical Association. March 2004. Retrieved March 21, 2016. In our view, the potential for GM foods to cause harmful health effects is very small and many of the concerns expressed apply with equal vigour to conventionally derived foods. However, safety concerns cannot, as yet, be dismissed completely on the basis of information currently available.*
64. When seeking to optimise the balance between benefits and risks, it is prudent to err on the side of caution and, above all, learn from accumulating knowledge and experience. Any new technology such as genetic modification must be examined for possible benefits and risks to human health and the environment. As with all novel foods, safety assessments in relation to GM foods must be

made on a case-by-case basis.

65. Members of the GM jury project were briefed on various aspects of genetic modification by a diverse group of acknowledged experts in the relevant subjects. The GM jury reached the conclusion that the sale of GM foods currently available should be halted and the moratorium on commercial growth of GM crops should be continued. These conclusions were based on the precautionary principle and lack of evidence of any benefit. The Jury expressed concern over the impact of GM crops on farming, the environment, food safety and other potential health effects.

66. The Royal Society review (2002) concluded that the risks to human health associated with the use of specific viral DNA sequences in GM plants are negligible, and while calling for caution in the introduction of potential allergens into food crops, stressed the absence of evidence that commercially available GM foods cause clinical allergic manifestations. The BMA shares the view that that there is no robust evidence to prove that GM foods are unsafe but we endorse the call for further research and surveillance to provide convincing evidence of safety and benefit.

67. *Funk, Cary; Rainie, Lee (January 29, 2015). "Public and Scientists' Views on Science and Society". Pew Research Center. Retrieved February 24, 2016. The largest differences between the public and the AAAS scientists are found in beliefs about the safety of eating genetically modified (GM) foods. Nearly nine-in-ten (88%) scientists say it is generally safe to eat GM foods compared with 37% of the general public, a difference of 51 percentage points.*

68. *Marris, Claire (2001). "Public views on GMOs: deconstructing the myths"(PDF). EMBO Reports.* ***2****: 545–548. doi:10.1093/embo-reports/kve142.*

69. *Final Report of the PABE research project (December 2001). "Public Perceptions of Agricultural Biotechnologies in Europe". Commission of European Communities. Retrieved February 24, 2016.*

70. *Scott, Sydney E.; Inbar, Yoel; Rozin, Paul (2016). "Evidence for Absolute Moral Opposition to Genetically Modified Food in the United States" (PDF). Perspectives on Psychological Science.* ***11*** *(3): 315–324. doi:10.1177/1745691615621275.*

71. *"Restrictions on Genetically Modified Organisms". Library of Congress. June 9, 2015. Retrieved February 24, 2016.*

72. *Bashshur, Ramona (February 2013). "FDA and Regulation of GMOs". American Bar Association. Retrieved February 24, 2016.*

73. *Sifferlin, Alexandra (October 3, 2015). "Over Half of E.U. Countries Are Opting Out of GMOs". Time.*

74. *Lynch, Diahanna; Vogel, David (April 5, 2001). "The Regulation of GMOs in Europe and the United States: A Case-Study of Contemporary European Regulatory Politics". Council on Foreign Relations. Retrieved February 24,2016.*

75. Andrew Pollack for the New York Times. April 13, 2010 Study Says Overuse Threatens Gains From Modified Crops

76. Industrial Biotechnology and Biomass Utilisation

77. Industrial biotechnology, A powerful, innovative technology to mitigate climate change

78. Daniel A. Vallero, *Environmental Biotechnology: A Biosystems Approach*, Academic Press, Amsterdam, NV; ISBN 978-0-12-375089-1; 2010.

79. *Gaskell, G.; Bauer, M. W.; Durant, J.; Allum, N. C. (1999). "Worlds Apart? The Reception of Genetically Modified Foods in Europe and the U.S". Science.* ***285*** *(5426): 384–387. doi:10.1126/science.285.5426.384. PMID 10411496.*

80. PotatoPro

81. Wesseler, J. and Kalaitzandonakes, N. (2011) "Present and Future EU GMO policy", pp. 23–323 to 23–332 in Arie Oskam, Gerrit Meesters and Huib Silvis (eds.) *EU Policy for Agriculture, Food and Rural Areas*. 2nd ed. Wageningen: Wageningen Academic Publishers

82. Beckmann, V., C. Soregaroli, J. Wesseler (2011): Coexistence of genetically modified (GM) and non-modified (non GM) crops: Are the two main property rights regimes equivalent with respect to the coexistence value? In "Genetically modified food and global welfare" edited by Colin Carter, GianCarlo Moschini and Ian Sheldon, pp 201–224. Volume 10 in Frontiers of Economics and Globalization Series. Bingley, UK: Emerald Group Publishing

83. "Biotechnology Predoctoral Training Program". National Institute of General Medical Sciences. 18 December 2013. Retrieved 28 October 2014.

CHAPTER 2

History of Biotechnology

Chapter Contents

It is not known precisely when people began using the word biotechnology to refer to technologies based on living things. Many believe the phrase was first coined in the early 1900s by a Hungarian engineer named Károly Ereky, but the practice of biotechnology began long before that and dates back almost to the beginning of human existence. The following are examples of biotechnologies over the course of history.

INTRODUCTION

Biotechnology is the application of scientific and engineering principles to the processing of materials by biological agents to provide goods and services. From its inception, biotechnology has maintained a close relationship with society. Although now most often associated with the development of drugs, historically biotechnology has been principally associated with food, addressing such issues as malnutrition and famine. The history of biotechnology begins with zymotechnology, which commenced with a focus on brewing techniques for beer. By World War I, however, zymotechnology would expand to tackle larger industrial issues, and the potential of industrial fermentation gave rise to biotechnology. However, both the single-cell protein and gasohol projects failed to progress due to varying issues including public resistance, a changing economic scene, and shifts in political power.

Yet the formation of a new field, genetic engineering, would soon bring biotechnology to the forefront of science in society, and the intimate relationship between the scientific community, the public, and the government would ensue. These debates gained exposure in 1975 at the Asilomar Conference, where Joshua Lederberg was the most outspoken supporter for this emerging field in biotechnology. By as early as 1978, with the development of synthetic human insulin, Lederberg's claims would prove valid, and the biotechnology industry grew rapidly. Each new scientific advance became a media event designed to capture public support, and by the 1980s, biotechnology grew into a promising real industry. In 1988, only five proteins from genetically engineered cells had been approved as drugs by the United States Food and Drug Administration (FDA), but this number would skyrocket to over 125 by the end of the 1990s.

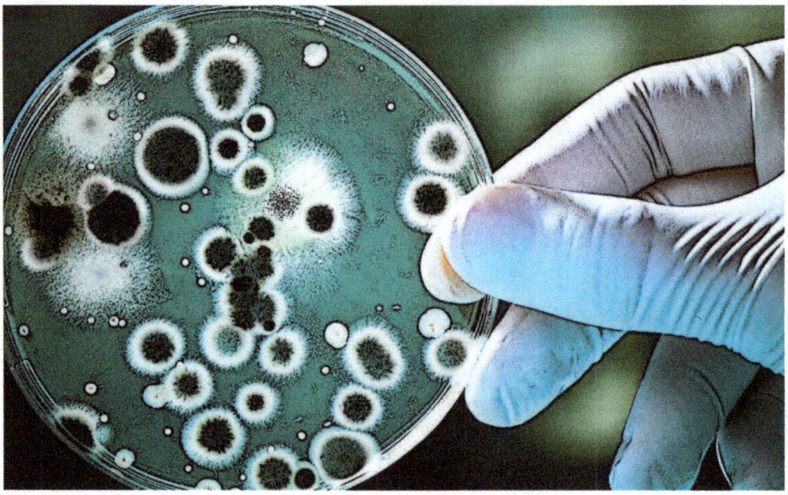

The field of genetic engineering remains a heated topic of discussion in today's society with the advent of gene therapy, stem cell research, cloning, and genetically modified food. While it seems only natural nowadays to link pharmaceutical drugs as solutions to health and societal problems, this relationship of biotechnology serving social needs began centuries ago.

ORIGINS OF BIOTECHNOLOGY

Biotechnology arose from the field of zymotechnology or zymurgy, which began as a search for a better understanding of industrial fermentation, particularly beer. Beer was an important industrial, and not just social, commodity. In late 19th-century Germany, brewing contributed as much to the gross national product as steel, and taxes on alcohol proved to be significant sources of revenue to the government. In the 1860s, institutes and remunerative consultancies were dedicated to the technology of brewing. The most famous was the private Carlsberg Institute, founded in 1875, which employed Emil Christian Hansen, who pioneered the pure yeast process for the reliable production of consistent beer. Less well known were private consultancies that advised the brewing industry. One of these, the Zymotechnic Institute, was established in Chicago by the German-born chemist John Ewald Siebel.

The heyday and expansion of zymotechnology came in World War I in response to industrial needs to support the war. Max Delbrück grew yeast on an immense scale during the war to meet 60 percent of Germany's animal feed needs. Compounds of another fermentation product, lactic acid, made up for a lack of hydraulic fluid, glycerol. On the Allied side the Russian chemist Chaim Weizmann used starch to eliminate Britain's shortage of acetone, a key raw material for cordite, by fermenting maize to acetone. The industrial potential of fermentation was outgrowing its traditional home in brewing, and "zymotechnology" soon gave way to "biotechnology."

With food shortages spreading and resources fading, some dreamed of a new industrial solution. The Hungarian Károly Ereky coined the word "biotechnology" in Hungary during 1919 to describe a technology based on converting raw materials into a more useful product. He built a slaughterhouse for a thousand pigs and also a fattening farm with space for 50,000 pigs, raising over 100,000 pigs a year. The enterprise was enormous, becoming one of the largest and most profitable meat and fat operations in the world. In a book entitled *Biotechnologie*, Ereky further developed a theme that would be reiterated through the 20th century: biotechnology could provide solutions to societal crises, such as food and energy shortages. For Ereky, the term "biotechnologie" indicated the process by which raw materials could be biologically upgraded into socially useful products.

This catchword spread quickly after the First World War, as "biotechnology" entered German dictionaries and was taken up abroad by business-hungry private consultancies as far away as the United States. In Chicago, for example, the coming of prohibition at the end of World War I encouraged biological industries to create opportunities for new fermentation products, in particular a market for nonalcoholic drinks. Emil Siebel, the son of the founder of the Zymotechnic Institute, broke away from his father's company to establish his own called the "Bureau of Biotechnology," which specifically offered expertise in fermented nonalcoholic drinks.

The belief that the needs of an industrial society could be met by fermenting agricultural waste was an important ingredient of the "chemurgic movement."Fermentation-based processes generated products of ever-growing utility. In the 1940s, penicillin was the most dramatic. While it was discovered in England, it was produced industrially in the U.S. using a deep fermentation process originally developed in Peoria, Illinois. The enormous profits and the public expectations penicillin engendered caused a radical shift in the standing of the pharmaceutical industry. Doctors used the phrase "miracle drug", and the historian of its wartime use, David Adams, has suggested that to the public penicillin represented the perfect health that went together with the car and the dream house of wartime American advertising. Beginning in the 1950s, fermentation technology also became advanced enough to produce steroids on industrially significant scales. Of particular importance was the improved semisynthesis of cortisone which simplified the old 31 step synthesis to 11 steps.This advance was estimated to reduce the cost of the drug by 70%, making the medicine inexpensive and available. Today biotechnology still plays a central role in the production of these compounds and likely will for years to come.

SINGLE-CELL PROTEIN AND GASOHOL PROJECTS

Even greater expectations of biotechnology were raised during the 1960s by a process that grew single-cell protein. When the so-called protein gap threatened world hunger, producing food locally by growing it from waste seemed to offer a solution. It was the possibilities of growing microorganisms on oil that captured the imagination of scientists, policy makers, and commerce. Major companies such as British Petroleum (BP) staked their futures on it. In 1962, BP built a pilot plant at Cap de Lavera in Southern France to publicize its product, Toprina. Initial research work at Lavera was done by Alfred Champagnat, In 1963, construction started on BP's second pilot plant at Grangemouth Oil Refinery in Britain.

As there was no well-accepted term to describe the new foods, in 1966 the term "single-cell protein" (SCP) was coined at MIT to provide an acceptable and exciting new title, avoiding the unpleasant connotations of microbial or bacterial.

The "food from oil" idea became quite popular by the 1970s, when facilities for growing yeast fed by n-paraffins were built in a number of countries. The Soviets were particularly enthusiastic, opening large "BVK" (*belkovo-vitaminny kontsentrat*, i.e., "protein-vitamin concentrate") plants next to their oil refineries in Kstovo (1973) and Kirishi (1974).

By the late 1970s, however, the cultural climate had completely changed, as the growth in SCP interest had taken place against a shifting economic and cultural scene (136). First, the price of oil rose catastrophically in 1974, so that its cost per barrel was five times greater than it had been two years earlier. Second, despite continuing hunger around the world, anticipated demand also began to shift from humans to animals. The program had begun with the vision of growing food for Third World people, yet the product was instead launched as an animal food for the developed world. The rapidly rising demand for animal feed made that market appear economically more attractive. The ultimate downfall of the SCP project, however, came from public resistance.

This was particularly vocal in Japan, where production came closest to fruition. For all their enthusiasm for innovation and traditional interest in microbiologically produced foods, the Japanese were the first to ban the production of single-cell proteins. The Japanese ultimately were unable to separate the idea of their new "natural" foods from the far from natural connotation of oil. These arguments were made against a background of suspicion of heavy industry in which anxiety over minute traces of petroleum was expressed. Thus, public resistance to an unnatural product led to the end of the SCP project as an attempt to solve world hunger.

Also, in 1989 in the USSR, the public environmental concerns made the government decide to close down (or convert to different technologies) all 8 paraffin-fed-yeast plants that the Soviet Ministry of Microbiological Industry had by that time.

In the late 1970s, biotechnology offered another possible solution to a societal crisis. The escalation in the price of oil in 1974 increased the cost of the Western world's energy tenfold. In response, the U.S. government promoted the production of gasohol, gasoline with 10 percent alcohol added, as an answer to the energy crisis. In 1979, when the Soviet Union sent troops to Afghanistan, the Carter administration cut off its supplies to agricultural produce in retaliation, creating a surplus of agriculture in the U.S. As a result, fermenting the agricultural surpluses to synthesize fuel seemed to be an economical solution to the shortage of oil threatened by the Iran-Iraq war. Before the new direction could be taken, however, the political wind changed again: the Reaganadministration came to power in January 1981 and, with the declining oil prices of the 1980s, ended support for the gasohol industry before it was born.

Biotechnology seemed to be the solution for major social problems, including world hunger and energy crises. In the 1960s, radical measures would be needed to meet world starvation, and biotechnology seemed to provide an answer. However, the solutions proved to be too expensive and socially unacceptable, and solving world hunger through SCP food was dismissed. In the 1970s, the food crisis was succeeded by the energy crisis, and here too, biotechnology seemed to provide an answer. But once again, costs proved prohibitive as oil prices slumped in the 1980s. Thus, in practice, the implications of biotechnology were not fully realized in these situations. But this would soon change with the rise of genetic engineering.

GENETIC ENGINEERING

The origins of biotechnology culminated with the birth of genetic engineering. There were two key events that have come to be seen as scientific breakthroughs beginning the era that would unite genetics with biotechnology. One was the 1953 discovery of the structure of DNA, by Watson and Crick, and the other was the 1973 discovery by Cohen and Boyer of a recombinant DNA technique by which a section of DNA was cut from the plasmid of an E. coli bacterium and transferred into the DNA of another. This approach could, in principle, enable bacteria to adopt the genes and produce proteins of other organisms, including humans. Popularly referred to as "genetic engineering," it came to be defined as the basis of new biotechnology.

Genetic engineering proved to be a topic that thrust biotechnology into the public scene, and the interaction between scientists, politicians, and the public defined the work that was accomplished in this area. Technical developments during this time were revolutionary and at times frightening. In December 1967, the first heart transplant by Christian Barnard reminded the public that the physical identity of a person was becoming increasingly problematic. While poetic imagination had always seen the heart at the center of the soul, now there was the prospect of individuals being defined by other people's hearts. During the same month, Arthur Kornberg announced that he had managed to biochemically replicate a viral gene. "Life had been synthesized," said the head of the National Institutes of Health. Genetic engineering was now on the scientific agenda, as it was becoming possible to identify genetic characteristics with diseases such as beta thalassemia and sickle-cell anemia.

Responses to scientific achievements were colored by cultural skepticism. Scientists and their expertise were looked upon with suspicion. In 1968, an immensely popular work, *The Biological Time Bomb*, was written by the British journalist Gordon Rattray Taylor. The author's preface saw Kornberg's discovery of replicating a viral gene as a route to lethal doomsday bugs. The publisher's blurb for the book warned that within ten years, "You may marry a semi-artificial man or woman…choose your children's sex…tune out pain…change your memories…and live to be 150 if the scientific revolution doesn't destroy us first."The book ended with a chapter called "The Future – If Any." While it is rare for current science to be represented in the movies, in this period of "Star Trek", science fiction and science fact seemed to be converging. "Cloning" became a popular word in the media. Woody Allen satirized the cloning of a person from a nose in his 1973 movie *Sleeper*, and cloning Adolf Hitler from surviving cells was the theme of the 1976 novel by Ira Levin, *The Boys from Brazil*.

In response to these public concerns, scientists, industry, and governments increasingly linked the power of recombinant DNA to the immensely practical functions that biotechnology promised. One of the key scientific figures that attempted to highlight the promising aspects of genetic engineering was Joshua Lederberg, a Stanford professor and Nobel laureate. While in the 1960s "genetic engineering" described eugenics and work involving the manipulation of the human genome, Lederberg stressed research that would involve microbes instead. Lederberg emphasized the importance of focusing on curing living people. Lederberg's 1963 paper, "Biological Future of Man" suggested that, while molecular biology might one day make it possible to change the human genotype, "what we have overlooked is euphenics, the engineering of human development." Lederberg constructed the word "euphenics" to emphasize changing the phenotype after conception rather than the genotype which would affect future generations.

With the discovery of recombinant DNA by Cohen and Boyer in 1973, the idea that genetic engineering would have major human and societal consequences was born. In July 1974, a group of eminent molecular biologists headed by Paul Berg wrote to *Science* suggesting that the consequences of this work were so potentially destructive that there should be a pause until its implications had been thought through. This suggestion was explored at a meeting in February 1975 at California's Monterey Peninsula, forever immortalized by the location, Asilomar. Its historic outcome was an unprecedented call for a halt in research until it could be regulated in such a way that the public need not be anxious, and it led to a 16-month moratorium until National Institutes of Health (NIH) guidelines were established.

Joshua Lederberg was the leading exception in emphasizing, as he had for years, the potential benefits. At Asilomar, in an atmosphere favoring control and regulation, he circulated a paper countering the pessimism and fears of misuses with the benefits conferred by successful use. He described "an early chance for a technology of untold importance for diagnostic and therapeutic medicine: the ready production of an unlimited variety of human proteins. Analogous applications may be foreseen in fermentation process for cheaply manufacturing essential nutrients, and in the improvement of microbes for the production of antibiotics and of special industrial chemicals." In June 1976, the 16-month moratorium on research expired with the Director's Advisory Committee (DAC) publication of the NIH guidelines of good practice. They defined the risks of certain kinds of experiments and the appropriate physical conditions for their pursuit, as well as a list of things too dangerous to perform at all. Moreover, modified organisms were not to be tested outside the confines of a laboratory or allowed into the environment.

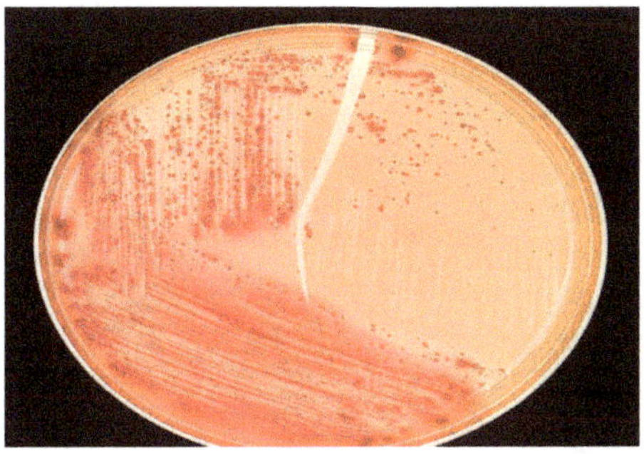

Synthetic insulin crystals synthesized using recombinant DNA technology

Atypical as Lederberg was at Asilomar, his optimistic vision of genetic engineering would soon lead to the development of the biotechnology industry. Over the next two years, as public concern over the dangers of recombinant DNAresearch grew, so too did interest in its technical and practical applications. Curing genetic diseases remained in the realms of science fiction, but it appeared that producing human simple proteins could be good business. Insulin, one of the smaller, best characterized and understood proteins, had been used in treating type 1 diabetes for a half century. It had been extracted from animals in a chemically slightly different form from the human product. Yet, if one could produce synthetic human insulin, one could meet an existing demand with a product whose approval would be relatively easy to obtain from regulators. In the period 1975 to 1977, synthetic "human" insulin represented the aspirations for new products that could be made with the new biotechnology. Microbial production of synthetic human insulin was finally announced in September 1978 and was produced by a startup company, Genentech. Although that company did not commercialize the product themselves, instead, it licensed the production method to Eli Lilly and Company. 1978 also saw the first application for a patent on a gene, the gene which produces human growth hormone, by the University of California, thus introducing the legal principle that genes could be patented. Since that filing, almost 20% of the more than 20,000 genes in the human DNA have been patented.

The radical shift in the connotation of "genetic engineering" from an emphasis on the inherited characteristics of people to the commercial production of proteins and therapeutic drugs was nurtured by Joshua Lederberg. His broad concerns since the 1960s had been stimulated by enthusiasm for science and its potential medical benefits. Countering calls for strict regulation, he expressed a vision of potential utility. Against a belief that new techniques would entail unmentionable and uncontrollable consequences for humanity and the environment, a growing consensus on the economic value of recombinant DNA emerged.

BIOTECHNOLOGY AND INDUSTRY

With ancestral roots in industrial microbiology that date back centuries, the new biotechnology industry grew rapidly beginning in the mid-1970s. Each new scientific advance became a media event designed to capture investment confidence and public support. Although market expectations and social benefits of new products were frequently overstated, many people were prepared to see genetic engineering as the next great advance in technological progress. By the 1980s, biotechnology characterized a nascent real industry, providing titles for emerging trade organizations such as the Biotechnology Industry Organization (BIO).

A Genentech-sponsored sign declaring South San Francisco to be "The Birthplace of Biotechnology."

The main focus of attention after insulin were the potential profit makers in the pharmaceutical industry: human growth hormone and what promised to be a miraculous cure for viral diseases, interferon. Cancer was a central target in the 1970s because increasingly the disease was linked to viruses. By 1980, a new company, Biogen, had produced interferon through recombinant DNA. The emergence of interferon and the possibility of curing cancer raised money in the community for research and increased the enthusiasm of an otherwise uncertain and tentative society. Moreover, to the 1970s plight of cancer was added AIDS in the 1980s, offering an enormous potential market for a successful therapy, and more immediately, a market for diagnostic tests based on monoclonal antibodies. By 1988, only five proteins from genetically engineered cells had been approved as drugs by the United States Food and Drug Administration (FDA): synthetic insulin, human growth hormone, hepatitis B vaccine, alpha-interferon, and tissue plasminogen activator (TPa), for lysis of blood clots. By the end of the 1990s, however, 125 more genetically engineered drugs would be approved.

Genetic engineering also reached the agricultural front as well. There was tremendous progress since the market introduction of the genetically engineered Flavr Savr tomato in 1994. Ernst and Young reported that in 1998, 30% of the U.S. soybean crop was expected to be from genetically engineered seeds. In 1998, about 30% of the US cotton and corn crops were also expected to be products of genetic engineering.

Genetic engineering in biotechnology stimulated hopes for both therapeutic proteins, drugs and biological organisms themselves, such as seeds, pesticides, engineered yeasts, and modified human cells for treating genetic diseases. From the perspective of its commercial promoters, scientific breakthroughs, industrial commitment, and official support were finally coming together, and biotechnology became a normal part of business. No longer were the proponents for the economic and technological significance of biotechnology the iconoclasts. Their message had finally become accepted and incorporated into the policies of governments and industry.

GLOBAL TRENDS

According to Burrill and Company, an industry investment bank, over $350 billion has been invested in biotech since the emergence of the industry, and global revenues rose from $23 billion in 2000 to more than $50 billion in 2005. The greatest growth has been in Latin America but all regions of the world have shown strong growth trends. By 2007 and into 2008, though, a downturn in the fortunes of biotech emerged, at least in the United Kingdom, as the result of declining investment in the face of failure of biotech pipelines to deliver and a consequent downturn in return on investment.

TIMELINE OF BIOTECHNOLOGY

The historical application of Biotechnology throughout is provided below in chronological order. These discoveries, inventions and modifications are evidence of the evolution of biotechnology since before the common era.

Before Common Era

- 7000 BCE – Chinese discover fermentation through beer making.
- 6000 BCE – Yogurt and cheese made with lactic acid-producing bacteria by various peoples.
- 4000 BCE – Egyptians bake leavened bread using yeast.
- 500 BCE – Moldy soybean curds used as an antibiotic.
- 250 BCE – The Greeks practice crop rotation for maximum soil fertility.
- 100 CE – Chinese use chrysanthemum as a natural insecticide.

Pre-20th Century

- 1663 – First recorded description of living cells by Robert Hooke.
- 1675 – Antonie van Leeuwenhoek discovers and describes bacteria and protozoa.
- 1798 – Edward Jenner uses first viral vaccine to inoculate a child from smallpox.
- 1802 – The first recorded use of the word biology.
- 1824 – Henri Dutrochet discovers that tissues are composed of living cells.
- 1838 – Protein discovered, named and recorded by Gerardus Johannes Mulder and Jöns Jacob Berzelius.
- 1862 – Louis Pasteur discovers the bacterial origin of fermentation.
- 1863 – Gregor Mendel discovers the laws of inheritance.
- 1864 – Antonin Prandtl invents first centrifuge to separate cream from milk.

- 1869 – Friedrich Miescher identifies DNA in the sperm of a trout.
- 1871 – Ernst Hoppe-Seyler discovers invertase, which is still used for making artificial sweeteners.
- 1877 – Robert Koch develops a technique for staining bacteria for identification.
- 1878 – Walther Flemming discovers chromatin leading to the discovery of chromosomes.
- 1881 – Louis Pasteur develops vaccines against bacteria that cause cholera and anthrax in chickens.
- 1885 – Louis Pasteur and Emile Roux develop the first rabies vaccine and use it on Joseph Meister.

20th century"

- 1919 – Károly Ereky, a Hungarian agricultural engineer, first uses the word biotechnology.
- 1928 – Alexander Fleming notices that a certain mould could stop the duplication of bacteria, leading to the first antibiotic: penicillin.
- 1933 – Hybrid corn is commercialized.
- 1942 – Penicillin is mass-produced in microbes for the first time.
- 1950 – The first synthetic antibiotic is created.
- 1951 – Artificial insemination of livestock is accomplished using frozen semen.
- 1952 – L.V. Radushkevich and V.M. Lukyanovich publish clear images of 50 nanometer diameter tubes made of carbon, in the Soviet Journal of Physical Chemistry.
- 1953 – James D. Watson and Francis Crick describe the structure of DNA.
- 1958 – The term bionics is coined by Jack E. Steele.
- 1964 – The first commercial myoelectric arm is developed by the Central Prosthetic Research Institute of the USSR, and distributed by the Hangar Limb Factory of the UK.
- 1972 – The DNA composition of chimpanzees and gorillas is discovered to be 99% similar to that of humans.
- 1973 – Stanley Norman Cohen and Herbert Boyer perform the first successful recombinant DNA experiment, using bacterial genes.
- 1974 – Scientist invent the first biocement for industrial applications.
- 1975 – Method for producing monoclonal antibodies developed by Köhler and César Milstein.
- 1978 – North Carolina scientists Clyde Hutchison and Marshall Edgell show it is possible to introduce specific mutations at specific sites in a DNA molecule.
- 1980 – The U.S. patent for gene cloning is awarded to Cohen and Boyer.
- 1982 – Humulin, Genentech's human insulin drug produced by genetically

engineered bacteria for the treatment of diabetes, is the first biotech drug to be approved by the Food and Drug Administration.

- 1983 – The Polymerase Chain Reaction (PCR) technique is conceived.
- 1990 – First federally approved gene therapy treatment is performed successfully on a young girl who suffered from an immune disorder.
- 1994 – The United States Food and Drug Administration approves the first GM food: the "Flavr Savr" tomato.
- 1997 – British scientists, led by Ian Wilmut from the Roslin Institute, report cloning Dolly the sheep using DNA from two adult sheep cells.
- 1999 – Discovery of the gene responsible for developing cystic fibrosis.
- 2000 – Completion of a "rough draft" of the human genome in the Human Genome Project.

21st Century

- 2001 – Celera Genomics and the Human Genome Project create a draft of the human genome sequence. It is published by *Science and Nature*Magazine.
- 2002 – Rice becomes the first crop to have its genome decoded.
- 2003 – The Human Genome Project is completed, providing information on the locations and sequence of human genes on all 46 chromosomes.
- 2008 – Japanese astronomers launch the first Medical Experiment Module called "Kibo", to be used on the International Space Station.
- 2009 – Cedars-Sinai Heart Institute uses modified SAN heart genes to create the first *viral pacemaker* in guinea pigs, now known as iSANs.
- 2012 – Thirty-one-year-old Zac Vawter successfully uses a nervous system-controlled bionic leg to climb the Chicago Willis Tower.

COMPETITIONS AND PRIZES IN BIOTECHNOLOGY

There exist a number of competitions and prizes to reward distinguished contributions and to encourage developments in biotechnology.

Inducement Prizes

- The Archon X Prize for Genomics of US$10,000,000 is to be awarded to "the first Team that can build a device and use it to sequence 100 human genomes within 10 days or less, with an accuracy of no more than one error in every 100,000 bases sequenced, with sequences accurately covering at least 98% of the genome, and at a recurring cost of no more than $10,000 (US) per genome."
- The Prize4Life ALS biomarker prize is a US$1,000,000 award for a reliable way of tracking progression of amyotrophic lateral sclerosis (ALS).

- The Prize4Life ALS treatment prize is a US$1,000,000 award for a therapy that reliably and effectively extends the life of ALS mice by 25%.
- People for Ethical Treatment of Animals (PETA) is offering a US$1,000,000 reward for a method of producing enough meat to be marketed in 10 U.S. states at a price competitive with chicken prices.
- Illumina iDEA Challenge to develop new visualization and data analysis techniques.

Recognition Prizes

- The Gotham Prize for Cancer Research is a US$1,000,000 prize awarded annually to "encourage new and innovative approaches to cancer research by fostering collaboration among top thinkers in the field--with the goal of leading to progress in the prevention, diagnosis, etiology and treatment of cancer."
- Gruber Prize in Genetics is a US$500,000 prize awarded annually for distinguished contributions in any realm of genetics research.
- The Nobel Prize in Physiology or Medicine is an annual grant worth approximately 10 million SEK. It is routinely awarded for contributions to biotechnology.

Inducement Prize Contest

An inducement prize contest (IPC) is a competition that awards a cash prize for the accomplishment of a feat, usually of engineering. IPCs are typically designed to extend the limits of human ability. Some of the most famous IPCs include the Longitude prize (1714–1765), the Orteig Prize (1919–1927) and the prizes from the X Prize Foundation.

IPCs are distinct from recognition prizes, such as the Nobel Prize, in that IPCs have prospectively defined criteria for what feat is to be achieved for winning the prize, while recognition prizes may be based on the beneficial effects of the feat.

Research has shown that IPCs can be extremely effective in pushing the advancement of technology.

Throughout history, there have been instances where IPCs were successfully utilized to push the boundaries of what would have been considered state-of-the-art at the time.

The Longitude Prize was a reward offered by the British government for a simple and practical method for the precise determination of a ship's longitude. The prize, established through an Act of Parliament (the Longitude Act) in 1714, was administered by the Board of Longitude. Another example happened during the first years of the Napoleonic Wars. The French government offered a hefty cash award of 12,000 francs to any inventor who could devise a cheap and effective method of preserving large amounts of food. The larger armies of the period required increased, regular supplies of quality food. Limited food availability was among the factors limiting military campaigns to the summer and autumn months. In 1809, a French confectioner and brewer, Nicolas Appert, observed that food cooked inside a jar did not spoil unless the seals leaked, and developed a method of sealing food in glass jars. The reason for lack of spoilage was unknown at the time, since it would be another 50 years before Louis Pasteur demonstrated the role of microbes in food spoilage. Yet another example is the Orteig Prize which was a $25,000 reward offered on May 19, 1919, by New York hotel owner Raymond Orteig to the first allied aviator(s) to fly non-stop from New York City to Paris or vice versa. On offer for five years, it attracted no competitors. Orteig renewed the offer for another five years in 1924 when the state of aviation technology had advanced to the point that numerous competitors vied for the prize. Several famous aviators made unsuccessful attempts at the New York–Paris flight before relatively unknown American Charles Lindbergh won the prize in 1927 in his aircraft *Spirit of St. Louis*.

In recent years, there has been increasing interest in the development of IPCs to push technical boundaries in several key areas such as aerospace, life sciences, and sustainable energy among others. Some now predict that global recession may make prizes more prevalent rather than less as prizes are argued to be a cost effective tool for advancing markets. Big prizes for big challenges - 13 predictions for 2013

A leading organization in development and managing IPCs is the X PRIZE Foundation. Its mission is to bring about "radical breakthroughs for the benefit of humanity" through incentivized competition. It fosters high-profile competitions that motivate individuals, companies and organizations across all disciplines to develop innovative ideas and technologies that help solve the grand challenges that restrict humanity's progress. The most high-profile X PRIZE to date was the Ansari X PRIZE relating to spacecraft development awarded in 2004. This prize was intended to inspire research and development into technology for space exploration. Indeed, the X Prize has inspired other "letter" named inducement prize competitions such as the H-Prize, N-Prize, and so forth. In 2006, there was much interest in prizes for automotive achievement, such as the 250 mpg car.

In Europe there has been a re-emergence of challenge prizes that following in the tradition of the Longitude Prize for solutions which impact on social problems. The Centre for Challenge Prizes based in London is an example of this running prizes for innovations that for example reduce social isolation or make renewable energy generators accessible to off the grid refugees and returnees.

Economics of IPCs

In some literature on the subject, it has been stated that well-designed IPCs can garner economic activity on the order of 10 to 20 times the amount of the prize face value.

List of IPCs

Ansari X PRIZE for Suborbital Spaceflight, won in 2004 by Scaled Composites SpaceShipOne

- Automotive X PRIZE
- Alkali Prize for a process to turn salt into soda ash, won by Nicolas Leblanc posthumously in the early 19th century
- Brexit Prize - prize for British exit from the European Union
- NASA's Centennial Challenges

- The Clay Mathematics Institute has a Millennium Prize it will award to anyone who provides a solution to one of seven important mathematics problems.
- Cornell Cup USA, presented by Intel
- DARPA Grand Challenge
- The Foresight Institute offers a $250,000 Feynman Grand Prize for the first persons to design and build two nanotechnology devices - a nano-scale robotic arm and a computing device that demonstrates the feasibility of building a nanotechnology computer.
- Global Security Challenge
- Google Lunar X Prize
- Intelligent Ground Vehicle Competition
- Kremer prize for man-powered aircraft, won in 1977 by the MacCready Gossamer Condor.
- L Prize is a US Department of Energy competition to increase efficiency of solid-state lighting.
- Longitude prize won in the 18th century by John Harrison
- Methuselah Mouse Prize or also known as the "M-Prize"
- Montyon Prizes established in 1820, a series of prizes awarded annually by the Académie française
- Orteig Prize for a non-stop flight between New York and Paris, won by Charles Lindbergh in 1927
- Peugeot Concours Design
- Prize4Life offers between $15,000 and $5 million in prize awards for medical discoveries that remove the largest barriers to finding a cure for ALS (Lou Gehrig's disease)
- Tricorder X Prize
- Virgin Earth Challenge
- Wolfskehl Prize for proving Fermat's Last Theorem, won by Andrew Wiles in 1997

REFERENCES

1. Bud, Robert; Cantley, Mark F. (1994). The Uses of Life: A History of Biotechnology (1st ed.). London: Cambridge University Press. pp. 1, 6, 7, 30, 133, 135, 138, 141–142, 155, 171–173, 165, 167, 174, 177, & 191. ISBN 9780521476997.
2. Thackray, Arnold (1998). Private Science: Biotechnology and the Rise of the Molecular Sciences. Philadelphia: University of Pennsylvania Press. pp. 6–8. ISBN 9780812234282.
3. Sifniades, Stylianos; Levy, Alan B. (2000). "Acetone". Ullmann's Encyclopedia of Industrial Chemistry. doi:10.1002/14356007.a01_079.

4. Fiechter, A.; Beppu, T. (2000). History of Modern Biotechnology I (1st ed.). Berlin: Springer Science & Business Media. pp. 153 & 170. ISBN 9783540677932.
5. Gordon, J. J.; Grenfell, E.; Legge, B. J.; Mcallister, R. C. A.; White, T. (1947). "Methods of Penicillin Production in Submerged Culture on a Pilot-Plant Scale". Microbiology. 1: 187–202. doi:10.1099/00221287-1-2-187.
6. Capek, Milantadra; Oldrich, Hanc; Alois, Capek (1966). Microbial Transformations of Steroids. Prague: Academia Publishing House of Czechoslovak Academy of Sciences. doi:10.1007/978-94-011-7603-3. ISBN 9789401176057.
7. Leigh, H. M.; Meister, P. D.; Weintraub,, A.; Reineke, L. M.; Eppstein, S. H.; Murray, H. C.; Peterson, D. H. (1952). "Microbiological Transformations of Steroids.1 I. Introduction of Oxygen at Carbon-11 of Progesterone". Journal of the American Chemical Society. 73 (23): 5933–5936. doi:10.1021/ja01143a033.
8. *Liese, Andreas; Seelbach, Karsten; Wandrey, Christian (2006). History of Industrial Biotransformations – Dreams and Realities (2nd ed.). New York: Wiley. doi:10.1002/9783527608188.ch1. ISBN 9783527310012.*
9. *Ohno, Masaji; Otsuka, Masami; Yagisawa, Morimasa; Kondo, Shinichi; Öppinger, Heinz; Hoffmann, Hinrich; Sukatsch, Dieter; Hepner, Leo; Male, Celia (2000). "Antibiotics". Ullmann's Encyclopedia of Industrial Chemistry. doi:10.1002/14356007.a02_467.*
10. *Sandow, Jürgen; Scheiffele, Ekkehard; Haring, Michael; Neef, Günter; Prezewowsky, Klaus; Stache, Ulrich (2000). "Hormones". Ullmann's Encyclopedia of Industrial Chemistry. doi:10.1002/14356007.a13_089.*
11. *Bamberg, J. H. (2000). British Petroleum and global oil, 1950-1975: the challenge of nationalism. Volume 3 of British Petroleum and Global Oil 1950-1975: The Challenge of Nationalism, J. H. Bamberg British Petroleum series. Cambridge University Press. pp. 426–428. ISBN 0-521-78515-4.*
12. Soviet Plant to Convert Oil to Protein for Feed; Use of Yeast Involved, By THEODORE SHABAD. the New York Times, November 10, 1973.
13. Первенец микробиологической промышленности (Microbiological industry's first plant), in: Станислав Марков (Stanislav Markov) "Кстово - молодой город России› (*Kstovo, Russia's Young City*)
14. *Grace, Eric S. (2006). Biotechnology Unzipped:: Promises and Realities (2nd ed.). Washington, D.C.: Joseph Henry Press. pp. 78 & 155. ISBN 9780309096218.*
15. *Krimsky, Sheldon (1991). Biotechnics & society: the rise of industrial genetics (1st ed.). Westport, CT: Praeger. pp. 18, & 21. ISBN 9780275938598.*
16. *Rita R, Colwell (2002). "Fulfilling the promise of biotechnology". Biotechnology Advances. 20 (3-4): 215–228. doi:10.1016/S0734-9750(02)00011-3. PMID 14550029.*
17. *Pearson, Sue (2008-08-01). "Is the U.K. Biotech Industry on the Skids?". Genetic Engineering & Biotechnology News. Mary Ann Liebert. 28 (14): 12–13. ISSN 1935-472X. Retrieved 2008-09-20.*

18. ‹Highlights in the History of Biotechnology" (PDF). St Louis Science Center. Retrieved 27 December 2012.

19. "Agriculture in Ancient Greece". The Ancient History Encyclopedia. Retrieved 27 December 2012.

20. "Biotechnology Timeline". Biotechnology Institute of Washington DC. Retrieved 27 December 2012.

21. "1973_Boyer". Genome News Network. Retrieved 19 August 2015.

22. C A Hutchison, 3rd, S Phillips, M H Edgell, S Gillam, P Jahnke and M Smith (1978). "Mutagenesis at a specific position in a DNA sequence". J Biol Chem. 253: 6551–6560. PMID 681366.

23. Wadhwa, Vivek (21 May 2014). "The powerful role of incentive competitions to spur innovation". Washington Post. Retrieved 14 January 2015.

24. "Association Internationale Nicolas Appert" (in French).

25. "Big prizes for big challenges - 13 predictions for 2013 - Nesta".

26. Centre for Challenge Prizes

27. "Brexit": IEA offers prize for UK exit plan from EU". British Broadcasting Corporation. 15 July 2013. Retrieved 17 February 2014.

CHAPTER 3

Cell: The Fundamental Unit of Life

Chapter Contents

- **Introduction**
- **Anatomy**
- **Subcellular Components**
- **Structures Outside the Cell Membrane**
- **Cellular Processes**
- **Multicellularity**
- **Origins**
- **History of Research**

Cell biology is the study of cell structure and function, and it revolves around the concept that the cell is the fundamental unit of life. Focusing on the cell permits a detailed understanding of the tissues and organisms that cells compose. Some organisms have only one cell, while others are organized into cooperative groups with huge numbers of cells. On the whole, cell biology focuses on the structure and function of a cell, from the most general properties shared by all cells, to the unique, highly intricate functions particular to specialized cells.

INTRODUCTION

The **cell** (from Latin *cella*, meaning "small room") is the basic structural, functional, and biological unit of all known living organisms. A cell is the smallest unit of life that can replicate independently, and cells are often called the "building blocks of life". The study of cells is called cell biology.

Cells consist of cytoplasm enclosed within a membrane, which contains many biomolecules such as proteins and nucleic acids. Organisms can be classified as unicellular (consisting of a single cell; including bacteria) or multicellular (including plants and animals). While the number of cells in plants and animals varies from species to species, humans contain more than 10 trillion (10^{12}) cells. Most plant and animal cells are visible only under a microscope, with dimensions between 1 and 100 micrometres.

The cell was discovered by Robert Hooke in 1665, who named the biological unit for its resemblance to cells inhabited by Christian monks in a monastery. Cell theory, first developed in 1839 by Matthias Jakob Schleiden and Theodor Schwann, states that all organisms are composed of one or more cells, that cells are the fundamental unit of structure and function in all living organisms, that all cells come from preexisting cells, and that all cells contain the hereditary information necessary for regulating cell functions and for transmitting information to the next generation of cells. Cells emerged on Earth at least 3.5 billion years ago.

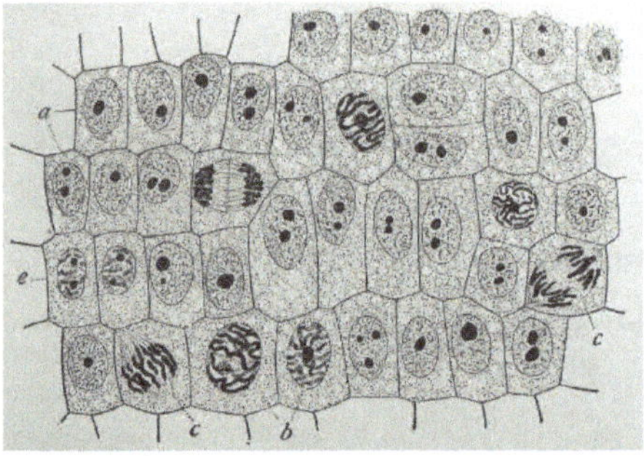

Onion (Allium) cells in different phases of the cell cycle

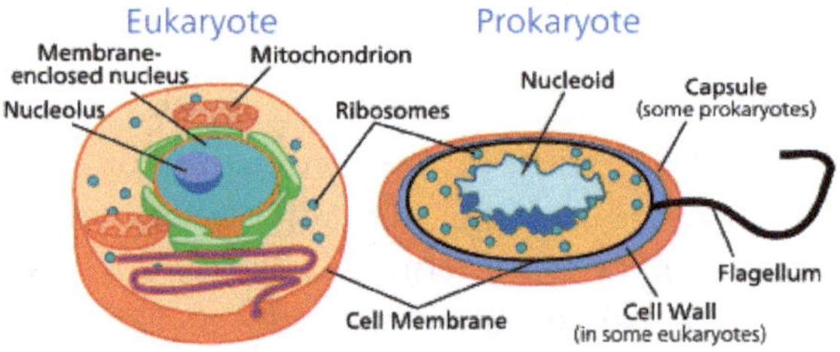

A eukaryotic cell (left) and a prokaryotic cell (right)

ANATOMY

Comparison of features of prokaryotic and eukaryotic cells		
	Prokaryotes	**Eukaryotes**
Typical organisms	bacteria, archaea	protists, fungi, plants, animals
Typical size	~ 1–5 µm	~ 10–100 µm

Type of nucleus	nucleoid region; no true nucleus	true nucleus with double membrane
DNA	circular (usually)	linear molecules (chromosomes) with histone proteins
RNA/protein synthesis	coupled in the cytoplasm	RNA synthesis in the nucleus protein synthesis in the cytoplasm
Ribosomes	50S and 30S	60S and 40S
Cytoplasmic structure	very few structures	highly structured by endomembranes and a cytoskeleton
Cell movement	flagella made of flagellin	flagella and cilia containing microtubules; lamellipodia and filopodia containing actin
Mitochondria	none	one to several thousand
Chloroplasts	none	in algae and plants
Organization	usually single cells	single cells, colonies, higher multicellular organisms with specialized cells
Cell division	binary fission (simple division)	mitosis (fission or budding) meiosis
Chromosomes	single chromosome	more than one chromosome
Membranes	cell membrane	Cell membrane and membrane-bound organelles

Cells are of two types, eukaryotic, which contain a nucleus, and prokaryotic, which do not. Prokaryotes are single-celled organisms, while eukaryotes can be either single-celled or multicellular.

Prokaryotic Cells

Prokaryotic cells were the first form of life on Earth, characterised by having vital biological processesincluding cell signaling and being self-sustaining. They are simpler and smaller than eukaryotic cells, and lack membrane-bound organelles such as the nucleus. Prokaryotes include two of the domains of life, bacteria and archaea. The DNA of a prokaryotic cell consists of a single chromosome that is in direct contact with the cytoplasm. The nuclear region in the cytoplasm is called the nucleoid. Most prokaryotesare the smallest of all organisms ranging from 0.5 to 2.0 μm in diameter.

A prokaryotic cell has three architectural regions:

- Enclosing the cell is the cell envelope – generally consisting of a plasma membrane covered by a cell wall which, for some bacteria, may be further covered by a third layer called a capsule. Though most prokaryotes have both a cell membrane and a cell wall, there are exceptions such as *Mycoplasma*(bacteria) and *Thermoplasma* (archaea) which only possess the cell membrane layer. The envelope gives rigidity to the cell and separates the interior of the cell from its environment, serving as a protective filter. The cell wall consists of peptidoglycan in bacteria, and acts as an additional barrier against exterior forces. It also prevents the cell from expanding and bursting (cytolysis) from osmotic pressure due to a hypotonic environment. Some eukaryotic cells (plant cells and fungal cells) also have a cell wall.

- Inside the cell is the cytoplasmic region that contains the genome (DNA), ribosomes and various sorts of inclusions. The genetic material is freely found in the cytoplasm. Prokaryotes can carry extrachromosomal DNA elements called plasmids, which are usually circular. Linear bacterial plasmids have been identified in several species of spirochete bacteria, including members of the genus Borrelia notably *Borrelia burgdorferi*, which causes Lyme disease.Though not forming a *nucleus*, the DNA is condensed in a *nucleoid*. Plasmids encode additional genes, such as antibiotic resistance genes.
- On the outside, flagella and pili project from the cell's surface. These are structures (not present in all prokaryotes) made of proteins that facilitate movement and communication between cells.

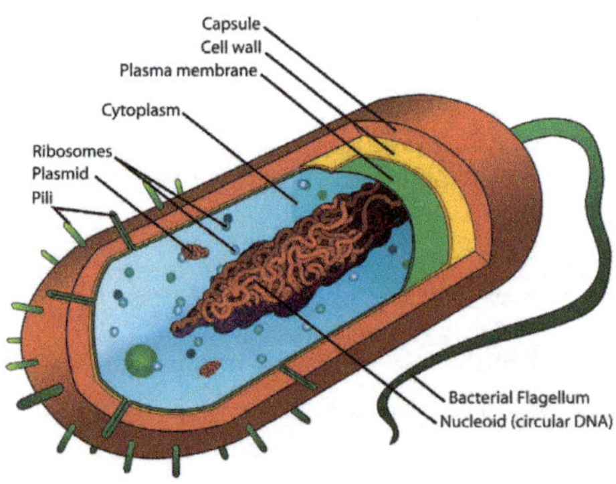

Structure of a typical prokaryotic cell

Eukaryotic Cells

Plants, animals, fungi, slime moulds, protozoa, and algae are all eukaryotic. These cells are about fifteen times wider than a typical prokaryote and can be as much as a thousand times greater in volume. The main distinguishing feature of eukaryotes as compared to prokaryotes is compartmentalization: the presence of membrane-bound organelles(compartments) in which specific metabolic activities take place. Most important among these is a cell nucleus, an organelle that houses the cell's DNA. This nucleus gives the eukaryote its name, which means "true kernel (nucleus)". Other differences include:

- The plasma membrane resembles that of prokaryotes in function, with minor differences in the setup. Cell walls may or may not be present.
- The eukaryotic DNA is organized in one or more linear molecules, called chromosomes, which are associated with histone proteins. All chromosomal DNA is stored in the *cell nucleus*, separated from the cytoplasm by a membrane. Some eukaryotic organelles such as mitochondria also contain some DNA.

- Many eukaryotic cells are ciliated with *primary cilia*. Primary cilia play important roles in chemosensation, mechanosensation, and thermosensation. Cilia may thus be "viewed as a sensory cellular antennae that coordinates a large number of cellular signaling pathways, sometimes coupling the signaling to ciliary motility or alternatively to cell division and differentiation."
- Motile cells of eukaryotes can move using *motile cilia* or *flagella*. Motile cells are absent in conifers and flowering plants. Eukaryotic flagella are less complex than those of prokaryotes.

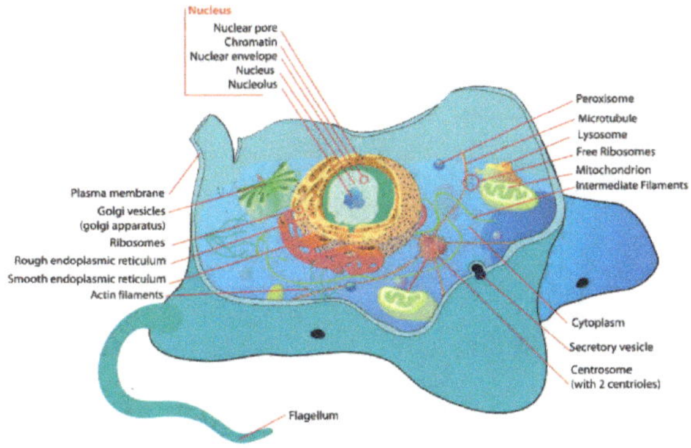

Structure of a typical animal cell

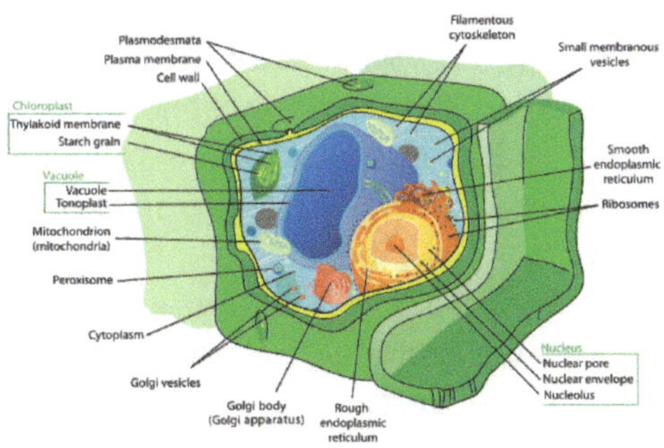

Structure of a typical plant cell

SUBCELLULAR COMPONENTS

All cells, whether prokaryotic or eukaryotic, have a membrane that envelops the cell, regulates what moves in and out (selectively permeable), and maintains the electric potential of the cell. Inside the membrane, the cytoplasm takes up most of the cell's volume. All cells (except red blood cells which lack a cell nucleus and most organelles to accommodate maximum space for hemoglobin) possess DNA, the hereditary material of genes, and RNA, containing the information necessary to build various proteins such as enzymes, the cell's primary machinery. There are also other kinds of biomolecules in cells. This article lists these primary components of the cell, then briefly describes their function.

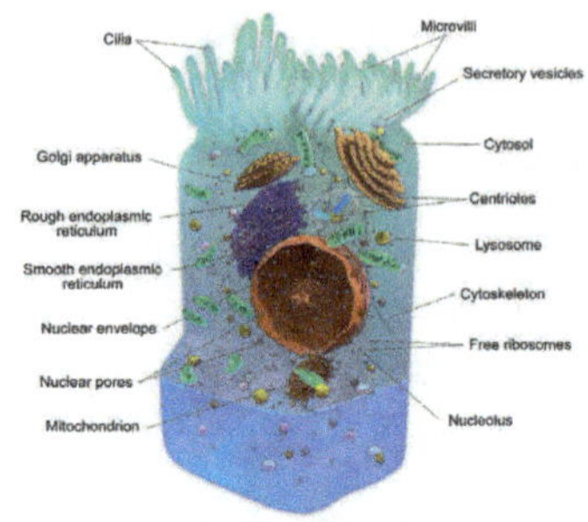

Anatomy of a Cell

Illustration depicting major structures inside a eukaryotic animal cell

Membrane

The cell membrane, or plasma membrane, is a biological membrane that surrounds the cytoplasm of a cell. In animals, the plasma membrane is the outer boundary of the cell, while in plants and prokaryotes it is usually covered by a cell wall. This membrane serves to separate and protect a cell from its surrounding environment and is made mostly from a double layer of phospholipids, which are amphiphilic (partly hydrophobic and partly hydrophilic). Hence, the layer is called a phospholipid bilayer, or sometimes a fluid mosaic membrane. Embedded within this membrane is a variety of protein molecules that act as channels and pumps that move different molecules into and out of the cell. The membrane is said to be 'semi-permeable', in that it can either let a substance (molecule or ion) pass through freely, pass through to a limited extent or not pass through at all. Cell surface membranes also contain receptor proteins that allow cells to detect external signaling molecules such as hormones.

Cytoskeleton

The cytoskeleton acts to organize and maintain the cell's shape; anchors organelles in place; helps during endocytosis, the uptake of external materials by a cell, and cytokinesis, the separation of daughter cells after cell division; and moves parts of the cell in processes of growth and mobility. The eukaryotic cytoskeleton is composed of microfilaments, intermediate filaments and microtubules. There are a great number of proteins associated with them, each controlling a cell's structure by directing, bundling, and aligning filaments. The prokaryotic cytoskeleton is less well-studied but is involved in the maintenance of cell shape, polarity and cytokinesis. The subunit protein of microfilaments is a small, monomeric protein called actin. The subunit of microtubules is a dimeric molecule called tubulin. Intermediate filaments are heteropolymers whose subunits vary among the cell types in different tissues. But some of the subunit protein of intermediate filaments include vimentin, desmin, lamin (lamins A, B and C), keratin(multiple acidic and basic keratins), neurofilament proteins (NF - L, NF - M).

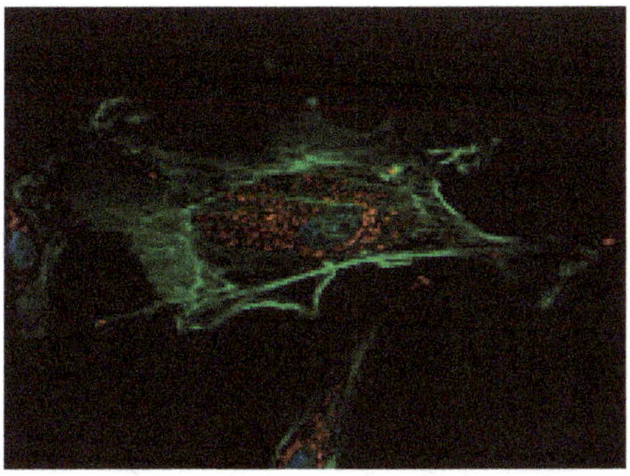

A fluorescent image of an endothelial cell. Nuclei are stained blue, mitochondria are stained red, and microfilaments are stained green.

Genetic Material

Two different kinds of genetic material exist: deoxyribonucleic acid (DNA) and ribonucleic acid (RNA). Cells use DNA for their long-term information storage. The biological information contained in an organism is encoded in its DNA sequence. RNA is used for information transport (e.g., mRNA) and enzymatic functions (e.g., ribosomal RNA). Transfer RNA (tRNA) molecules are used to add amino acids during protein translation.

Prokaryotic genetic material is organized in a simple circular DNA molecule (the bacterial chromosome) in the nucleoid region of the cytoplasm. Eukaryotic genetic material is divided into different, linear molecules called chromosomes inside a discrete nucleus, usually with additional genetic material in some organelles like mitochondria and chloroplasts (see endosymbiotic theory).

A human cell has genetic material contained in the cell nucleus (the nuclear genome) and in the mitochondria (the mitochondrial genome). In humans the nuclear genome is divided into 46 linear DNA molecules called chromosomes, including 22 homologous chromosome pairs and a pair of sex chromosomes. The mitochondrial genome is a circular DNA molecule distinct from the nuclear DNA. Although the mitochondrial DNA is very small compared to nuclear chromosomes, it codes for 13 proteins involved in mitochondrial energy production and specific tRNAs.

Foreign genetic material (most commonly DNA) can also be artificially introduced into the cell by a process called transfection. This can be transient, if the DNA is not inserted into the cell's genome, or stable, if it is. Certain viruses also insert their genetic material into the genome.

Organelles

Organelles are parts of the cell which are adapted and/or specialized for carrying out one or more vital functions, analogous to the organs of the human body (such as the heart, lung, and kidney, with each organ performing a different function). Both eukaryotic and prokaryotic cells have organelles, but prokaryotic organelles are generally simpler and are not membrane-bound.

There are several types of organelles in a cell. Some (such as the nucleus and golgi apparatus) are typically solitary, while others (such as mitochondria, chloroplasts, peroxisomes and lysosomes) can be numerous (hundreds to thousands). The cytosol is the gelatinous fluid that fills the cell and surrounds the organelles.

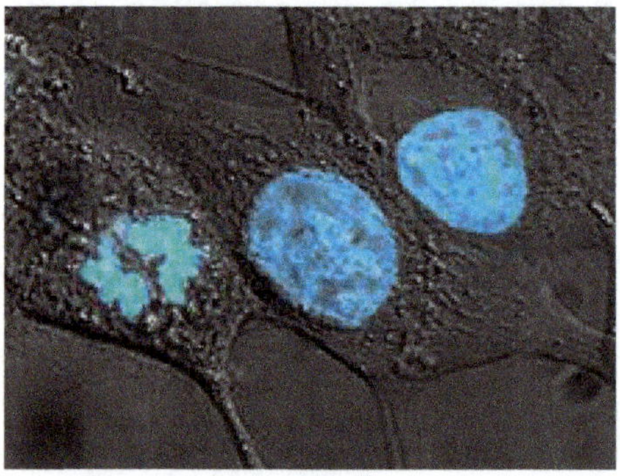

Human cancer cells with nuclei (specifically the DNA) stained blue. The central and rightmost cell are in interphase, so the entire nuclei are labeled. The cell on the left is going through mitosis and its DNA has condensed.

Eukaryotic

- **Cell nucleus**: A cell's information center, the cell nucleus is the most conspicuous organelle found in a eukaryoticcell. It houses the cell's chromosomes, and is the place where almost all DNA replication and RNA synthesis (transcription) occur. The nucleus is spherical and separated from the cytoplasm by a double membrane called the nuclear envelope. The nuclear envelope isolates and protects a cell's DNA from various molecules that could accidentally damage its structure or interfere with its processing. During processing, DNA is transcribed, or copied into a special RNA, called messenger RNA (mRNA). This mRNA is then transported out of the nucleus, where it is translated into a specific protein molecule. The nucleolus is a specialized region within the nucleus where ribosome subunits are assembled. In prokaryotes, DNA processing takes place in the cytoplasm.
- **Mitochondria and Chloroplasts**: generate energy for the cell. Mitochondria are self-replicating organelles that occur in various numbers, shapes, and sizes in the cytoplasm of all eukaryotic cells. Respiration occurs in the cell mitochondria,

which generate the cell's energy by oxidative phosphorylation, using oxygen to release energy stored in cellular nutrients (typically pertaining to glucose) to generate ATP. Mitochondria multiply by binary fission, like prokaryotes. Chloroplasts can only be found in plants and algae, and they capture the sun's energy to make carbohydrates through photosynthesis.

- **Endoplasmic reticulum**: The endoplasmic reticulum (ER) is a transport network for molecules targeted for certain modifications and specific destinations, as compared to molecules that float freely in the cytoplasm. The ER has two forms: the rough ER, which has ribosomes on its surface that secrete proteins into the ER, and the smooth ER, which lacks ribosomes. The smooth ER plays a role in calcium sequestration and release.
- **Golgi apparatus**: The primary function of the Golgi apparatus is to process and package the macromolecules such as proteins and lipids that are synthesized by the cell.
- **Lysosomes and Peroxisomes**: Lysosomes contain digestive enzymes (acid hydrolases). They digest excess or worn-out organelles, food particles, and engulfed viruses or bacteria. Peroxisomes have enzymes that rid the cell of toxic peroxides. The cell could not house these destructive enzymes if they were not contained in a membrane-bound system.
- **Centrosome**: the cytoskeleton organiser: The centrosome produces the microtubules of a cell – a key component of the cytoskeleton. It directs the transport through the ER and the Golgi apparatus. Centrosomes are composed of two centrioles, which separate during cell division and help in the formation of the mitotic spindle. A single centrosome is present in the animal cells. They are also found in some fungi and algae cells.
- **Vacuoles**: Vacuoles sequester waste products and in plant cells store water. They are often described as liquid filled space and are surrounded by a membrane. Some cells, most notably *Amoeba*, have contractile vacuoles, which can pump water out of the cell if there is too much water. The vacuoles of plant cells and fungal cells are usually larger than those of animal cells.

Diagram of an endomembrane system

Eukaryotic and prokaryotic

Ribosomes: The ribosome is a large complex of RNA and protein molecules. They each consist of two subunits, and act as an assembly line where RNA from the nucleus is used to synthesise proteins from amino acids. Ribosomes can be found either floating freely or bound to a membrane (the rough endoplasmatic reticulum in eukaryotes, or the cell membrane in prokaryotes).

STRUCTURES OUTSIDE THE CELL MEMBRANE

Many cells also have structures which exist wholly or partially outside the cell membrane. These structures are notable because they are not protected from the external environment by the semipermeable cell membrane. In order to assemble these structures, their components must be carried across the cell membrane by export processes.

Cell wall

Many types of prokaryotic and eukaryotic cells have a cell wall. The cell wall acts to protect the cell mechanically and chemically from its environment, and is an additional layer of protection to the cell membrane. Different types of cell have cell walls made up of different materials; plant cell walls are primarily made up of cellulose, fungi cell walls are made up of chitin and bacteria cell walls are made up of peptidoglycan.

Prokaryotic

Capsule

A gelatinous capsule is present in some bacteria outside the cell membrane and cell wall. The capsule may be polysaccharide as in pneumococci, meningococci or polypeptide as *Bacillus anthracis* or hyaluronic acidas in streptococci. Capsules are not marked by normal staining protocols and can be detected by India inkor methyl blue; which allows for higher contrast between the cells for observation.[87]

Flagella

Flagella are organelles for cellular mobility. The bacterial flagellum stretches from cytoplasm through the cell membrane(s) and extrudes through the cell wall. They are long and thick thread-like appendages, protein in nature. A different type of flagellum is found in archaea and a different type is found in eukaryotes.

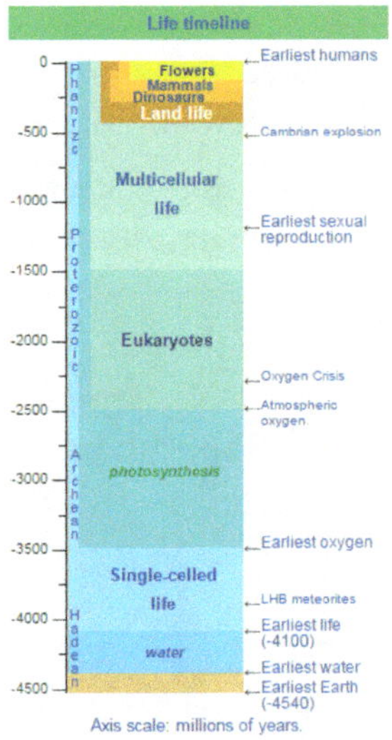

Fimbria

A fimbria also known as a pilus is a short, thin, hair-like filament found on the surface of bacteria. Fimbriae, or pili are formed of a protein called pilin (antigenic) and are responsible for attachment of bacteria to specific receptors of human cell (cell adhesion). There are special types of specific pili involved in bacterial conjugation.

CELLULAR PROCESSES

Growth and metabolism

Between successive cell divisions, cells grow through the functioning of cellular metabolism. Cell metabolism is the process by which individual cells process nutrient molecules. Metabolism has two distinct divisions: catabolism, in which the cell breaks down complex molecules to produce energy and reducing power, and anabolism, in which the cell uses energy and reducing power to construct complex molecules and perform other biological functions. Complex sugars consumed by the organism can be broken down into simpler sugar molecules called monosaccharides such as glucose. Once inside the cell, glucose is broken down to make adenosine triphosphate (ATP), a molecule that possesses readily available energy, through two different pathways.

Replication

Cell division involves a single cell (called a *mother cell*) dividing into two daughter cells. This leads to growth in multicellular organisms(the growth of tissue) and to procreation (vegetative reproduction) in unicellular organisms. Prokaryotic cells divide by binary fission, while eukaryotic cells usually undergo a process of nuclear division, called mitosis, followed by division of the cell, called cytokinesis. A diploidcell may also undergo meiosis to produce haploid cells, usually four. Haploid cells serve as gametes in multicellular organisms, fusing to form new diploid cells.

DNA replication, or the process of duplicating a cell's genome, always happens when a cell divides through mitosis or binary fission. This occurs during the S phase of the cell cycle.

In meiosis, the DNA is replicated only once, while the cell divides twice. DNA replication only occurs before meiosis I. DNA replication does not occur when the cells divide the second time, in meiosis II. Replication, like all cellular activities, requires specialized proteins for carrying out the job.

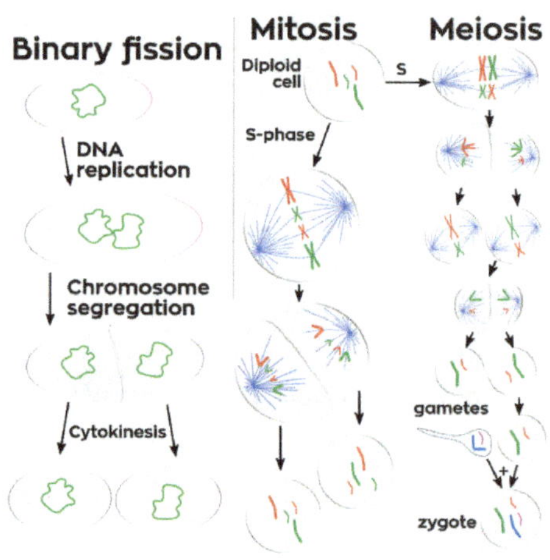

Bacteria divide by binary fission, while eukaryotes divide by mitosis or meiosis.

Protein Synthesis

Cells are capable of synthesizing new proteins, which are essential for the modulation and maintenance of cellular activities. This process involves the formation of new protein molecules from amino acid building blocks based on information encoded in DNA/RNA. Protein synthesis generally consists of two major steps: transcription and translation.

Transcription is the process where genetic information in DNA is used to produce a complementary RNA strand. This RNA strand is then processed to give messenger RNA (mRNA), which is free to migrate through the cell. mRNA molecules bind to protein-RNA complexes called ribosomes located in the cytosol, where they are translated into polypeptide sequences. The ribosome mediates the formation of a polypeptide sequence based on the mRNA sequence. The mRNA sequence directly relates to the polypeptide sequence by binding to transfer RNA (tRNA) adapter molecules in binding pockets within the ribosome. The new polypeptide then folds into a functional three-dimensional protein molecule.

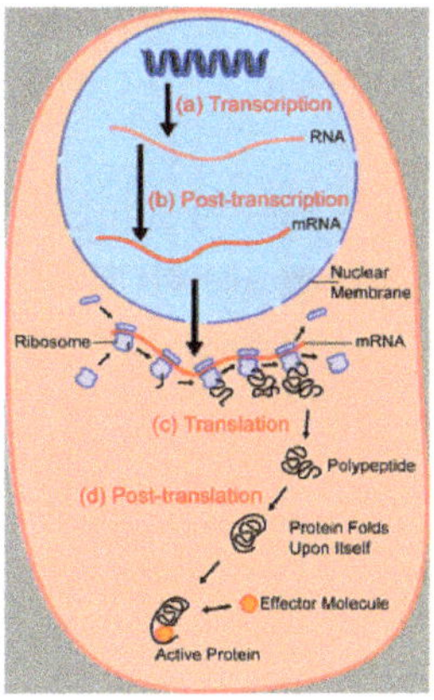

An overview of protein synthesis. Within the nucleus of the cell (*light blue*), genes (DNA, *dark blue*) are transcribed into RNA. This RNA is then subject to post-transcriptional modification and control, resulting in a mature mRNA (*red*) that is then transported out of the nucleus and into the cytoplasm (*peach*), where it undergoes translation into a protein. mRNA is translated by ribosomes(*purple*) that match the three-base codons of the mRNA to the three-base anti-codons of the appropriate tRNA. Newly synthesized proteins (*black*) are often further modified, such as by binding to an effector molecule (*orange*), to become fully active.

Movement or Motility

Unicellular organisms can move in order to find food or escape predators. Common mechanisms of motion include flagella and cilia.

In multicellular organisms, cells can move during processes such as wound healing, the immune response and cancer metastasis. For example, in wound healing in animals, white blood cells move to the wound site to kill the microorganisms that cause infection. Cell motility involves many receptors, crosslinking, bundling, binding, adhesion, motor and other proteins. The process is divided into three steps – protrusion of the leading edge of the cell, adhesion of the leading edge and de-adhesion at the cell body and rear, and cytoskeletal contraction to pull the cell forward. Each step is driven by physical forces generated by unique segments of the cytoskeleton.

MULTICELLULARITY

Cell specialization

Multicellular organisms are organisms that consist of more than one cell, in contrast to single-celled organisms.

In complex multicellular organisms, cells specialize into different cell types that are adapted to particular functions. In mammals, major cell types include skin cells, muscle cells, neurons, blood cells, fibroblasts, stem cells, and others. Cell types differ both in appearance and function, yet are genetically identical. Cells are able to be of the same genotype but of different cell type due to the differential expression of the genes they contain.

Most distinct cell types arise from a single totipotent cell, called a zygote, that differentiates into hundreds of different cell types during the course of development. Differentiation of cells is driven by different environmental cues (such as cell–cell interaction) and intrinsic differences (such as those caused by the uneven distribution of molecules during division).

Origin of Multicellularity

Multicellularity has evolved independently at least 25 times, including in some prokaryotes, like cyanobacteria, myxobacteria, actinomycetes, *Magnetoglobus multicellularis* or *Methanosarcina*. However, complex multicellular organisms evolved only in six eukaryotic groups: animals, fungi, brown algae, red algae, green algae, and plants. It evolved repeatedly for plants (Chloroplastida), once or twice for animals, once for brown algae, and perhaps several times for fungi, slime molds, and red algae. Multicellularity may have evolved from colonies of interdependent organisms, from cellularization, or from organisms in symbiotic relationships.

The first evidence of multicellularity is from cyanobacteria-like organisms that lived between 3 and 3.5 billion years ago. Other early fossils of multicellular organisms include the contested Grypania spiralis and the fossils of the black shales of the PalaeoproterozoicFrancevillian Group Fossil B Formation in Gabon.

The evolution of multicellularity from unicellular ancestors has been replicated in the laboratory, in evolution experiments using predation as the selective pressure.

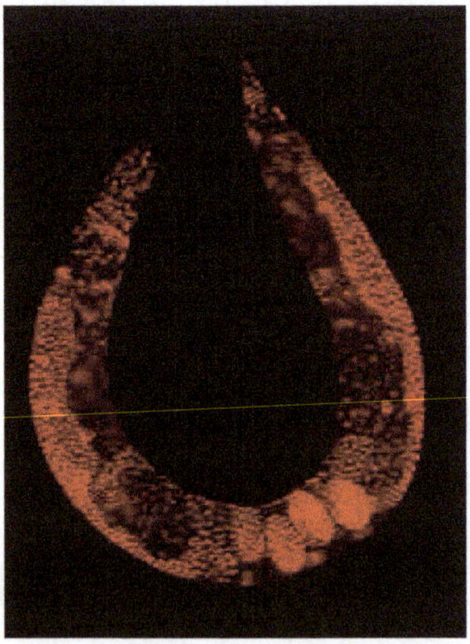

Staining of a *Caenorhabditis elegans* which highlights the nuclei of its cells.

ORIGINS

The origin of cells has to do with the origin of life, which began the history of life on Earth.

Origin of the First Cell

There are several theories about the origin of small molecules that led to life on the early Earth. They may have been carried to Earth on meteorites, created at deep-sea vents, or synthesized by lightning in a reducing atmosphere. There is little experimental data defining what the first self-replicating forms were. RNA is thought to be the earliest self-replicating molecule, as it is capable of both storing genetic information and catalyzing chemical reactions, but some other entity with the potential to self-replicate could have preceded RNA, such as clay or peptide nucleic acid.

Cells emerged at least 3.5 billion years ago. The current belief is that these cells were heterotrophs. The early cell membranes were probably more simple and permeable than modern ones, with only a single fatty acid chain per lipid. Lipids are known to spontaneously form bilayered vesicles in water, and could have preceded RNA, but the first cell membranes could also have been produced by catalytic RNA, or even have required structural proteins before they could form.

Stromatolites are left behind by cyanobacteria, also called blue-green algae. They are the oldest known fossils of life on Earth. This one-billion-year-old fossil is from Glacier National Park in the United States.

Origin of Eukaryotic Cells

The eukaryotic cell seems to have evolved from a symbiotic community of prokaryotic cells. DNA-bearing organelles like the mitochondria and the chloroplasts are descended from ancient symbiotic oxygen-breathing proteobacteria and cyanobacteria, respectively, which were endosymbiosed by an ancestral archaean prokaryote.

There is still considerable debate about whether organelles like the hydrogenosome predated the origin of mitochondria, or vice versa: see the hydrogen hypothesis for the origin of eukaryotic cells.

HISTORY OF RESEARCH

- 1632–1723: Antonie van Leeuwenhoek teaches himself to make lenses, constructs basic optical microscopes and draws protozoa, such as *Vorticella* from rain water, and bacteria from his own mouth.
- 1665: Robert Hooke discovers cells in cork, then in living plant tissue using an early compound microscope. He coins the term *cell* (from Latin *cella*, meaning "small room") in his book *Micrographia* (1665).
- 1839: Theodor Schwann and Matthias Jakob Schleiden elucidate the principle that plants and animals are made of cells, concluding that cells are a common unit of structure and development, and thus founding the cell theory.
- 1855: Rudolf Virchow states that new cells come from pre-existing cells by cell division (*omnis cellula ex cellula*).
- 1859: The belief that life forms can occur spontaneously (*generatio spontanea*) is contradicted by Louis Pasteur (1822–1895) (although Francesco Redi had performed an experiment in 1668 that suggested the same conclusion).
- 1931: Ernst Ruska builds the first transmission electron microscope (TEM) at the University of Berlin. By 1935, he has built an EM with twice the resolution of a light microscope, revealing previously unresolvable organelles.
- 1953: Watson and Crick made their first announcement on the double helix structure of DNA on February 28.
- 1981: Lynn Margulis published *Symbiosis in Cell Evolution* detailing the endosymbiotic theory.

REFERENCES

1. *"Cell". Online Etymology Dictionary. Retrieved 31 December 2012.*
2. Cell Movements and the Shaping of the Vertebrate Body in Chapter 21 of *Molecular Biology of the Cell* fourth edition, edited by Bruce Alberts (2002) published by Garland Science. The Alberts text discusses how the "cellular building blocks" move to shape developing embryos. It is also common to describe small molecules such as amino acids as "molecular building blocks".
3. Alberts, p. 2.
4. *Campbell, Neil A.; Brad Williamson; Robin J. Heyden (2006). Biology: Exploring Life. Boston, Massachusetts: Pearson Prentice Hall. ISBN 0-13-250882-6.*
5. *Karp, Gerald (19 October 2009). Cell and Molecular Biology: Concepts and Experiments. John Wiley & Sons. p. 2. ISBN 9780470483374. Hooke called the pores cells because they re- minded him of the cells inhabited by monks living in a monastery.*
6. *Alan Chong Tero (1990). Achiever's Biology. Allied Publishers. p. 36. ISBN 9788184243697. In 1665, an Englishman, Robert Hooke observed a thin slice of" cork under a simple microscope. (A simple microscope is a*

microscope with only one biconvex lens, rather like a magnifying glass). He saw many small box like structures. These reminded him of small rooms called "cells" in which Christian monks lived and meditated.

7. *Maton, Anthea (1997). Cells Building Blocks of Life. New Jersey: Prentice Hall. ISBN 0-13-423476-6.*
8. Schopf, JW, Kudryavtsev, AB, Czaja, AD, and Tripathi, AB. (2007). *Evidence of Archean life: Stromatolites and microfossils.* Precambrian Research 158:141-155.
9. Schopf, JW (2006). *Fossil evidence of Archaean life.* Philos Trans R Soc Lond B Biol Sci 29;361(1470):869-85.
10. *Peter Hamilton Raven; George Brooks Johnson (2002). Biology. McGraw-Hill Education. p. 68. ISBN 978-0-07-112261-0. Retrieved 7 July 2013.*
11. *Campbell Biology—Concepts and Connections. Pearson Education. 2009. p. 320.*
12. *Microbiology : Principles and Explorations* By Jacquelyn G. Black
13. European Bioinformatics Institute, Karyn's Genomes: Borrelia burgdorferi, part of 2can on the EBI-EMBL database. Retrieved 5 August 2012
14. *Satir, Peter; Christensen, ST; Søren T. Christensen (2008-03-26). "Structure and function of mammalian cilia". Histochemistry and Cell Biology. Springer Berlin/Heidelberg.* ***129*** *(6): 687–693. doi:10.1007/s00418-008-0416-9. PMC 2386530. PMID 18365235. 1432-119X. Retrieved 2009-09-12.*
15. PH Raven , Evert RF, Eichhorm SE (1999) Biology of Plants, 6th edition. WH Freeman, New York
16. *Michie K, Löwe J (2006). "Dynamic filaments of the bacterial cytoskeleton". Annu Rev Biochem.* ***75****: 467–92. doi:10.1146/annurev.biochem.75.103004.142452. PMID 16756499.*
17. *Ménétret JF, Schaletzky J, Clemons WM, et al. (December 2007). "Ribosome binding of a single copy of the SecY complex: implications for protein translocation". Mol. Cell.* ***28****(6): 1083–92. doi:10.1016/j.molcel.2007.10.034. PMID 18158904.*
18. *Prokaryotes. Newnes. Apr 11, 1996. ISBN 9780080984735.*
19. *Campbell Biology—Concepts and Connections. Pearson Education. 2009. p. 138.*
20. *Revathi Ananthakrishnan; Allen Ehrlicher. "The Forces Behind Cell Movement". Biolsci.org. Retrieved 2009-04-17.*
21. Alberts B, Johnson A, Lewis J. et al. Molecular Biology of the Cell, 4e. Garland Science. 2002
22. *Ananthakrishnan, R; Ehrlicher, A (2007). "The Forces Behind Cell Movement". Int J Biol Sci.* ***3*** *(5): 303–317. doi:10.7150/ijbs.3.303. PMC 1893118. PMID 17589565.*
23. *Becker, Wayne M.; et al. (2009). The world of the cell. Pearson Benjamin Cummings. p. 480. ISBN 978-0-321-55418-5.*
24. [c] Grosberg RK, Strathmann RR. The evolution of multicellularity: A minor major transition? Annu Rev Ecol Evol Syst. 2007;38:621–654.
25. http://public.wsu.edu/~lange-m/Documnets/Teaching2011/Popper2011.pdf

26. *Bonner, John Tyler (1998). "The Origins of Multicellularity" (PDF). Integrative Biology: Issues, News, and Reviews.* ***1*** *(1): 27–36. doi:10.1002/(SICI)1520-6602(1998)1:1<27::AID-INBI4>3.0.CO;2-6. ISSN 1093-4391. Archived from the original (PDF, 0.2 MB) on March 8, 2012.*

27. *El Albani A, Bengtson S, Canfield DE, Bekker A, Macchiarelli R, Mazurier A, Hammarlund EU, Boulvais P, Dupuy JJ, Fontaine C, Fürsich FT, Gauthier-Lafaye F, Janvier P, Javaux E, Ossa FO, Pierson-Wickmann AC, Riboulleau A, Sardini P, Vachard D, Whitehouse M, Meunier A (1 July 2010). "Large colonial organisms with coordinated growth in oxygenated environments 2.1 Gyr ago". Nature.* ***466*** *(7302): 100–104. Bibcode:2010Natur.466..100A. doi:10.1038/nature09166. ISSN 0028-0836. PMID 20596019.*

28. *Orgel LE (1998). "The origin of life--a review of facts and speculations". Trends Biochem Sci.* ***23*** *(12): 491–5. doi:10.1016/S0968-0004(98)01300-0. PMID 9868373.*

29. *Griffiths G (December 2007). "Cell evolution and the problem of membrane topology". Nature reviews. Molecular cell biology.* ***8*** *(12): 1018–24. doi:10.1038/nrm2287. PMID 17971839.*

30. "*... I could exceedingly plainly perceive it to be all perforated and porous, much like a Honey-comb, but that the pores of it were not regular [..] these pores, or cells, [..] were indeed the first microscopical pores I ever saw, and perhaps, that were ever seen, for I had not met with any Writer or Person, that had made any mention of them before this. . .*" – Hooke describing his observations on a thin slice of cork. Robert Hooke

CHAPTER 4

Allele and Selective Breeding

Chapter Contents

- **Introduction**
- **Dominant and Recessive Alleles**
- **Multiple Alleles**
- **Allele and Genotype Frequencies**
- **Allelic Dominance in Genetic Disorders**
- **Epialleles**
- **Animal Breeding**
- **Plant Breeding**
- **Selective Breeding in Aquaculture**
- **Advantages and Disadvantages**

Selective breeding is evolution by human selection. As nineteenth-century British naturalist Charles Darwin noted in *Variation of Animals and Plants under Domestication,* selective breeding may be methodical or unconscious. Methodical selection is oriented toward a predetermined standard, whereas unconscious selection is the result of biases in the preservation of valuable individuals. Methodical selection requires great care in discriminating among organisms and is capable of rapid change in specific traits, such as milk production or silk color.

INTRODUCTION

An allele is the variant form of a given gene. Sometimes, different alleles can result in different observable phenotypic traits, such as different pigmentation. However, most genetic variations result in little or no observable variation.

The word "allele" is a short form of allelomorph ("other form", a word coined by William Bateson), which was used in the early days of genetics to describe variant forms of a gene detected as different phenotypes. It derives from the Greek prefix ἀλλήλ, *allel*, meaning "reciprocal" or "each other", which itself is related to the Greek adjective ἄ λλος (allos; cognate with Latin "alius"), meaning "other".

Most multicellular organisms have two sets of chromosomes; that is, they are diploid. These chromosomes are referred to as homologous chromosomes. If both alleles at a gene (or locus) on the homologous chromosomes are the same, they and the organism are homozygous with respect to that gene (or locus). If the alleles are different, they and the organism are heterozygous with respect to that gene.

Selective breeding (also called artificial selection) is the process by which humans use animal breeding and plant breeding to selectively develop particular phenotypic traits (characteristics) by choosing which typically animal or plant males and females will sexually reproduce and have offspring together. Domesticated animals are known as breeds, normally bred by a professional breeder, while domesticated plants are known as varieties, cultigens, or cultivars. Two purebred animals of different breeds produce a crossbreed, and crossbred plants are called hybrids. Flowers, vegetables and fruit-trees may be bred by amateurs and commercial or non-commercial professionals: major crops are usually the provenance of the professionals.

There are two approaches or types of artificial selection, or selective breeding. First is the traditional "breeder's approach" in which the breeder or experimenter applies "a known amount of selection to a single phenotypic trait" by examining the chosen trait and choosing to breed only those that exhibit higher or "extreme values" of that trait. The second is called "controlled natural selection," which is essentially natural selection in a controlled environment. In this, the breeder does not choose which individuals being tested "survive or reproduce," as he or she could in the traditional approach. There are also "selection experiments," which is a third approach and these are conducted in order to determine the "strength of natural selection in the wild." However, this is more often an observational approach as opposed to an experimental approach.

In animal breeding, techniques such as inbreeding, linebreeding, and outcrossing are utilized. In plant breeding, similar methods are used. Charles Darwin discussed how selective breeding had been successful in producing change over time in his 1859 book, *On the Origin of Species.* Its first chapter discusses selective breeding and domestication of such animals as pigeons, cats, cattle, and dogs. Darwin used artificial selection as a springboard to introduce and support the theory of natural selection.

The deliberate exploitation of selective breeding to produce desired results has become very common in agriculture and experimental biology.

Selective breeding can be unintentional, e.g., resulting from the process of human cultivation; and it may also produce unintended – desirable or undesirable – results. For example, in some grains, an increase in seed size may have resulted from certain ploughing practices rather than from the intentional selection of larger seeds. Most likely, there has been an interdependence between natural and artificial factors that have resulted in plant domestication.

A Belgian Blue cow. The defect in the breed's myostatin gene is maintained through linebreeding and is responsible for its accelerated lean muscle growth.

This Chihuahua mix and Great Dane shows the wide range of dog breed sizes created using selective breeding.

Selective breeding of both plants and animals has been practiced since early prehistory; key species such as wheat, rice, and dogs have been significantly different from their wild ancestors for millennia, and maize, which required especially large changes from teosinte, its wild form, was selectively bred in Mesoamerica. Selective breeding was practiced by the Romans. Treatises as much as 2,000 years old give advice on selecting animals for different purposes, and these ancient works cite still older authorities, such as Mago the Carthaginian. The notion of selective breeding was later expressed by the Persian Muslim polymath Abu Rayhan Biruni in the 11th century. He noted the idea in his book titled *India*, which included various examples.

The agriculturist selects his corn, letting grow as much as he requires, and tearing out the remainder. The forester leaves those branches which he perceives to be excellent, whilst he cuts away all others. The bees kill those of their kind who only eat, but do not work in their beehive.

—Abu Rayhan Biruni, India

Selective breeding transformed teosinte's few fruitcases (left) into modern maize's rows of exposed kernels (right).

Selective breeding was established as a scientific practice by Robert Bakewell during the British Agricultural Revolutionin the 18th century. Arguably, his most important breeding program was with sheep. Using native stock, he was able to quickly select for large, yet fine-boned sheep, with long, lustrous wool. The Lincoln Longwool was improved by Bakewell, and in turn the Lincoln was used to develop the subsequent breed, named the New (or Dishley) Leicester. It was hornless and had a square, meaty body with straight top lines.

These sheep were exported widely, including to Australia and North America, and have contributed to numerous modern breeds, despite the fact that they fell quickly out of favor as market preferences in meat and textiles changed. Bloodlines of these original New Leicesters survive today as the English Leicester (or Leicester Longwool), which is primarily kept for wool production.

Bakewell was also the first to breed cattle to be used primarily for beef. Previously, cattle were first and foremost kept for pulling ploughs as oxen, but he crossed long-horned heifers and a Westmoreland bull to eventually create the Dishley Longhorn. As more and more farmers followed his lead, farm animals increased dramatically in size and quality. In 1700, the average weight of a bull sold for slaughter was 370 pounds (168 kg). By 1786, that weight had more than doubled to 840 pounds (381 kg). However, after his death, the Dishley Longhorn was replaced with short-horn versions.

He also bred the Improved Black Cart horse, which later became the Shire horse.

Charles Darwin coined the term 'selective breeding'; he was interested in the process as an illustration of his proposed wider process of natural selection. Darwin noted that many domesticated animals and plants had special properties that were developed by intentional animal and plant breeding from individuals that showed desirable characteristics, and discouraging the breeding of individuals with less desirable characteristics.

Darwin used the term "artificial selection" twice in the 1859 first edition of his work *On the Origin of Species*, in Chapter IV: Natural Selection, and in Chapter VI: Difficulties on Theory:

Slow though the process of selection may be, if feeble man can do much by his powers of artificial selection, I can see no limit to the amount of change, to the beauty and infinite complexity of the co-adaptations between all organic beings, one with another and with their physical conditions of life, which may be effected in the long course of time by nature's power of selection.

—Charles Darwin, On the Origin of Species

We are profoundly ignorant of the causes producing slight and unimportant variations; and we are immediately made conscious of this by reflecting on the differences in the breeds of our domesticated animals in different countries,—more especially in the less civilized countries where there has been but little artificial selection.

DOMINANT AND RECESSIVE ALLELES

In many cases, genotypic interactions between the two alleles at a locus can be described as dominant or recessive, according to which of the two homozygous phenotypes the heterozygote most resembles. Where the heterozygote is indistinguishable from one of the homozygotes, the allele involved is said to be dominant to the other, which is said to be recessive to the former. The degree and pattern of dominance varies among loci. This type of interaction was first formally described by Gregor Mendel. However, many traits defy this simple categorization and the phenotypes are modeled by co-dominance and polygenic inheritance.

The term "wild type" allele is sometimes used to describe an allele that is thought to contribute to the typical phenotypic character as seen in "wild" populations of organisms, such as fruit flies (*Drosophila melanogaster*). Such a "wild type" allele was historically regarded as dominant (overpowering - always expressed), common, and normal, in contrast to "mutant" alleles regarded as recessive, rare, and frequently deleterious. It was formerly thought that most individuals were homozygous for the "wild type" allele at most gene loci, and that any alternative "mutant" allele was found in homozygous form in a small minority of "affected" individuals, often as genetic diseases, and more frequently in heterozygous form in "carriers" for the mutant allele. It is now appreciated that most or all gene loci are highly polymorphic, with multiple alleles, whose frequencies vary from population to population, and that a great deal of genetic variation is hidden in the form of alleles that do not produce obvious phenotypic differences.

MULTIPLE ALLELES

A population or species of organisms typically includes multiple alleles at each locus among various individuals. Allelic variation at a locus is measurable as the number of alleles (polymorphism) present, or the proportion of heterozygotes in the population. A null allele is a gene variant that lacks the gene's normal function because it either is not expressed, or the expressed protein is inactive.

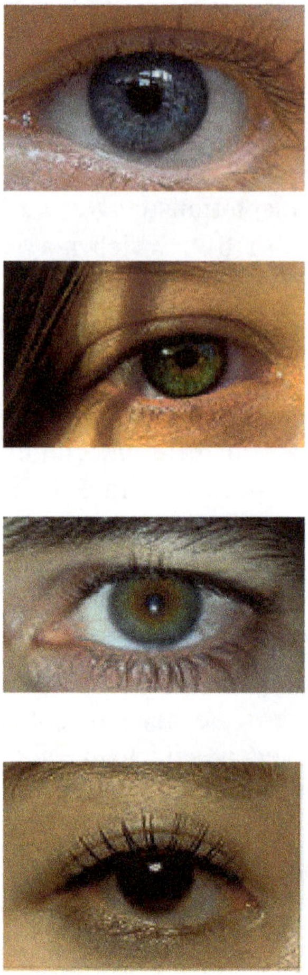

Eye color is an inherited trait influenced by more than one gene, including OCA2 and HERC2. The interaction of multiple genes—and the variation in these genes ("alleles") between individuals—help to determine a person's eye color phenotype. Eye color is influenced by pigmentation of the iris and the frequency-dependence of the light scattering by the turbid medium within the stroma of the iris.

For example, at the gene locus for the ABO blood typecarbohydrate antigens in humans, classical genetics recognizes three alleles, I^A, I^B, and i, that determine compatibility of blood transfusions. Any individual has one of six possible genotypes (I^AI^A, I^Ai, I^BI^B, I^Bi, I^AI^B, and ii) that produce one of four possible phenotypes: "Type A" (produced by I^AI^A homozygous and I^Ai heterozygous genotypes), "Type B" (produced by I^BI^B homozygous and I^Bi heterozygous genotypes), "Type AB" produced by I^AI^B heterozygous genotype, and "Type O" produced by ii homozygous genotype. (It is now known that each of the A, B, and O alleles is actually a class of multiple alleles with different DNA sequences that produce proteins with identical properties: more than 70 alleles are known at the ABO locus. Hence an individual with "Type A" blood may be an AO heterozygote, an AA homozygote, or an AA heterozygote with two different "A" alleles.)

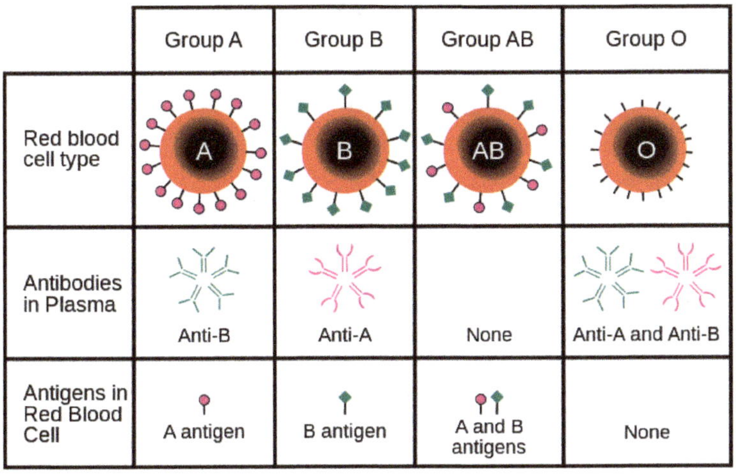

In the ABO blood group system, a person with Type A blood displays A-antigens and may have a genotype I^AI^A or I^Ai. A person with Type B blood displays B-antigens and may have the genotype I^BI^B or I^Bi. A person with Type AB blood displays both A- and B-antigens and has the genotype $I^AI^B$and a person with Type O blood, displaying neither antigen, has the genotype ii.

ALLELE AND GENOTYPE FREQUENCIES

The frequency of alleles in a diploid population can be used to predict the frequencies of the corresponding genotypes (see Hardy-Weinberg principle). For a simple model, with two alleles:

$$p+q=1$$

$$p^2+2pq+q^2=1$$

where p is the frequency of one allele and q is the frequency of the alternative allele, which necessarily sum to unity. Then, p^2 is the fraction of the population homozygous for the first allele, $2pq$ is the fraction of heterozygotes, and q^2 is the fraction homozygous for the alternative allele. If the first allele is dominant to the second then the fraction of the population that will show the dominant phenotype is $p^2 + 2pq$, and the fraction with the recessive phenotype is q^2.

With three alleles:

$$p+q+r=1 \text{ and}$$

$$p^2+q^2+r^2+2pq+2pr+2qr=1.$$

In the case of multiple alleles at a diploid locus, the number of possible genotypes (G) with a number of alleles (a) is given by the expression:

$$G=\frac{a(a+1)}{2}.$$

ALLELIC DOMINANCE IN GENETIC DISORDERS

A number of genetic disorders are caused when an individual inherits two recessive alleles for a single-gene trait. Recessive genetic disorders include albinism, cystic fibrosis, galactosemia, phenylketonuria (PKU), and Tay–Sachs disease. Other disorders are also due to recessive alleles, but because the gene locus is located on the X chromosome, so that males have only one copy (that is, they are hemizygous), they are more frequent in males than in females. Examples include red-green color blindness and fragile X syndrome.

Other disorders, such as Huntington disease, occur when an individual inherits only one dominant allele.

EPIALLELES

While heritable traits are typically studied in terms of genetic alleles, epigenetic marks such as DNA methylation can be inherited at specific genomic regions in certain species, a process termed transgenerational epigenetic inheritance. The term *epiallele* is used to distinguish these heritable marks from traditional alleles, which are defined by nucleotide sequence. A specific class of epiallele, the metastable epialleles, has been discovered in mice and in humans which is characterized by stochastic (probabilistic) establishment of epigenetic state that can be mitotically inherited.

ANIMAL BREEDING

Animals with homogeneous appearance, behavior, and other characteristics are known as particular breeds, and they are bred through culling animals with particular traits and selecting for further breeding those with other traits. Purebredanimals have a single, recognizable breed, and purebreds with recorded lineage are called pedigreed. Crossbreedsare a mix of two purebreds, whereas mixed breeds are a mix of several breeds, often unknown. Animal breeding begins with breeding stock, a group of animals used for the purpose of planned breeding. When individuals are looking to breed animals, they look for certain valuable traits in purebred stock for a certain purpose, or may intend to use some type of crossbreeding to produce a new type of stock with different, and, it is presumed, superior abilities in a given area of endeavor. For example, to breed chickens, a breeder typically intends to receive eggs, meat, and new, young birds for further reproduction. Thus, the breeder has to study different breeds and types of chickens and analyze what can be expected from a certain set of characteristics before he or she starts breeding them. Therefore, when purchasing initial breeding stock, the breeder seeks a group of birds that will most closely fit the purpose intended.

Three generations of "Westies" in a village in Fife, Scotland

Purebred breeding aims to establish and maintain stable traits, that animals will pass to the next generation. By "breeding the best to the best, employing a certain degree of inbreeding, considerable culling, and selection for "superior" qualities, one could develop a bloodline superior in certain respects to the original base stock. Such animals can be recorded with a breed registry, the organization that maintains pedigrees and/or stud books. However, single-trait breeding, breeding for only one trait over all others, can be problematic. In one case mentioned by animal behaviorist Temple Grandin, roosters bred for fast growth or heavy muscles did not know how to perform typical rooster courtship dances, which alienated the roosters from hens and led the roosters to kill the hens after reproducing with them.

The observable phenomenon of hybrid vigor stands in contrast to the notion of breed purity. However, on the other hand, indiscriminate breeding of crossbred or hybrid animals may also result in degradation of quality. Studies in evolutionary physiology, behavioral genetics, and other areas of organismal biology have also made use of deliberate selective breeding, though longer generation times and greater difficulty in breeding can make such projects challenging in vertebrates.

PLANT BREEDING

Plant breeding has been used for thousands of years, and began with the domestication of wild plants into uniform and predictable agricultural cultigens. High-yielding varieties have been particularly important in agriculture.

Selective plant breeding is also used in research to produce transgenic animals that breed "true" (i.e., are homozygous) for artificially inserted or deleted genes.

Researchers at the USDA have selectively bred carrots with a variety of colors.

SELECTIVE BREEDING IN AQUACULTURE

Selective breeding in aquaculture holds high potential for the genetic improvement of fish and shellfish. Unlike terrestrial livestock, the potential benefits of selective breeding in aquaculture were not realized until recently. This is because high mortality led to the selection of only a few broodstock, causing inbreeding depression, which then forced the use of wild broodstock. This was evident in selective breeding programs for growth rate, which resulted in slow growth and high mortality.

Control of the reproduction cycle was one of the main reasons as it is a requisite for selective breeding programmes. Artificial reproduction was not achieved because of the difficulties in hatching or feeding some farmed species such as eel and yellowtail farming. A suspected reason associated with the late realisation of success in selective breeding programs in aquaculture was the education of the concerned people – researchers, advisory personnel and fish farmers. The education of fish biologists paid less attention to quantitative genetics and breeding plans.

Another was the failure of documentation of the genetic gains in successive generations. This in turn led to failure in quantifying economic benefits that successful selective breeding programs produce. Documentation of the genetic changes was considered important as they help in fine tuning further selection schemes.

Quality Traits in Aquaculture

Aquaculture species are reared for particular traits such as growth rate, survival rate, meat quality, resistance to diseases, age at sexual maturation, fecundity, shell traits like shell size, shell colour, etc.

- **Growth rate** – growth rate is normally measured as either body weight or body length. This trait is of great economic importance for all aquaculture species as faster growth rate speeds up the turnover of production. Improved growth rates show that farmed animals utilize their feed more efficiently through a correlated response.
- **Survival rate** – survival rate may take into account the degrees of resistance to diseases. This may also see the stress response as fish under stress are highly vulnerable to diseases. The stress fish experience could be of biological, chemical or environmental influence.
- **Meat quality** – the quality of fish is of great economic importance in the market. Fish quality usually takes into account size, meatiness, and percentage of fat, colour of flesh, taste, shape of the body, ideal oil and omega-3 content.
- **Age at sexual maturation** – The age of maturity in aquaculture species is another very important attribute for farmers as during early maturation the species divert all their energy to gonad production affecting growth and meat production and are more susceptible to health problems (Gjerde 1986).
- **Fecundity** – As the fecundity in fish and shellfish is usually high it is not considered as a major trait for improvement. However, selective breeding practices may consider the size of the egg and correlate it with survival and early growth rate.

Finfish Response to Selection

Salmonids

Gjedrem (1979) showed that selection of Atlantic salmon (*Salmo salar*) led to an increase in body weight by 30% per generation. A comparative study on the performance of select Atlantic salmon with wild fish was conducted by AKVAFORSK Genetics Centre in Norway. The traits, for which the selection was done included growth rate, feed consumption, protein retention, energy retention, and feed conversion efficiency. Selected fish had a twice better growth rate, a 40% higher feed intake, and an increased protein and energy retention. This led to an overall 20% better Fed Conversion Efficiency as compared to the wild stock. Atlantic salmon have also been selected for resistance to bacterial and viral diseases. Selection was done to check resistance to Infectious Pancreatic Necrosis Virus (IPNV). The results showed 66.6% mortality for low-resistant species whereas the high-resistant species showed 29.3% mortality compared to wild species.

Rainbow trout (*S. gairdneri*) was reported to show large improvements in growth rate after 7–10 generations of selection. Kincaid et al. (1977) showed that growth gains by 30% could be achieved by selectively breeding rainbow trout for three generations. A 7% increase in growth was recorded per generation for rainbow trout by Kause et al. (2005).

In Japan, high resistance to IPNV in rainbow trout has been achieved by selectively breeding the stock. Resistant strains were found to have an average mortality of 4.3% whereas 96.1% mortality was observed in a highly sensitive strain.

Coho salmon (*Oncorhynchus kisutch*) increase in weight was found to be more than 60% after four generations of selective breeding. In Chile, Neira et al. (2006) conducted experiments on early spawning dates in coho salmon. After selectively breeding the fish for four generations, spawning dates were 13–15 days earlier.

Cyprinids

Selective breeding programs for the Common carp (*Cyprinus carpio*) include improvement in growth, shape and resistance to disease. Experiments carried out in the USSR used crossings of broodstocks to increase genetic diversity and then selected the species for traits like growth rate, exterior traits and viability, and/or adaptation to environmental conditions like variations in temperature. Kirpichnikov *et al.* (1974) and Babouchkine (1987) selected carp for fast growth and tolerance to cold, the Ropsha carp. The results showed a 30–40% to 77.4% improvement of cold tolerance but did not provide any data for growth rate. An increase in growth rate was observed in the second generation in Vietnam. Moav and Wohlfarth (1976) showed positive results when selecting for slower growth for three generations compared to selecting for faster growth. Schaperclaus (1962) showed resistance to the dropsy disease wherein selected lines suffered low mortality (11.5%) compared to unselected (57%).

Channel Catfish

Growth was seen to increase by 12–20% in selectively bred *Iictalurus punctatus*. More recently, the response of the Channel Catfish to selection for improved growth rate was found to be approximately 80%, i.e., an average of 13% per generation.

Shellfish response to selection

Oysters

Selection for live weight of Pacific oysters showed improvements ranging from 0.4% to 25.6% compared to the wild stock. Sydney-rock oysters (*Saccostrea commercialis*) showed a 4% increase after one generation and a 15% increase after two generations. Chilean oysters (*Ostrea chilensis*), selected for improvement in live weight and shell length showed a 10–13% gain in one generation. Bonamia ostrea is a protistan parasite that causes catastrophic losses (nearly 98%) in European flat oyster *Ostrea edulis* L. This protistan parasite is endemic to three oyster-regions in Europe. Selective breeding programs show that *O. edulis* susceptibility to the infection differs across oyster strains in Europe. A study carried out by Culloty et al. showed that 'Rossmore' oysters in Cork harbour, Ireland had better resistance compared to other Irish strains. A selective breeding program at Cork harbour uses broodstock from 3– to 4-year-old survivors and is further controlled until a viable percentage reaches market size. Over the years 'Rossmore' oysters have shown to develop lower prevalence to *B. ostreae* infection and percentage mortality. Ragone Calvo et al. (2003) selectively bred the eastern oyster, *Crassostrea virginica*, for resistance against co-occurring parasites *Haplosporidium nelson* (MSX) and *Perkinsus marinus* (Dermo). They achieved dual resistance to the disease in four generations of selective breeding. The oysters showed higher growth and survival rates and low susceptibility to the infections. At the end of the experiment, artificially selected *C. virginica* showed a 34–48% higher survival rate.

Penaeid shrimps

Selection for growth in Penaeid shrimps yielded successful results. A selective breeding program for *Litopenaeus stylirostris* saw an 18% increase in growth after the fourth generation and 21% growth after the fifth generation. *Marsupenaeus japonicas* showed a 10.7% increase in growth after the first generation. Argue et al. (2002) conducted a selective breeding program on the Pacific White Shrimp, *Litopenaeus vannamei* at The Oceanic Institute, Waimanalo, USA from 1995 to 1998. They reported significant responses to selection compared to the unselected control shrimps. After one generation, a 21% increase was observed in growth and 18.4% increase in survival to TSV. The Taura Syndrome Virus (TSV) causes mortalities of 70% or more in shrimps. C.I. Oceanos S.A. in Colombia selected the survivors of the disease from infected ponds and used them as parents for the next generation. They achieved satisfying results in two or three generations wherein survival rates approached levels before the outbreak of the disease. The resulting heavy losses (up to 90%) caused by Infectious hypodermal and haematopoietic necrosis virus (IHHNV) caused a number of shrimp farming industries started to selectively breed shrimps resistant to this disease. Successful outcomes led to development of Super Shrimp, a selected line of *L. stylirostris* that is resistant to IHHNV infection. Tang et al. (2000) confirmed this by showing no mortalities in IHHNV- challenged Super Shrimp post larvae and juveniles.

Aquatic species versus terrestrial livestock

Selective breeding programs for aquatic species provide better outcomes compared to terrestrial livestock. This higher response to selection of aquatic farmed species can be attributed to the following:

- High fecundity in both sexes fish and shellfish enabling higher selection intensity.
- Large phenotypic and genetic variation in the selected traits.

Selective breeding in aquaculture provide remarkable economic benefits to the industry, the primary one being that it reduces production costs due to faster turnover rates. This is because of faster growth rates, decreased maintenance rates, increased energy and protein retention, and better feed efficiency.Applying such genetic improvement program to aquaculture species will increase productivity to meet the increasing demands of growing populations.

ADVANTAGES AND DISADVANTAGES

Selective breeding is a direct way to determine if a specific trait can "evolve in response to selection." A single-generation method of breeding is not as accurate or direct. The process is also more practical and easier to understand than sibling analysis. The former tests "differences between line means" while the latter is dependent upon "variance and covariance components." Essentially, selective breeding is better for traits such as physiology and behavior that are hard to measure because it requires fewer individuals to test than single-generation testing.

However, there are disadvantages to this process. Because a single experiment done in selective breeding cannot be used to assess an entire group of "genetic variances and covariances," individual experiments must be done for every individual trait. Also, because of the necessity of selective breeding experiments to require maintaining the organisms tested in a lab or greenhouse, it is impractical to use this breeding method on many organisms. Controlled mating instances are difficult to carry out in this case and this is a necessary component of selective breeding.

REFERENCES AND NOTES

1. https://www.oxfordlearnersdictionaries.com/definition/english/allele
2. https://dictionary.cambridge.org/dictionary/english/allele
3. Wood, E.J. (1995). "The encyclopedia of molecular biology". Biochemical Education. 23 *(2): 1165.* doi:10.1016/0307-4412(95)90659-2.
4. Craft, Jude (2013). "Genes and genetics: the language of scientific discovery". Genes and genetics. Oxford English Dictionary. Retrieved 2016-01-14.
5. Hartl, Daniel L.; Elizabeth W. Jones (2005). Essential genetics: A genomics perspective (4th ed.). Jones & Bartlett Publishers. p. 600. ISBN 978-0-7637-3527-2.
6. Victor A. McKusick; Cassandra L. Kniffin; Paul J. Converse; Ada Hamosh (10 November 2009). "ABO Glycosyltransferase; ABO". Online Mendelian Inheritance in Man. National Library of Medicine. Archived from the original on 16 November 2010. Retrieved 24 March 2010.
7. Yip SP (January 2002). "Sequence variation at the human ABO locus". Annals of Human Genetics. 66 *(1): 1–27.* doi:10.1017/S0003480001008995. PMID 12014997.
8. Daxinger, Lucia; Whitelaw, Emma (31 January 2012). "Understanding transgenerational epigenetic inheritance via the gametes in mammals". Nature Reviews Genetics. 13 *(3): 155.* doi:10.1038/nrg3188.
9. Rakyan, Vardhman K; Blewitt, Marnie E; Druker, Riki; Preis, Jost I; Whitelaw, Emma (July 2002). "Metastable epialleles in mammals". Trends in Genetics. 18 *(7): 348–351.* doi:10.1016/S0168-9525(02)02709-9.
10. Waterland, RA; Dolinoy, DC; Lin, JR; Smith, CA; Shi, X; Tahiliani, KG (September 2006). "Maternal methyl supplements increase offspring DNA methylation at Axin Fused.". Genesis (New York, N.Y. : 2000). 44 *(9): 401–6. doi:10.1002/dvg.20230. PMID 16868943.*
11. Conner, J. K. (2003). "Artificial Selection: A Powerful Tool for Ecologists". Ecology. 84 *(7): 1650–1660.* doi:10.1890/0012-9658(2003)084[1650:asaptf]2.0.co;2.

12. Darwin Purugganan, M. D.; Fuller, D. Q. (2009). "The nature of selection during plant domestication". Nature. 457 *(7231): 843–8.* doi:10.1038/nature07895. PMID 19212403.

13. Buffum, Burt C. (2008). Arid Agriculture; A Hand-Book for the Western Farmer and Stockman. Read Books. p. 232. ISBN 978-1-4086-6710-1.

14. Lush, Jay L. (2008). Animal Breeding Plans. Orchard Press. p. 21. ISBN 978-1-4437-8451-1.

15. Wilczynski, J. Z. (1959). "On the Presumed Darwinism of Alberuni Eight Hundred Years before Darwin". Isis. 50 *(4): 459.* doi:10.1086/348801.

16. «Robert Bakewell (1725–1795)". BBC History. Retrieved 20 July 2012.

17. Darwin, p. 109

18. Darwin, pp. 197–198

19. Grandin, Temple; Johnson, Catherine (2005). Animals in Translation. New York, New York: Scribner. pp. 69–71. ISBN 0-7432-4769-8.

20. Swallow, JG; Garland; Jr (2005). "Selection experiments as a tool in evolutionary and comparative physiology: insights into complex traits—an introduction to the symposium" (PDF). Integr Comp Biol. 45 *(3): 387–390. doi:10.1093/* icb/45.3.387. PMID 21676784.

21. Garland T, Jr. (2003). Selection experiments: an under-utilized tool in biomechanics and organismal biology. Ch. 3, Vertebrate Biomechanics and Evolution ed. Bels VL, Gasc JP, Casinos A. PDF

22. Garland T, Jr., Rose MR, eds. (2009). Experimental Evolution: Concepts, Methods, and Applications of Selection Experiments. University of California Press, Berkeley, California.

23. Jain, H. K.; Kharkwal, M. C. (2004). Plant breeding - Mendelian to molecular approaches. Boston, London, Dordecht: Kluwer Academic Publishers. ISBN 1-4020-1981-5.

24. Gjedrem, T & Baranski, M. (2009). Selective breeding in Aquaculture: An Introduction. 1st Edition. Springer. ISBN 978-90-481-2772-6

25. Gjedrem, T. (1985). "Improvement of productivity through breeding schemes". GeoJournal. 10 *(3): 233–241.* doi:10.1007/BF00462124.

26. Gjedrem, T. (1983). "Genetic variation in quantitative traits and selective breeding in fish and shellfish". Aquaculture. 33: 51–72. doi:10.1016/0044-8486(83)90386-1.

27. Thodesen, J. R.; Grisdale-Helland, B.; Helland, S. L. J.; Gjerde, B. (1999). "Feed intake, growth and feed utilization of offspring from wild and selected Atlantic salmon (Salmo salar)". Aquaculture. 180 *(3–4): 237–246.* doi:10.1016/ s0044-8486(99)00204-5.

28. Storset, A.; Strand, C.; Wetten, M.; Kjøglum, S.; Ramstad, A. (2007). "Response to selection for resistance against infectious pancreatic necrosis in Atlantic salmon (Salmo salar *L.)". Aquaculture.* 272: S62. doi:10.1016/j.aquaculture.2007.08.011.

29. Donaldson, L. R.; Olson, P. R. (1957). "Development of Rainbow Trout Brood Stock by Selective Breeding". Transactions of the American Fisheries Society. 85: 93–101. doi:10.1577/1548-8659(1955)85[93:dortbs]2.0.co;2.

30. Kincaid, H. L.; Bridges, W. R.; von Limbach, B. (1977). "Three Generations of Selection for Growth Rate in Fall-Spawning Rainbow Trout". Transactions of the American Fisheries Society. 106 *(6): 621–628.* doi:10.1577/1548-8659(1977)106<621:tgosfg>2.0.co;2.

31. Kause, A.; Ritola, O.; Paananen, T.; Wahlroos, H.; Mäntysaari, E. A. (2005). "Genetic trends in growth, sexual maturity and skeletal deformations, and rate of inbreeding in a breeding programme for rainbow trout (Oncorhynchus mykiss)". Aquaculture. 247: 177–187. doi:10.1016/j.aquaculture.2005.02.023.

32. Okamoto, N.; Tayama, T.; Kawanobe, M.; Fujiki, N.; Yasuda, Y.; Sano, T. (1993). "Resistance of a rainbow trout strain to infectious pancreatic necrosis". Aquaculture. 117: 71–76. doi:10.1016/0044-8486(93)90124-h.

33. Hershberger, W. K.; Myers, J. M.; Iwamoto, R. N.; McAuley, W. C.; Saxton, A. M. (1990). "Genetic changes in the growth of coho salmon (Oncorhynchus kisutch) in marine net-pens, produced by ten years of selection". Aquaculture. 85: 187–197. doi:10.1016/0044-8486(90)90018-i.

34. Neira, R.; Díaz, N. F.; Gall, G. A. E.; Gallardo, J. A.; Lhorente, J. P.; Alert, A. (2006). "Genetic improvement in coho salmon (Oncorhynchus kisutch). II: Selection response for early spawning date" (PDF). Aquaculture. 257: 1–8. doi:10.1016/j.aquaculture.2006.03.001.

35. Kirpichnikov, V. S.; Ilyasov, I.; Shart, L. A.; Vikhman, A. A.; Ganchenko, M. V.; Ostashevsky, A. L.; Simonov, V. M.; Tikhonov, G. F.; Tjurin, V. V. (1993). "Selection of Krasnodar common carp (Cyprinus carpio *L.)* for resistance to dropsy: Principal results and prospects". Genetics in Aquaculture. p. 7. doi:10.1016/b978-0-444-81527-9.50006-3. ISBN 9780444815279.

36. Babouchkine, Y.P., 1987. La sélection d'une carpe résistant à l'hiver. In: Tiews, K. (Ed.), Proceedings ofWorld Symposium on Selection,Hybridization, and Genetic Engineering in Aquaculture, Bordeaux 27–30 May 1986, vol. 1. HeenemannVerlagsgesellschaft mbH, Berlin, pp. 447–454.

37. Mai Thien Tran; Cong Thang Nguyen (1993). "Selection of common carp (Cyprinus carpio *L.)* in Vietnam". Aquaculture. 111: 301–302. doi:10.1016/0044-8486(93)90064-6.

38. Moav, R; Wohlfarth, G (1976). "Two-way selection for growth rate in the common carp (Cyprinus carpio *L.)". Genetics.* 82 *(1): 83–101. PMC 1213447.* PMID 1248737.

39. Bondari, K. (1983). "Response to bidirectional selection for body weight in channel catfish". Aquaculture. 33: 73–81. doi:10.1016/0044-8486(83)90387-3.

40. Langdon, C.; Evans, F.; Jacobson, D.; Blouin, M. (2003). "Yields of cultured Pacific oysters Crassostrea gigas Thunberg improved after one generation of selection". Aquaculture. 220: 227–244. doi:10.1016/s0044-8486(02)00621-x.

41. Nell, J. A.; Sheridan, A. K.; Smith, I. R. (1996). "Progress in a Sydney rock oyster, Saccostrea commercialis (Iredale and Roughley), breeding program". Aquaculture. 144 *(4): 295–302.* doi:10.1016/0044-8486(96)01328-2.

42. Nell, J. A.; Smith, I. R.; Sheridan, A. K. (1999). "Third generation evaluation of Sydney rock oyster Saccostrea commercialis (Iredale and Roughley) breeding lines". Aquaculture. 170 *(3–4): 195–203.* doi:10.1016/s0044-8486(98)00408-6.

43. Culloty, S. C.; Cronin, M. A.; Mulcahy, M. I. F. (2001). "An investigation into the relative resistance of Irish flat oysters Ostrea edulis L. to the parasite Bonamia ostreae". Aquaculture. 199 *(3–4): 229–244.* doi:10.1016/s0044-8486(01)00569-5.

44. Culloty, S. C.; Cronin, M. A.; Mulcahy, M. F. (2004). "Potential resistance of a number of populations of the oyster Ostrea edulis to the parasite Bonamia ostreae". Aquaculture. 237: 41–58. doi:10.1016/j.aquaculture.2004.04.007.

45. Ragone Calvo, L. M.; Calvo, G. W.; Burreson, E. M. (2003). "Dual disease resistance in a selectively bred eastern oyster, Crassostrea virginica, strain tested in Chesapeake Bay". Aquaculture. 220: 69–87. doi:10.1016/s0044-8486(02)00399-x.

46. Goyard, E.; Patrois, J.; Reignon, J.-M.; Vanaa, V.; Dufour, R; Be (1999). "IFREMER's shrimp genetics program". Global Aquaculture Advocate. 2 *(6): 26–28.*

47. Hetzel, D. J. S.; Crocos, P. J.; Davis, G. P.; Moore, S. S.; Preston, N. C. (2000). "Response to selection and heritability for growth in the Kuruma prawn, Penaeus japonicus". Aquaculture. 181 *(3–4): 215–223.* doi:10.1016/S0044-8486(99)00237-9.

48. Argue, B. J.; Arce, S. M.; Lotz, J. M.; Moss, S. M. (2002). "Selective breeding of Pacific white shrimp (Litopenaeus vannamei) for growth and resistance to Taura Syndrome Virus". Aquaculture. 204 *(3–4): 447–460.* doi:10.1016/s0044-8486(01)00830-4.

49. Cock, J.; Gitterle, T.; Salazar, M.; Rye, M. (2009). "Breeding for disease resistance of Penaeid shrimps". Aquaculture. 286: 1–11. doi:10.1016/j.aquaculture.2008.09.011.

50. Tang, K. F. J.; Durand, S. V.; White, B. L.; Redman, R. M.; Pantoja, C. R.; Lightner, D. V. (2000). "Postlarvae and juveniles of a selected line of Penaeus stylirostris *are resistant to infectious hypodermal and hematopoietic necrosis virus infection". Aquaculture.* 190 *(3–4): 203–210.* doi:10.1016/s0044-8486(00)00407-5.

CHAPTER 5

DNA Molecule

Chapter Contents

DNA, or deoxyribonucleic acid, is the hereditary material in humans and almost all other organisms. Nearly every cell in a person's body has the same DNA. Most DNA is located in the cell nucleus (where it is called nuclear DNA), but a small amount of DNA can also be found in the mitochondria (where it is called mitochondrial DNA or mtDNA).

INTRODUCTION

Deoxyribonucleic acid is a molecule that carries the genetic instructions used in the growth, development, functioning and reproduction of all known living organisms and many viruses. DNA and RNA are nucleic acids; alongside proteins, lipids and complex carbohydrates (polysaccharides), they are one of the four major types of macromolecules that are essential for all known forms of life. Most DNA molecules consist of two biopolymer strands coiled around each other to form a double helix.

The two DNA strands are termed polynucleotides since they are composed of simpler monomer units called nucleotides. Each nucleotide is composed of one of four nitrogen-containing nucleobases—either cytosine (C), guanine (G), adenine (A), or thymine (T)—and a sugar called deoxyribose and a phosphate group. The nucleotides are joined to one another in a chain by covalent bonds between the sugar of one nucleotide and the phosphate of the next, resulting in an alternating sugar-phosphate backbone. The nitrogenous bases of the two separate polynucleotide strands are bound together (according to base pairing rules (A with T, and C with G) with hydrogen bonds to make double-stranded DNA. The total amount of related DNA base pairs on Earth is estimated at 5.0×10^{37}, and weighs 50 billion tonnes. In comparison, the total mass of the biosphere has been estimated to be as much as 4 trillion tons of carbon (TtC).

DNA stores biological information. The DNA backbone is resistant to cleavage, and both strands of the double-stranded structure store the same biological information. This information is replicated as and when the two strands separate. A large part of DNA (more than 98% for humans) is non-coding, meaning that these sections do not serve as patterns for protein sequences.

The two strands of DNA run in opposite directions to each other and are thus antiparallel. Attached to each sugar is one of four types of nucleobases (informally, *bases*). It is the sequence of these four nucleobases along the backbone that encodes biological information. RNA strands are created using DNA strands as a template in a process called transcription. Under the genetic code, these RNAstrands are translated to specify the sequence of amino acids within proteins in a process called translation.

Within eukaryotic cells, DNA is organized into long structures called chromosomes. During cell division these chromosomes are duplicated in the process of DNA replication, providing each cell its own complete set of chromosomes. Eukaryotic organisms (animals, plants, fungi, and protists) store most of their DNA inside the cell nucleus and some of their DNA in organelles, such as mitochondria or chloroplasts. In contrast, prokaryotes (bacteria and archaea) store their DNA only in the cytoplasm. Within the eukaryotic chromosomes, chromatin proteins such as histones compact and organize DNA. These compact structures guide the interactions between DNA and other proteins, helping control which parts of the DNA are transcribed.

DNA was first isolated by Friedrich Miescher in 1869. Its molecular structure was identified by James Watson and Francis Crick in 1953, whose model-building efforts were guided by X-ray diffraction data acquired by Rosalind Franklin. DNA is used by researchers as a molecular tool to explore physical laws and theories, such as the ergodic theorem and the theory of elasticity. The unique material properties of DNA have made it an attractive molecule for material scientists and engineers interested in micro- and nano-fabrication. Among notable advances in this field are DNA origami and DNA-based hybrid materials.

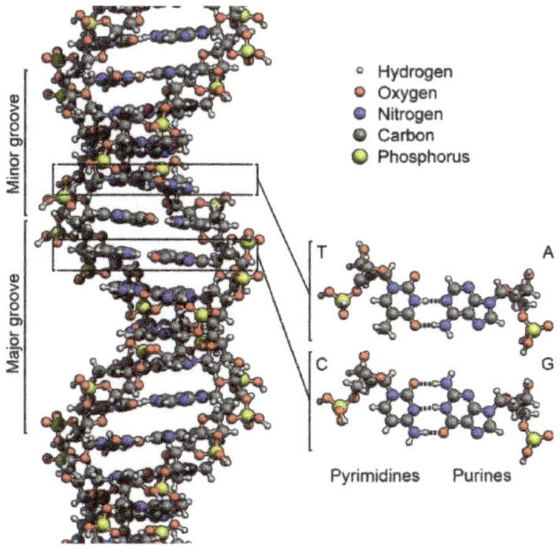

The structure of the DNA double helix. The atoms in the structure are colour-coded by element and the detailed structure of two base pairs are shown in the bottom right.

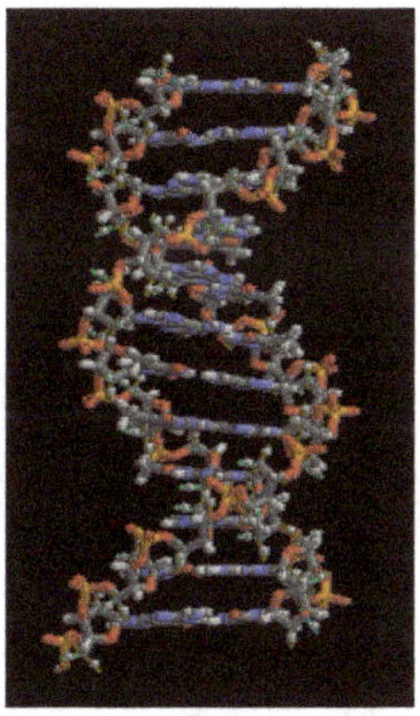

The structure of part of a DNA double helix

PROPERTIES

DNA is a long polymer made from repeating units called nucleotides. The structure of DNA is non-static, all species comprises two helical chains each coiled round the same axis, and each with a pitch of 34 ångströms (3.4 nanometres) and a radius of 10 ångströms (1.0 nanometre). According to another study, when measured in a particular solution, the DNA chain measured 22 to 26 ångströms wide (2.2 to 2.6 nanometres), and one nucleotide unit measured 3.3 Å (0.33 nm) long. Although each individual repeating unit is very small, DNA polymers can be very large molecules containing millions of nucleotides. For instance, the DNA in the largest human chromosome, chromosome number 1, consists of approximately 220 million base pairs and would be 85 mm long if straightened.

In living organisms DNA does not usually exist as a single molecule, but instead as a pair of molecules that are held tightly together. These two long strands entwine like vines, in the shape of a double helix. The nucleotide contains both a segment of the backbone of the molecule (which holds the chain together) and a nucleobase (which interacts with the other DNA strand in the helix). A nucleobase linked to a sugar is called a nucleoside and a base linked to a sugar and one or more phosphate groups is called a nucleotide. A polymer comprising multiple linked nucleotides (as in DNA) is called a polynucleotide. The backbone of the DNA strand is made from alternating phosphate and sugar residues. The sugar in DNA is 2-deoxyribose, which is a pentose (five-carbon) sugar. The sugars are joined together by phosphate groups that form phosphodiester bonds between the third and fifth carbon atoms of adjacent sugar rings. These asymmetric bonds mean a strand of DNA has a direction. In a double helix, the direction of the nucleotides in one strand is opposite to their direction in the other strand: the strands are *antiparallel*. The asymmetric ends of DNA strands are said to have a directionality of *five prime* (5′) and *three prime* (3′), with the 5′ end having a terminal phosphate group and the 3′ end a terminal hydroxyl group. One major difference between DNA and RNA is the sugar, with the 2-deoxyribose in DNA being replaced by the alternative pentose sugar ribose in RNA.

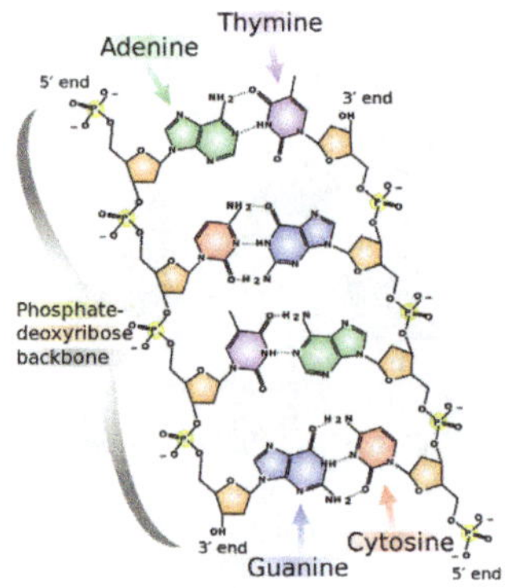

Chemical structure of DNA; hydrogen bonds shown as dotted lines

The DNA double helix is stabilized primarily by two forces: hydrogen bonds between nucleotides and base-stacking interactions among aromaticnucleobases. In the aqueous environment of the cell, the conjugated π bonds of nucleotide bases align perpendicular to the axis of the DNA molecule, minimizing their interaction with the solvation shell. The four bases found in DNA are adenine (A), cytosine (C), guanine (G) and thymine (T). These four bases are attached to the sugar-phosphate to form the complete nucleotide, as shown for adenosine monophosphate. Adenine pairs with thymine and guanine pairs with cytosine. It was represented by A-T base pairs and G-C base pairs.

Nucleobase Classification

The nucleobases are classified into two types: the purines, A and G, being fused five- and six-membered heterocyclic compounds, and the pyrimidines, the six-membered rings C and T. A fifth pyrimidine nucleobase, uracil (U), usually takes the place of thymine in RNA and differs from thymine by lacking a methyl group on its ring. In addition to

RNA and DNA, many artificial nucleic acid analogues have been created to study the properties of nucleic acids, or for use in biotechnology.

Uracil is not usually found in DNA, occurring only as a breakdown product of cytosine. However, in several bacteriophages, *Bacillus subtilis*bacteriophages PBS1 and PBS2 and *Yersinia* bacteriophage piR1-37, thymine has been replaced by uracil. Another phage - Staphylococcal phage S6 - has been identified with a genome where thymine has been replaced by uracil.

Base J (beta-d-glucopyranosyloxymethyluracil), a modified form of uracil, is also found in several organisms: the flagellates *Diplonema* and *Euglena*, and all the kinetoplastid genera. Biosynthesis of J occurs in two steps: in the first step a specific thymidine in DNA is converted into hydroxymethyldeoxyuridine; in the second HOMedU is glycosylated to form J. Proteins that bind specifically to this base have been identified. These proteins appear to be distant relatives of the Tet1 oncogene that is involved in the pathogenesis of acute myeloid leukemia. J appears to act as a termination signal for RNA polymerase II.

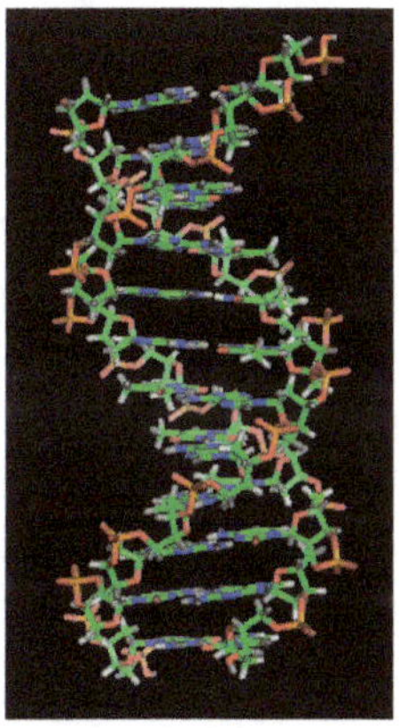

A section of DNA. The bases lie horizontally between the two spiraling strands.(animated version).

Grooves

Twin helical strands form the DNA backbone. Another double helix may be found tracing the spaces, or grooves, between the strands. These voids are adjacent to the base pairs and may provide a binding site. As the strands are not symmetrically located with respect to each other, the grooves are unequally sized. One groove, the major groove, is 22 Å wide and the other, the minor groove, is 12 Å wide. The width of the major groove means that the edges of the bases are more accessible in the major groove than in the minor groove. As a result, proteins such as transcription factors that can bind to specific sequences in double-stranded DNA usually make contact with the sides of the bases exposed in the major groove. This situation varies in unusual conformations of DNA within the cell *(see below)*, but the major and minor grooves are always named to reflect the differences in size that would be seen if the DNA is twisted back into the ordinary B form.

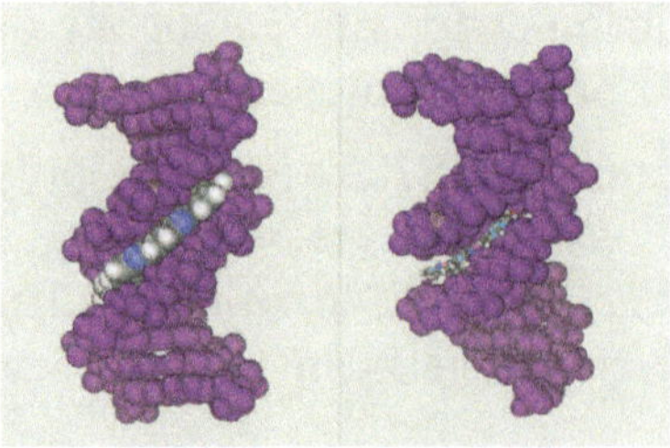

DNA major and minor grooves. The later is a binding site for the Hoechst stain dye 33258.

Base Pairing

In a DNA double helix, each type of nucleobase on one strand bonds with just one type of nucleobase on the other strand. This is called complementary base pairing. Here, purines form hydrogen bonds to pyrimidines, with adenine bonding only to thymine in two hydrogen bonds, and cytosine bonding only to guanine in three hydrogen bonds. This arrangement of two nucleotides binding together across the double helix is called a base pair. As hydrogen bonds are not covalent, they can be broken and rejoined relatively easily. The two strands of DNA in a double helix can thus be pulled apart like a zipper, either by a mechanical force or high temperature. As a result of this base pair complementarity, all the information in the double-stranded sequence of a DNA helix is duplicated on each strand, which is vital in DNA replication. This reversible and specific interaction between complementary base pairs is critical for all the functions of DNA in living organisms.

The two types of base pairs form different numbers of hydrogen bonds, AT forming two hydrogen bonds, and GC forming three hydrogen bonds. DNA with high GC-content is more stable than DNA with low GC-content.

As noted above, most DNA molecules are actually two polymer strands, bound together in a helical fashion by noncovalent bonds; this double stranded structure (**dsDNA**) is maintained largely by the intrastrand base stacking interactions, which are strongest for G,C stacks. The two strands can come apart – a process known as melting – to form two single-stranded DNA molecules (**ssDNA**) molecules. Melting occurs at high temperature, low salt and high pH (low pH also melts DNA, but since DNA is unstable due to acid depurination, low pH is rarely used).

The stability of the dsDNA form depends not only on the GC-content (% G,C basepairs) but also on sequence (since stacking is sequence specific) and also length (longer molecules are more stable). The stability can be measured in various ways; a common way is the "melting temperature", which is the temperature at which 50% of the ds molecules are converted to ss molecules; melting temperature is dependent on ionic strength and the concentration of DNA. As a result, it is both the percentage of GC base pairs and the overall length of a DNA double helix that determines the strength of the association between the two strands of DNA. Long DNA helices with a high GC-content have stronger-interacting strands, while short helices with high AT content have weaker-interacting strands. In biology, parts of the DNA double helix that need to separate easily, such as the TATAAT Pribnow box in some promoters, tend to have a high AT content, making the strands easier to pull apart.

In the laboratory, the strength of this interaction can be measured by finding the temperature necessary to break the hydrogen bonds, their melting temperature (also called T_m value). When all the base pairs in a DNA double helix melt, the strands separate and exist in solution as two entirely independent molecules. These single-stranded DNA molecules have no single common shape, but some conformations are more stable than others.

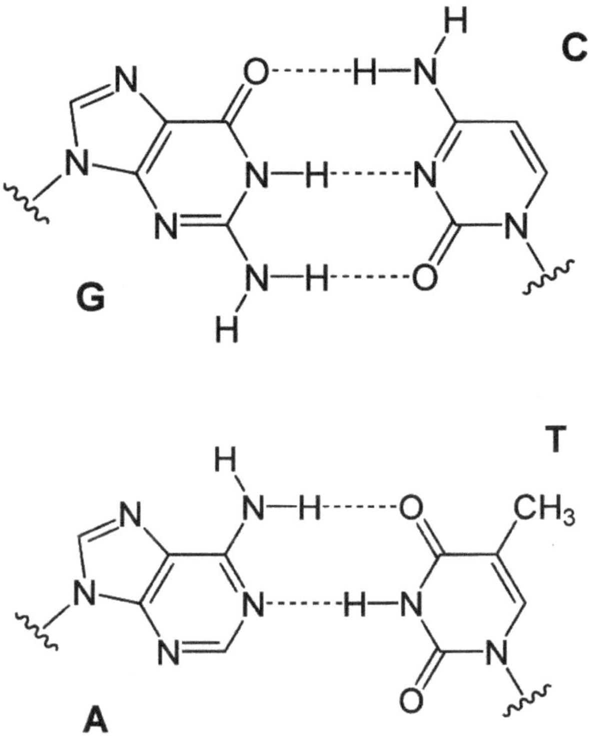

Top, a **GC** base pair with three hydrogen bonds. Bottom, an **AT** base pair with two hydrogen bonds. Non-covalent hydrogen bonds between the pairs are shown as dashed lines.

Sense and Antisense

A DNA sequence is called "sense" if its sequence is the same as that of a messenger RNA copy that is translated into protein. The sequence on the opposite strand is called the "antisense" sequence. Both sense and antisense sequences can exist on different parts of the same strand of DNA (i.e. both strands can contain both sense and antisense sequences). In both prokaryotes and eukaryotes, antisense RNA sequences are produced, but the functions of these RNAs are not entirely clear. One proposal is that antisense RNAs are involved in regulating gene expression through RNA-RNA base pairing.

A few DNA sequences in prokaryotes and eukaryotes, and more in plasmids and viruses, blur the distinction between sense and antisense strands by having overlapping genes. In these cases, some DNA sequences do double duty, encoding one protein when read along one strand, and a second protein when read in the opposite direction along the other strand. In bacteria, this overlap may be involved in the regulation of gene transcription, while in viruses, overlapping genes increase the amount of information that can be encoded within the small viral genome.

Supercoiling

DNA can be twisted like a rope in a process called DNA supercoiling. With DNA in its "relaxed" state, a strand usually circles the axis of the double helix once every 10.4 base pairs, but if the DNA is twisted the strands become more tightly or more loosely wound. If the DNA is twisted in the direction of the helix, this is positive supercoiling, and the bases are held more tightly together. If they are twisted in the opposite direction, this is negative supercoiling, and the bases come apart more easily. In nature, most DNA has slight negative supercoiling that is introduced by enzymes called topoisomerases. These enzymes are also needed to relieve the twisting stresses introduced into DNA strands during processes such as transcription and DNA replication.

Alternative DNA Structures

DNA exists in many possible conformations that include A-DNA, B-DNA, and Z-DNA forms, although, only B-DNA and Z-DNA have been directly observed in functional organisms. The conformation that DNA adopts depends on the hydration level, DNA sequence, the amount and direction of supercoiling, chemical modifications of the bases, the type and concentration of metal ions, and the presence of polyamines in solution.

The first published reports of A-DNA X-ray diffraction patterns—and also B-DNA—used analyses based on Patterson transforms that provided only a limited amount of structural information for oriented fibers of DNA. An alternative analysis was then proposed by Wilkins *et al.*, in 1953, for the *in vivo* B-DNA X-ray diffraction-scattering patterns of highly hydrated DNA fibers in terms of squares of Bessel functions. In the same journal, James Watson and Francis Crick presented their molecular modeling analysis of the DNA X-ray diffraction patterns to suggest that the structure was a double-helix.

Although the *B-DNA form* is most common under the conditions found in cells, it is not a well-defined conformation but a family of related DNA conformations that occur at the high hydration levels present in living cells. Their corresponding X-ray diffraction and scattering patterns are characteristic of molecular paracrystals with a significant degree of disorder.

Compared to B-DNA, the A-DNA form is a wider right-handed spiral, with a shallow, wide minor groove and a narrower, deeper major groove. The A form occurs under non-physiological conditions in partly dehydrated samples of DNA, while in the cell it may be produced in hybrid pairings of DNA and RNA strands, and in enzyme-DNA complexes. Segments of DNA where the bases have been chemically modified by methylation may undergo a larger change in conformation and adopt the Z form. Here, the strands turn about the helical axis in a left-handed spiral, the opposite of the more common B form. These unusual structures can be recognized by specific Z-DNA binding proteins and may be involved in the regulation of transcription.

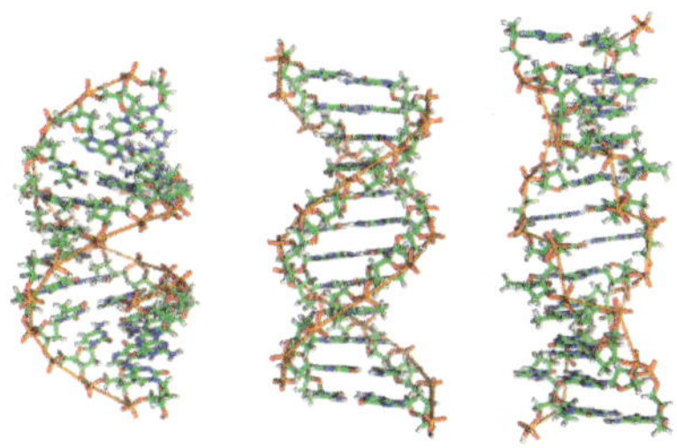

From left to right, the structures of A, B and Z DNA

Alternative DNA Chemistry

For many years exobiologists have proposed the existence of a shadow biosphere, a postulated microbial biosphere of Earth that uses radically different biochemical and molecular processes than currently known life. One of the proposals was the existence of lifeforms that use arsenic instead of phosphorus in DNA. A report in 2010 of the possibility in the bacterium GFAJ-1, was announced, though the research was disputed, and evidence suggests the bacterium actively prevents the incorporation of arsenic into the DNA backbone and other biomolecules.

Quadruplex Structures

At the ends of the linear chromosomes are specialized regions of DNA called telomeres. The main function of these regions is to allow the cell to replicate chromosome ends using the enzyme telomerase, as the enzymes that normally replicate DNA cannot copy the extreme 3′ ends of chromosomes. These specialized chromosome caps also help protect the DNA ends, and stop the DNA repair systems in the cell from treating them as damage to be corrected. In human cells, telomeres are usually lengths of single-stranded DNA containing several thousand repeats of a simple TTAGGG sequence.

These guanine-rich sequences may stabilize chromosome ends by forming structures of stacked sets of four-base units, rather than the usual base pairs found in other DNA molecules. Here, four guanine bases form a flat plate and these flat four-base units then stack on top of each other, to form a stable G-quadruplex structure. These structures are stabilized by hydrogen bonding between the edges of the bases and chelation of a metal ion in the centre of each four-base unit. Other structures can also be formed, with the central set of four bases coming from either a single strand folded around the bases, or several different parallel strands, each contributing one base to the central structure.

In addition to these stacked structures, telomeres also form large loop structures called telomere loops, or T-loops. Here, the single-stranded DNA curls around in a long circle stabilized by telomere-binding proteins. At the very end of the T-loop, the single-stranded telomere DNA is held onto a region of double-stranded DNA by the telomere strand disrupting the double-helical DNA and base pairing to one of the two strands. This triple-stranded structure is called a displacement loop or D-loop.

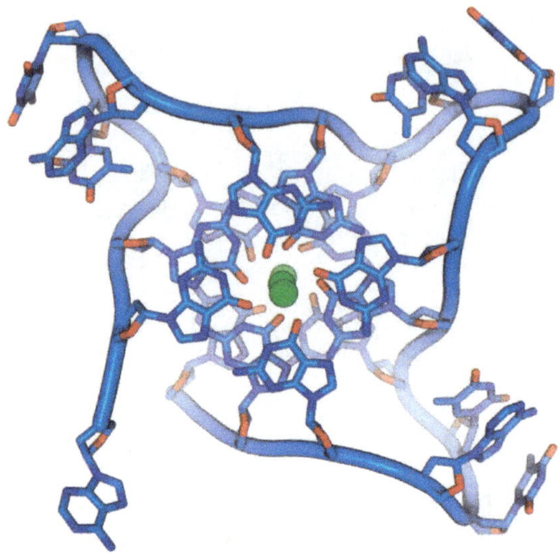

DNA quadruplex formed by telomere repeats. The looped conformation of the DNA backbone is very different from the typical DNA helix. The green spheres in the center represent potassium ions.

Branched DNA

In DNA, fraying occurs when non-complementary regions exist at the end of an otherwise complementary double-strand of DNA. However, branched DNA can occur if a third strand of DNA is introduced and contains adjoining regions able to hybridize with the frayed regions of the pre-existing double-strand. Although the simplest example of branched DNA involves only three strands of DNA, complexes involving additional strands and multiple branches are also possible. Branched DNA can be used in nanotechnology to construct geometric shapes, see the section on uses in technology below.

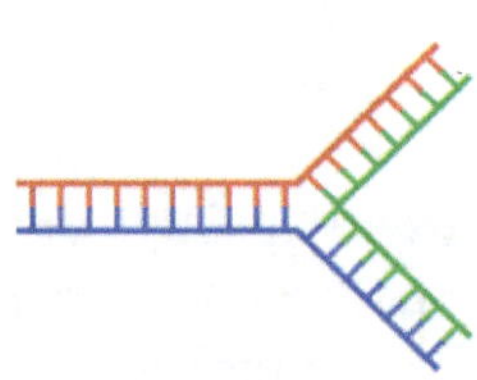

Branched DNA can form networks containing multiple branches.

CHEMICAL MODIFICATIONS AND ALTERED DNA PACKAGING

Base modifications and DNA packaging

The expression of genes is influenced by how the DNA is packaged in chromosomes, in a structure called chromatin. Base modifications can be involved in packaging, with regions that have low or no gene expression usually containing high levels of methylation of cytosine bases. DNA packaging and its influence on gene expression can also occur by covalent modifications of the histone protein core around which DNA is wrapped in the chromatin structure or else by remodeling carried out by chromatin remodeling complexes. There is, further, crosstalk between DNA methylation and histone modification, so they can coordinately affect chromatin and gene expression.

For one example, cytosine methylation, produces 5-methylcytosine, which is important for X-inactivation of chromosomes.The average level of methylation varies between organisms – the worm *Caenorhabditis elegans* lacks cytosine methylation, while vertebrates have higher levels, with up to 1% of their DNA containing 5-methylcytosine. Despite the importance of 5-methylcytosine, it can deaminate to leave a thymine base, so methylated cytosines are particularly prone to mutations.Other base modifications include adenine methylation in bacteria, the presence of 5-hydroxymethylcytosine in the brain, and the glycosylation of uracil to produce the "J-base" in kinetoplastids.

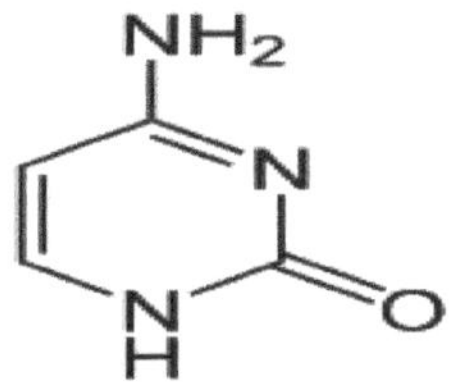

cytosine

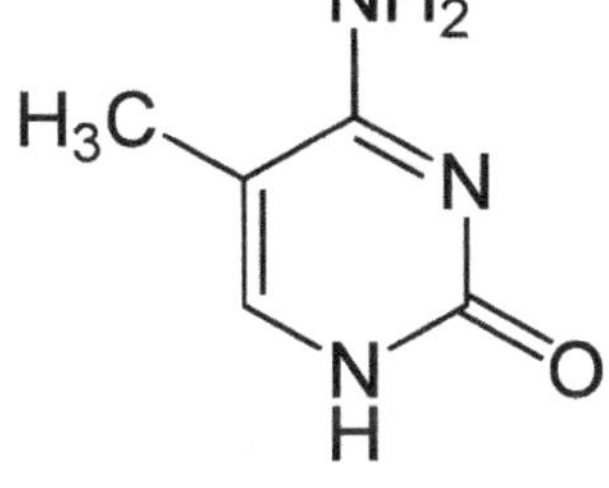

5-methylcytosine

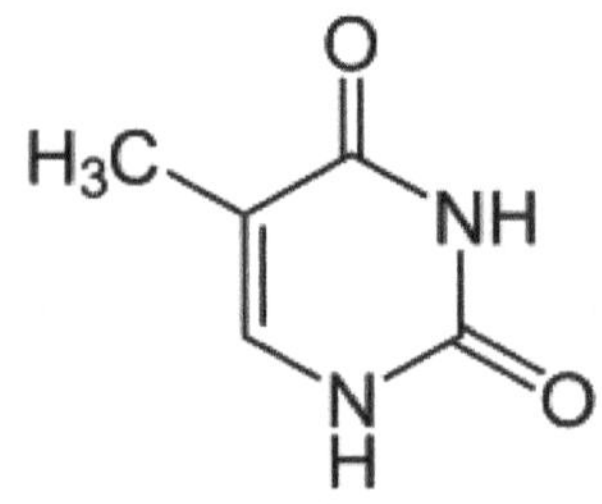

thymine

Structure of cytosine with and without the 5-methyl group. Deamination converts 5-methylcytosine into thymine.

Damage

DNA can be damaged by many sorts of mutagens, which change the DNA sequence. Mutagens include oxidizing agents, alkylating agents and also high-energy electromagnetic radiation such as ultraviolet light and X-rays. The type of DNA damage produced depends on the type of mutagen. For example, UV light can damage DNA by producing thymine dimers, which are cross-links between pyrimidine bases. On the other hand, oxidants such as free radicals or hydrogen peroxide produce multiple forms of damage, including base modifications, particularly of guanosine, and double-strand breaks. A typical human cell contains about 150,000 bases that have suffered oxidative damage. Of these oxidative lesions, the most dangerous are double-strand breaks, as these are difficult to repair and can produce point mutations, insertions, deletions from the DNA sequence, and chromosomal translocations. These mutations can cause cancer. Because of inherent limits in the DNA repair mechanisms, if humans lived long enough, they would all eventually develop cancer. DNA damages that are naturally occurring, due to normal cellular processes that produce reactive oxygen species, the hydrolytic activities of cellular water, etc., also occur frequently. Although most of these damages are repaired, in any cell some DNA damage may remain despite the action of repair processes. These remaining DNA damages accumulate with age in mammalian postmitotic tissues. This accumulation appears to be an important underlying cause of aging.

Many mutagens fit into the space between two adjacent base pairs, this is called *intercalation*. Most intercalators are aromatic and planar molecules; examples include ethidium bromide, acridines, daunomycin, and doxorubicin. For an intercalator to fit between base pairs, the bases must separate, distorting the DNA strands by unwinding of the double helix. This inhibits both transcription and DNA replication, causing toxicity and mutations. As a result, DNA intercalators may be carcinogens, and in the case of thalidomide, a teratogen.Others such as benzo[*a*]pyrene diol epoxide and aflatoxin form DNA adducts that induce errors in replication. Nevertheless, due to their ability to inhibit DNA transcription and replication, other similar toxins are also used in chemotherapy to inhibit rapidly growing cancercells.

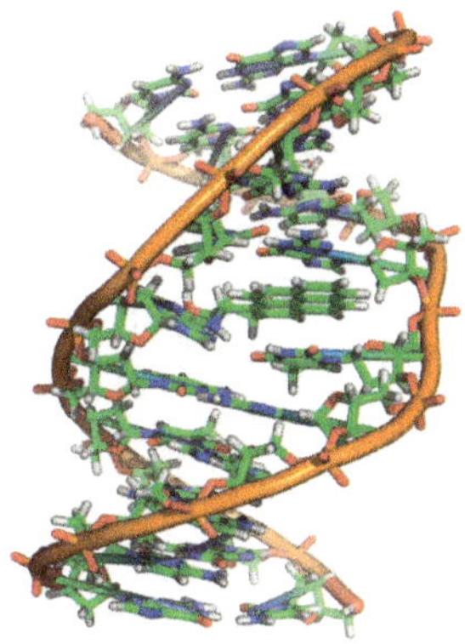

A covalent adduct between a metabolically activated form of benzo[*a*]pyrene, the major mutagen in tobacco smoke, and DNA

BIOLOGICAL FUNCTIONS

DNA usually occurs as linear chromosomes in eukaryotes, and circular chromosomes in prokaryotes. The set of chromosomes in a cell makes up its genome; the human genome has approximately 3 billion base pairs of DNA arranged into 46 chromosomes. The information carried by DNA is held in the sequence of pieces of DNA called genes. Transmission of genetic information in genes is achieved via complementary base pairing. For example, in transcription, when a cell uses the information in a gene, the DNA sequence is copied into a complementary RNA sequence through the attraction between the DNA and the correct RNA nucleotides. Usually, this RNA copy is then used to make a matching protein sequence in a process called translation, which depends on the same interaction between RNA nucleotides. In alternative fashion, a cell may simply copy its genetic information in a process called DNA replication. The details of these functions are covered in other articles; here the focus is on the interactions between DNA and other molecules that mediate the function of the genome.

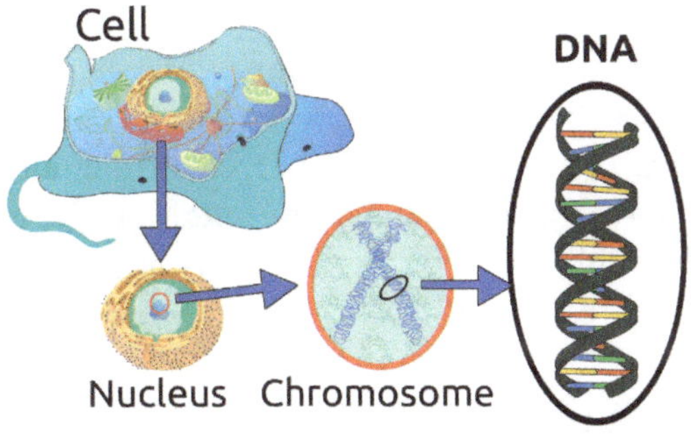

Location of eukaryote nuclear DNA within the chromosomes.

Genes and Genomes

Genomic DNA is tightly and orderly packed in the process called DNA condensation, to fit the small available volumes of the cell. In eukaryotes, DNA is located in the cell nucleus, with small amounts in mitochondria and chloroplasts. In prokaryotes, the DNA is held within an irregularly shaped body in the cytoplasm called the nucleoid. The genetic information in a genome is held within genes, and the complete set of this information in an organism is called its genotype. A gene is a unit of heredity and is a region of DNA that influences a particular characteristic in an organism. Genes contain an open reading frame that can be transcribed, and regulatory sequences such as promoters and enhancers, which control transcription of the open reading frame.

In many species, only a small fraction of the total sequence of the genome encodes protein. For example, only about 1.5% of the human genome consists of protein-coding exons, with over 50% of human DNA consisting of non-coding repetitive sequences. The reasons for the presence of so much noncoding DNA in eukaryotic genomes and the extraordinary differences in genome size, or *C-value*, among species represent a long-standing puzzle known as the “C-value enigma”. However, some DNA sequences that do not code protein may still encode functional non-coding RNA molecules, which are involved in the regulation of gene expression.

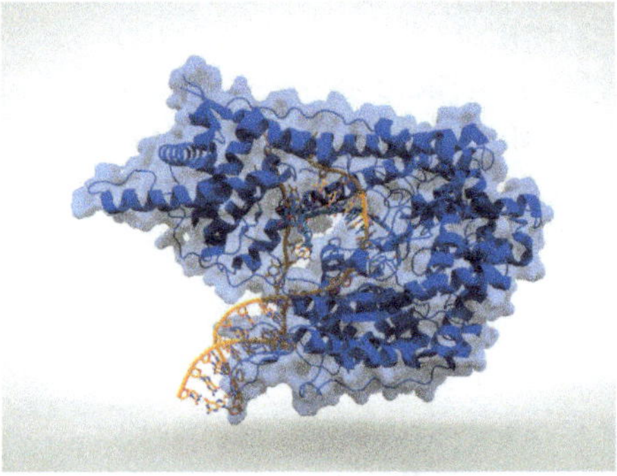

T7 RNA polymerase (blue) producing a mRNA (green) from a DNA template (orange).

Some noncoding DNA sequences play structural roles in chromosomes. Telomeres and centromeres typically contain few genes, but are important for the function and stability of chromosomes. An abundant form of noncoding DNA in humans are pseudogenes, which are copies of genes that have been disabled by mutation. These sequences are usually just molecular fossils, although they can occasionally serve as raw genetic material for the creation of new genes through the process of gene duplication and divergence.

Transcription and Translation

A gene is a sequence of DNA that contains genetic information and can influence the phenotype of an organism. Within a gene, the sequence of bases along a DNA strand defines a messenger RNA sequence, which then defines one or more protein sequences.

The relationship between the nucleotide sequences of genes and the amino-acid sequences of proteins is determined by the rules of translation, known collectively as the genetic code. The genetic code consists of three-letter ‹words’ called *codons* formed from a sequence of three nucleotides (e.g. ACT, CAG, TTT).

In transcription, the codons of a gene are copied into messenger RNA by RNA polymerase. This RNA copy is then decoded by a ribosome that reads the RNA sequence by base-pairing the messenger RNA to transfer RNA, which carries amino acids. Since there are 4 bases in 3-letter combinations, there are 64 possible codons (4^3 combinations). These encode the twenty standard amino acids, giving most amino acids more than one possible codon. There are also three ‹stop’ or ‹nonsense’ codons signifying the end of the coding region; these are the TAA, TGA, and TAG codons.

Replication

Cell division is essential for an organism to grow, but, when a cell divides, it must replicate the DNA in its genome so that the two daughter cells have the same genetic information as their parent. The double-stranded structure of DNA provides a simple mechanism for DNA replication. Here, the two strands are separated and then each strand's complementary DNA sequence is recreated by an enzyme called DNA polymerase. This enzyme makes the complementary strand by finding the correct base through complementary base pairing, and bonding it onto the original strand. As DNA polymerases can only extend a DNA strand in a 5′ to 3′ direction, different mechanisms are used to copy the antiparallel strands of the double helix. In this way, the base on the old strand dictates which base appears on the new strand, and the cell ends up with a perfect copy of its DNA.

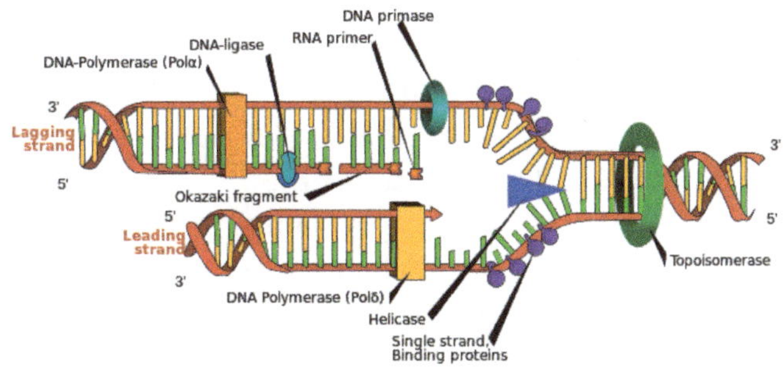

DNA replication. The double helix is unwound by a helicase and topoisomerase. Next, one DNA polymerase produces the leading strand copy. Another DNA polymerase binds to the lagging strand. This enzyme makes discontinuous segments (called Okazaki fragments) before DNA ligase joins them together.

Extracellular Nucleic Acids

Naked extracellular DNA (eDNA), most of it released by cell death, is nearly ubiquitous in the environment. Its concentration in soil may be as high as 2 μg/L, and its concentration in natural aquatic environments may be as high at 88 μg/L. Various possible functions have been proposed for eDNA: it may be involved in horizontal gene transfer; it may provide nutrients; and it may act as a buffer to recruit or titrate ions or antibiotics. Extracellular DNA acts as a functional extracellular matrix component in the biofilms of several bacterial species. It may act as a recognition factor to regulate the attachment and dispersal of specific cell types in the biofilm; it may contribute to biofilm formation; and it may contribute to the biofilm's physical strength and resistance to biological stress.

INTERACTIONS WITH PROTEINS

All the functions of DNA depend on interactions with proteins. These protein interactions can be non-specific, or the protein can bind specifically to a single DNA sequence. Enzymes can also bind to DNA and of these, the polymerases that copy the DNA base sequence in transcription and DNA replication are particularly important.

DNA-binding Proteins

Structural proteins that bind DNA are well-understood examples of non-specific DNA-protein interactions. Within chromosomes, DNA is held in complexes with structural proteins. These proteins organize the DNA into a compact structure called chromatin. In eukaryotes this structure involves DNA binding to a complex of small basic proteins called histones, while in prokaryotes multiple types of proteins are involved. The histones form a disk-shaped complex called a nucleosome, which contains two complete turns of double-stranded DNA wrapped around its surface. These non-specific interactions are formed through basic residues in the histones, making ionic bonds to the acidic sugar-phosphate backbone of the DNA, and are thus largely independent of the base sequence. Chemical modifications of these basic amino acid residues include methylation, phosphorylation and acetylation.These chemical changes alter the strength of the interaction between the DNA and the histones, making the DNA more or less accessible to transcription factors and changing the rate of transcription. Other non-specific DNA-binding proteins in chromatin include the high-mobility group proteins, which bind to bent or distorted DNA. These proteins are important in bending arrays of nucleosomes and arranging them into the larger structures that make up chromosomes.

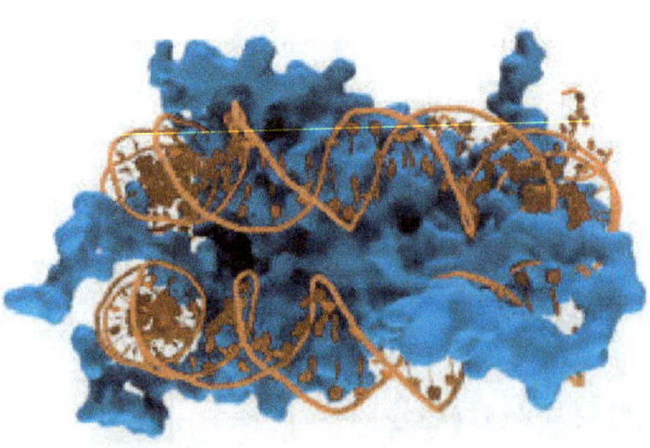

Interaction of DNA (in orange) with histones (in blue). These proteins' basic amino acids bind to the acidic phosphate groups on DNA.

A distinct group of DNA-binding proteins are the DNA-binding proteins that specifically bind single-stranded DNA. In humans, replication protein A is the best-understood member of this family and is used in processes where the double helix is separated, including DNA replication, recombination and DNA repair. These binding proteins seem to stabilize single-stranded DNA and protect it from forming stem-loops or being degraded by nucleases.

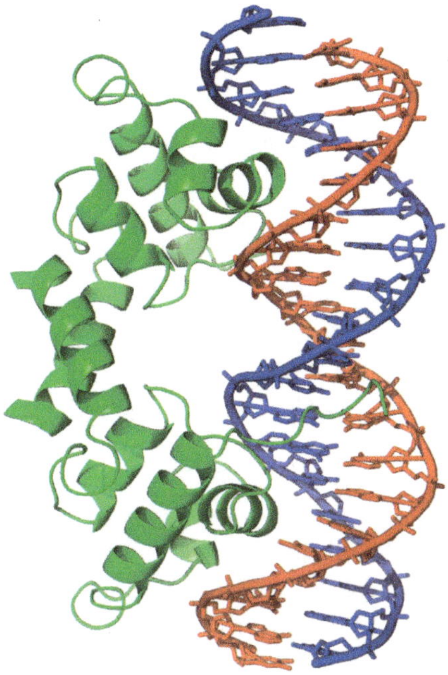

The lambda repressor helix-turn-helix transcription factor bound to its DNA target

In contrast, other proteins have evolved to bind to particular DNA sequences. The most intensively studied of these are the various transcription factors, which are proteins that regulate transcription. Each transcription factor binds to one particular set of DNA sequences and activates or inhibits the transcription of genes that have these sequences close to their promoters. The transcription factors do this in two ways. Firstly, they can bind the RNA polymerase responsible for transcription, either directly or through other mediator proteins; this locates the polymerase at the promoter and allows it to begin transcription. Alternatively, transcription factors can bind enzymes that modify the histones at the promoter. This changes the accessibility of the DNA template to the polymerase.

As these DNA targets can occur throughout an organism's genome, changes in the activity of one type of transcription factor can affect thousands of genes. Consequently, these proteins are often the targets of the signal transduction processes that control responses to environmental changes or cellular differentiation and development. The specificity of these transcription factors' interactions with DNA come from the proteins making multiple contacts to the edges of the DNA bases, allowing them to "read" the DNA sequence. Most of these base-interactions are made in the major groove, where the bases are most accessible.

DNA-modifying enzymes

Nucleases and ligases

Nucleases are enzymes that cut DNA strands by catalyzing the hydrolysis of the phosphodiester bonds. Nucleases that hydrolyse nucleotides from the ends of DNA strands are called exonucleases, while endonucleases cut within strands. The most frequently used nucleases in molecular biology are the restriction endonucleases, which cut DNA at specific sequences. For instance, the EcoRV enzyme shown to the left recognizes the 6-base sequence 5′-GATATC-3′ and makes a cut at the horizontal line. In nature, these enzymes protect bacteria against phage infection by digesting the phage DNA when it enters the bacterial cell, acting as part of the restriction modification system. In technology, these sequence-specific nucleases are used in molecular cloning and DNA fingerprinting.

Enzymes called DNA ligases can rejoin cut or broken DNA strands. Ligases are particularly important in lagging strand DNA replication, as they join together the short segments of DNA produced at the replication fork into a complete copy of the DNA template. They are also used in DNA repair and genetic recombination.

Topoisomerases and helicases

Topoisomerases are enzymes with both nuclease and ligase activity. These proteins change the amount of supercoiling in DNA. Some of these enzymes work by cutting the DNA helix and allowing one section to rotate, thereby reducing its level of supercoiling; the enzyme then seals the DNA break. Other types of these enzymes are capable of cutting one DNA helix and then passing a second strand of DNA through this break, before rejoining the helix. Topoisomerases are required for many processes involving DNA, such as DNA replication and transcription.

Helicases are proteins that are a type of molecular motor. They use the chemical energy in nucleoside triphosphates, predominantly adenosine triphosphate (ATP), to break hydrogen bonds between bases and unwind the DNA double helix into single strands. These enzymes are essential for most processes where enzymes need to access the DNA bases.

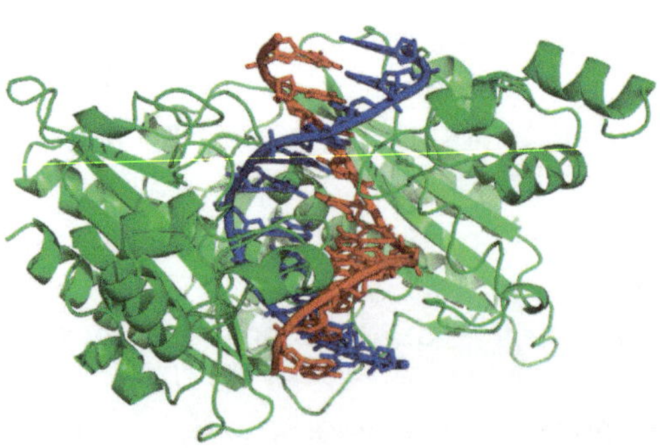

The restriction enzyme EcoRV(green) in a complex with its substrate DNA

Polymerases

Polymerases are enzymes that synthesize polynucleotide chains from nucleoside triphosphates. The sequence of their products are created based on existing polynucleotide chains—which are called *templates*. These enzymes function by repeatedly adding a nucleotide to the 3′ hydroxyl group at the end of the growing polynucleotide chain. As a consequence, all polymerases work in a 5′ to 3′ direction. In the active site of these enzymes, the incoming nucleoside triphosphate base-pairs to the template: this allows polymerases to accurately synthesize the complementary strand of their template. Polymerases are classified according to the type of template that they use.

In DNA replication, DNA-dependent DNA polymerases make copies of DNA polynucleotide chains. To preserve biological information, it is essential that the sequence of bases in each copy are precisely complementary to the sequence of bases in the template strand. Many DNA polymerases have a proofreading activity. Here, the polymerase recognizes the occasional mistakes in the synthesis reaction by the lack of base pairing between the mismatched nucleotides. If a mismatch is detected, a 3′ to 5′ exonuclease activity is activated and the incorrect base removed. In most organisms, DNA polymerases function in a large complex called the replisome that contains multiple accessory subunits, such as the DNA clamp or helicases.

RNA-dependent DNA polymerases are a specialized class of polymerases that copy the sequence of an RNA strand into DNA. They include reverse transcriptase, which is a viral enzyme involved in the infection of cells by retroviruses, and telomerase, which is required for the replication of telomeres. Telomerase is an unusual polymerase because it contains its own RNA template as part of its structure.

Transcription is carried out by a DNA-dependent RNA polymerase that copies the sequence of a DNA strand into RNA. To begin transcribing a gene, the RNA polymerase binds to a sequence of DNA called a promoter and separates the DNA strands. It then copies the gene sequence into a messenger RNA transcript until it reaches a region of DNA called the terminator, where it halts and detaches from the DNA. As with human DNA-dependent DNA polymerases, RNA polymerase II, the enzyme that transcribes most of the genes in the human genome, operates as part of a large protein complex with multiple regulatory and accessory subunits.

GENETIC RECOMBINATION

A DNA helix usually does not interact with other segments of DNA, and in human cells the different chromosomes even occupy separate areas in the nucleus called "chromosome territories". This physical separation of different chromosomes is important for the ability of DNA to function as a stable repository for information, as one of the few times chromosomes interact is in chromosomal crossover which occurs during sexual reproduction, when genetic recombination occurs. Chromosomal crossover is when two DNA helices break, swap a section and then rejoin.

Recombination allows chromosomes to exchange genetic information and produces new combinations of genes, which increases the efficiency of natural selection and can be important in the rapid evolution of new proteins. Genetic recombination can also be involved in DNA repair, particularly in the cell's response to double-strand breaks.

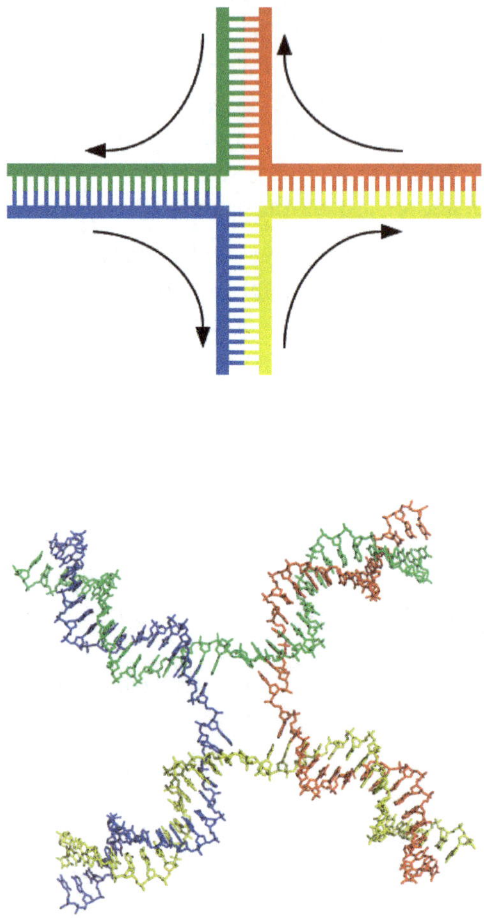

Structure of the Holliday junctionintermediate in genetic recombination. The four separate DNA strands are coloured red, blue, green and yellow.

The most common form of chromosomal crossover is homologous recombination, where the two chromosomes involved share very similar sequences. Non-homologous recombination can be damaging to cells, as it can produce chromosomal translocations and genetic abnormalities. The recombination reaction is catalyzed by enzymes known as recombinases, such as RAD51. The first step in recombination is a double-stranded break caused by either an endonuclease or damage to the DNA. A series of steps catalyzed in part by the recombinase then leads to joining of the two helices by at least one Holliday junction, in which a segment of a single strand in each helix is annealed to the complementary strand in the other helix. The Holliday junction is a tetrahedral junction structure that can be moved along the pair of chromosomes, swapping one strand for another. The recombination reaction is then halted by cleavage of the junction and re-ligation of the released DNA.

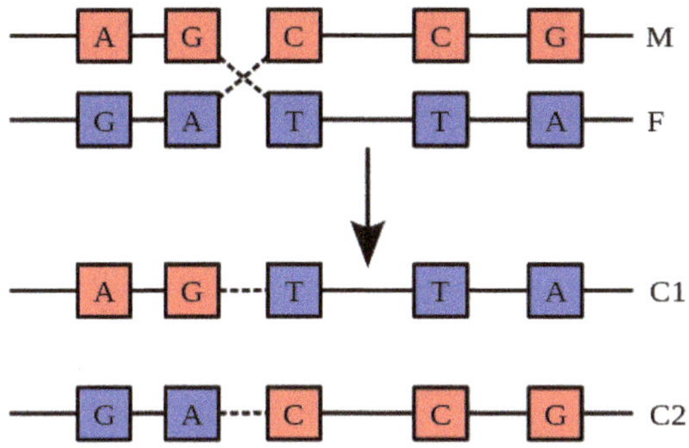

Recombination involves the breaking and rejoining of two chromosomes (M and F) to produce two rearranged chromosomes (C1 and C2).

EVOLUTION

DNA contains the genetic information that allows all modern living things to function, grow and reproduce. However, it is unclear how long in the 4-billion-year history of life DNA has performed this function, as it has been proposed that the earliest forms of life may have used RNA as their genetic material. RNA may have acted as the central part of early cell metabolism as it can both transmit genetic information and carry out catalysis as part of ribozymes.This ancient RNA world where nucleic acid would have been used for both catalysis and genetics may have influenced the evolution of the current genetic code based on four nucleotide bases. This would occur, since the number of different bases in such an organism is a trade-off between a small number of bases increasing replication accuracy and a large number of bases increasing the catalytic efficiency of ribozymes. However, there is no direct evidence of ancient genetic systems, as recovery of DNA from most fossils is impossible because DNA survives in the environment for less than one million years, and slowly degrades into short fragments in solution. Claims for older DNA have been made, most notably a report of the isolation of a viable bacterium from a salt crystal 250 million years old, but these claims are controversial.

Building blocks of DNA (adenine, guanine and related organic molecules) may have been formed extraterrestrially in outer space. Complex DNA and RNA organic compounds of life, including uracil, cytosine, and thymine, have also been formed in the laboratory under conditions mimicking those found in outer space, using starting chemicals, such as pyrimidine, found in meteorites. Pyrimidine, like polycyclic aromatic hydrocarbons (PAHs), the most carbon-rich chemical found in the universe, may have been formed in red giants or in interstellar cosmic dust and gas clouds.

USES IN TECHNOLOGY

Genetic Engineering

Methods have been developed to purify DNA from organisms, such as phenol-chloroform extraction, and to manipulate it in the laboratory, such as restriction digests and the polymerase chain reaction. Modern biology and biochemistry make intensive use of these techniques in recombinant DNA technology. Recombinant DNA is a man-made DNA sequence that has been assembled from other DNA sequences. They can be transformed into organisms in the form of plasmids or in the appropriate format, by using a viral vector. The genetically modified organisms produced can be used to produce products such as recombinant proteins, used in medical research, or be grown in agriculture.

DNA Profiling

Forensic scientists can use DNA in blood, semen, skin, saliva or hair found at a crime scene to identify a matching DNA of an individual, such as a perpetrator. This process is formally termed DNA profiling, but may also be called "genetic fingerprinting". In DNA profiling, the lengths of variable sections of repetitive DNA, such as short tandem repeatsand minisatellites, are compared between people. This method is usually an extremely reliable technique for identifying a matching DNA. However, identification can be complicated if the scene is contaminated with DNA from several people. DNA profiling was developed in 1984 by British geneticist Sir Alec Jeffreys, and first used in forensic science to convict Colin Pitchfork in the 1988 Enderby murders case.

The development of forensic science, and the ability to now obtain genetic matching on minute samples of blood, skin, saliva, or hair has led to re-examining many cases. Evidence can now be uncovered that was scientifically impossible at the time of the original examination. Combined with the removal of the double jeopardy law in some places, this can allow cases to be reopened where prior trials have failed to produce sufficient evidence to convince a jury. People charged with serious crimes may be required to provide a sample of DNA for matching purposes. The most obvious defence to DNA matches obtained forensically is to claim that cross-contamination of evidence has occurred. This has resulted in meticulous strict handling procedures with new cases of serious crime. DNA profiling is also used successfully to positively identify victims of mass casualty incidents, bodies or body parts in serious accidents, and individual victims in mass war graves, via matching to family members.

DNA profiling is also used in DNA paternity testing to determine if someone is the biological parent or grandparent of a child with the probability of parentage is typically 99.99% when the alleged parent is biologically related to the child. Normal DNA sequencing methods happen after birth but there are new methods to test paternity while a mother is still pregnant.

DNA Enzymes or Catalytic DNA

Deoxyribozymes, also called DNAzymes or catalytic DNA are first discovered in 1994. They are mostly single stranded DNA sequences isolated from a large pool of random DNA sequences through a combinatorial approach called in vitro selection or systematic evolution of ligands by exponential enrichment (SELEX). DNAzymes catalyze variety of chemical reactions including RNA-DNA cleavage, RNA-DNA ligation, amino acids phosphorylation-dephosphorylation, carbon-carbon bond formation, and etc. DNAzymes can enhance catalytic rate of chemical reactions up to 100,000,000,000-fold over the uncatalyzed reaction. The most extensively studied class of DNAzymes are RNA-cleaving types which have been used to detect different metal ions and designing therapeutic agents. Several metal-specific DNAzymes have been reported including the GR-5 DNAzyme (lead-specific), the CA1-3 DNAzymes (copper-specific), the 39E DNAzyme (uranyl-specific) and the NaA43 DNAzyme (sodium-specific). The NaA43 DNAzyme, which is reported to be more than 10,000-fold selective for sodium over other metal ions, was used to make a real-time sodium sensor in living cells.

Bioinformatics

Bioinformatics involves the development of techniques to store, data mine, search and manipulate biological data, including DNA nucleic acid sequence data. These have led to widely applied advances in computer science, especially string searching algorithms, machine learning and database theory. String searching or matching algorithms, which find an occurrence of a sequence of letters inside a larger sequence of letters, were developed to search for specific sequences of nucleotides. The DNA sequence may be aligned with other DNA sequences to identify homologous sequences and locate the specific mutations that make them distinct. These techniques, especially multiple sequence alignment, are used in studying phylogenetic relationships and protein function. Data sets representing entire genomes' worth of DNA sequences, such as those produced by the Human Genome Project, are difficult to use without the annotations that identify the locations of genes and regulatory elements on each chromosome. Regions of DNA sequence that have the characteristic patterns associated with protein- or RNA-coding genes can be identified by gene finding algorithms, which allow researchers to predict the presence of particular gene products and their possible functions in an organism even before they have been isolated experimentally. Entire genomes may also be compared, which can shed light on the evolutionary history of particular organism and permit the examination of complex evolutionary events.

DNA Nanotechnology

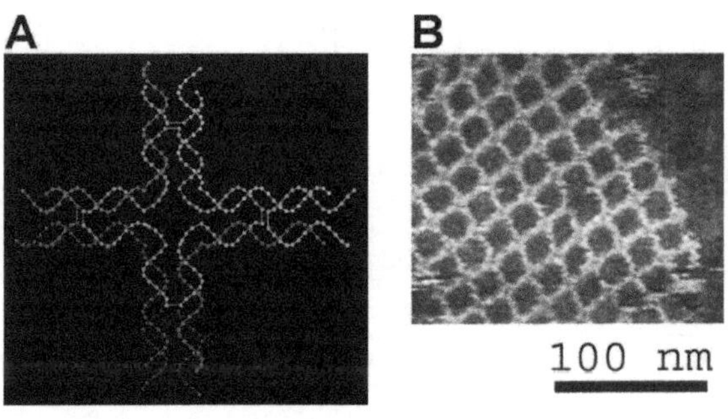

The DNA structure at left (schematic shown) will self-assemble into the structure visualized by atomic force microscopy at right. DNA nanotechnology is the field that seeks to design nanoscale structures using the molecular recognition properties of DNA molecules. Image from Strong, 2004.

DNA nanotechnology uses the unique molecular recognition properties of DNA and other nucleic acids to create self-assembling branched DNA complexes with useful properties. DNA is thus used as a structural material rather than as a carrier of biological information. This has led to the creation of two-dimensional periodic lattices (both tile-based and using the *DNA origami* method) and three-dimensional structures in the shapes of polyhedra. Nanomechanical devices and algorithmic self-assembly have also been demonstrated, and these DNA structures have been used to template the arrangement of other molecules such as gold nanoparticles and streptavidin proteins.

History and Anthropology

Because DNA collects mutations over time, which are then inherited, it contains historical information, and, by comparing DNA sequences, geneticists can infer the evolutionary history of organisms, their phylogeny. This field of phylogenetics is a powerful tool in evolutionary biology. If DNA sequences within a species are compared, population geneticists can learn the history of particular populations. This can be used in studies ranging from ecological genetics to anthropology; For example, DNA evidence is being used to try to identify the Ten Lost Tribes of Israel.

Information Storage

In a paper published in *Nature* in January 2013, scientists from the European Bioinformatics Institute and Agilent Technologies proposed a mechanism to use DNA's ability to code information as a means of digital data storage. The group was able to encode 739 kilobytes of data into DNA code, synthesize the actual DNA, then sequence the DNA and decode the information back to its original form, with a reported 100% accuracy. The encoded information consisted of text files and audio files. A prior experiment was published in August 2012. It was conducted by researchers at Harvard University, where the text of a 54,000-word book was encoded in DNA.

Moreover, in living cells the storage can be turned active by enzymes. Light-gated protein domains fused to DNA processing enzymes are suitable for that task *in vitro*. Fluorescent exonucleases can transmit the output according to the nucleotide they have read.

HISTORY OF DNA RESEARCH

DNA was first isolated by the Swiss physician Friedrich Miescher who, in 1869, discovered a microscopic substance in the pus of discarded surgical bandages. As it resided in the nuclei of cells, he called it "nuclein". In 1878, Albrecht Kossel isolated the non-protein component of "nuclein", nucleic acid, and later isolated its five primary nucleobases. In 1919, Phoebus Levene identified the base, sugar and phosphate nucleotide unit. Levene suggested that DNA consisted of a string of nucleotide units linked together through the phosphate groups. Levene thought the chain was short and the bases repeated in a fixed order. In 1937, William Astburyproduced the first X-ray diffraction patterns that showed that DNA had a regular structure.

In 1927, Nikolai Koltsov proposed that inherited traits would be inherited via a "giant hereditary molecule" made up of "two mirror strands that would replicate in a semi-conservative fashion using each strand as a template". In 1928, Frederick Griffith in his experimentdiscovered that traits of the "smooth" form of *Pneumococcus* could be transferred to the "rough" form of the same bacteria by mixing killed "smooth" bacteria with the live "rough" form. This system provided the first clear suggestion that DNA carries genetic information—the Avery–MacLeod–McCarty experiment—when Oswald Avery, along with coworkers Colin MacLeod and Maclyn McCarty, identified DNA as the transforming principle in 1943. DNA's role in heredity was confirmed in 1952, when Alfred Hershey and Martha Chase in the Hershey–Chase experiment showed that DNA is the genetic material of the T2 phage.

James Watson and Francis Crick (right), co-originators of the double-helix model, with Maclyn McCarty (left).

In 1953, James Watson and Francis Crick suggested what is now accepted as the first correct double-helix model of DNA structure in the journal *Nature*. Their double-helix, molecular model of DNA was then based on one X-ray diffraction image (labeled as "Photo 51")taken by Rosalind Franklin and Raymond Gosling in May 1952, and the information that the DNA bases are paired.

Pencil sketch of the DNA double helix by Francis Crick in 1953

Experimental evidence supporting the Watson and Crick model was published in a series of five articles in the same issue of *Nature*.Of these, Franklin and Gosling's paper was the first publication of their own X-ray diffraction data and original analysis method that partly supported the Watson and Crick model; this issue also contained an article on DNA structure by Maurice Wilkins and two of his colleagues, whose analysis and *in vivo* B-DNA X-ray patterns also supported the presence *in vivo* of the double-helical DNA configurations as proposed by Crick and Watson for their double-helix molecular model of DNA in the prior two pages of *Nature*. In 1962, after Franklin's death, Watson, Crick, and Wilkins jointly received the Nobel Prize in Physiology or Medicine. Nobel Prizes are awarded only to living recipients. A debate continues about who should receive credit for the discovery.

In an influential presentation in 1957, Crick laid out the central dogma of molecular biology, which foretold the relationship between DNA, RNA, and proteins, and articulated the "adaptor hypothesis". Final confirmation of the replication mechanism that was implied by the double-helical structure followed in 1958 through the Meselson–Stahl experiment. Further work by Crick and coworkers showed that the genetic code was based on non-overlapping triplets of bases, called codons, allowing Har Gobind Khorana, Robert W. Holley and Marshall Warren Nirenberg to decipher the genetic code. These findings represent the birth of molecular biology.

REFERENCES

1. "deoxyribonucleic acid". Merriam-Webster Dictionary.
2. Alberts B, Johnson A, Lewis J, Raff M, Roberts K, Walter P (2014). Molecular Biology of the Cell (6th ed.). Garland. p. Chapter 4: DNA, Chromosomes and Genomes. ISBN 9780815344322.
3. Purcell, Adam. "DNA". Basic Biology.
4. Nuwer, Rachel (18 July 2015). "Counting All the DNA on Earth". The New York Times. New York: The New York Times Company. ISSN 0362-4331. Retrieved 2015-07-18.
5. "The Biosphere: Diversity of Life". Aspen Global Change Institute. Basalt, CO. Retrieved 2015-07-19.
6. Russell, Peter (2001). iGenetics. New York: Benjamin Cummings. ISBN 0-8053-4553-1.
7. Mashaghi A, Katan A (2013). "A physicist's view of DNA". De Physicus. **24e** *(3): 59–61.* arXiv:1311.2545v1. Bibcode:2013arXiv1311.2545M.
8. Saenger, Wolfram (1984). Principles of Nucleic Acid Structure. New York: Springer-Verlag. ISBN 0-387-90762-9.
9. to:[a] [b] Alberts, Bruce; Johnson, Alexander; Lewis, Julian; Raff, Martin; Roberts, Keith; Walters, Peter (2002). Molecular Biology of the Cell; Fourth Edition. New York and London: Garland Science. ISBN 0-8153-3218-1. OCLC 145080076.
10. Irobalieva, Rossitza N.; Fogg, Jonathan M.; Catanese Jr, Daniel J.; Sutthibutpong, Thana; Chen, Muyuan; Barker, Anna K.; Ludtke, Steven J.; Harris, Sarah A.; Schmid, Michael F. (2015-10-12). "Structural diversity of supercoiled DNA". Nature Communications. **6**: 8440. doi:10.1038/ncomms9440. PMC 4608029. PMID 26455586.
11. to:[a] [b] [c] Watson JD, Crick FH (1953). "A Structure for Deoxyribose Nucleic Acid" (PDF).

Nature. **171** *(4356): 737–738. Bibcode*:1953Natur.171..737W. doi:10.1038/171737a0. PMID 13054692.

12. Mandelkern M, Elias JG, Eden D, Crothers DM (1981). "The dimensions of DNA in solution". J Mol Biol. **152** *(1): 153–61.* doi:10.1016/0022-2836(81)90099-1. PMID 7338906.
13. Gregory SG, Barlow KF, McLay KE, Kaul R, Swarbreck D, Dunham A, et al. (2006). "The DNA sequence and biological annotation of human chromosome 1". Nature. **441** *(7091): 315–21. Bibcode*:2006Natur.441..315G. doi:10.1038/nature04727. PMID 16710414.
14. Watson JD, Crick FH (1953). "A Structure for Deoxyribose Nucleic Acid" (PDF). Nature. **171** *(4356): 737–738. Bibcode*:1953Natur.171..737W. doi:10.1038/171737a0. PMID 13054692. Retrieved 4 May 2009.
15. to:[a] [b] [c] Berg J., Tymoczko J. and Stryer L. (2002) Biochemistry. W. H. Freeman and Company ISBN 0-7167-4955-6
16. Abbreviations and Symbols for Nucleic Acids, Polynucleotides and their ConstituentsIUPAC-IUB Commission on Biochemical Nomenclature (CBN). Retrieved 3 January 2006.
17. to:[a] [b] Ghosh A, Bansal M (2003). "A glossary of DNA structures from A to Z". Acta Crystallogr D. **59** *(4): 620–6.* doi:10.1107/S0907444903003251. PMID 12657780.
18. Created from PDB 1D65
19. Yakovchuk P, Protozanova E, Frank-Kamenetskii MD (2006). "Base-stacking and base-pairing contributions into thermal stability of the DNA double helix". Nucleic Acids Res. **34** *(2): 564–74. doi:10.1093/*nar/gkj454. PMC 1360284. PMID 16449200.
20. Burton E. Tropp - «Molecular Biology"- Jones and Barlett Learning, ISBN 978-0-7637-8663-2
21. "Watson-Crick Structure of DNA - 1953". Steven Carr. Memorial University of Newfoundland. Retrieved 13 July 2016.
22. Verma S, Eckstein F (1998). "Modified oligonucleotides: synthesis and strategy for users". Annu. Rev. Biochem. **67**: 99–134. doi:10.1146/annurev.biochem.67.1.99. PMID 9759484.
23. Kiljunen S, Hakala K, Pinta E, Huttunen S, Pluta P, Gador A, Lönnberg H, Skurnik M (2005). "Yersiniophage phiR1-37 is a tailed bacteriophage having a 270 kb DNA genome with thymidine replaced by deoxyuridine". Microbiology. **151** *(12): 4093–4102. doi:10.1099/mic.0.28265-0. PMID 16339954.*
24. Uchiyama J, Takemura-Uchiyama I, Sakaguchi Y, Gamoh K, Kato SI, Daibata M, Ujihara T, Misawa N, Matsuzaki S (Mar 2014). "Intragenus generalized transduction in Staphylococcus spp. by a novel giant phage". ISME J. **8**: 1949–1952. doi:10.1038/ismej.2014.29.
25. Simpson L (1998). "A base called J". Proc Natl Acad Sci USA. **95** *(5): 2037–2038. Bibcode*:1998PNAS...95.2037S. doi:10.1073/pnas.95.5.2037. PMC 33841. PMID 9482833.
26. Borst P, Sabatini R (2008). "Base J: discovery, biosynthesis, and possible functions". Annual Review of Microbiology. **62**: 235–51. doi:10.1146/annurev.micro.62.081307.162750. PMID 18729733.
27. Cross M, Kieft R, Sabatini R, Wilm M, de Kort M, van der Marel GA, van Boom JH, van Leeuwen F, Borst P (1999). "The modified base J is the target for a novel DNA-binding protein in kinetoplastid protozoans". The EMBO Journal. **18** *(22): 6573–6581. doi:10.1093/*emboj/18.22.6573. PMC 1171720. PMID 10562569.

28. DiPaolo C, Kieft R, Cross M, Sabatini R (2005). "Regulation of trypanosome DNA glycosylation by a SWI2/SNF2-like protein". Mol Cell. **17** *(3): 441–451. doi:10.1016/j.molcel.2004.12.022. PMID 15694344.*

29. Vainio S, Genest PA, ter Riet B, van Luenen H, Borst P (2009). "Evidence that J-binding protein 2 is a thymidine hydroxylase catalyzing the first step in the biosynthesis of DNA base J". Molecular and biochemical parasitology. **164** *(2): 157–61. doi:10.1016/j.molbiopara.2008.12.001. PMID 19114062.*

30. Iyer LM, Tahiliani M, Rao A, Aravind L (2009). "Prediction of novel families of enzymes involved in oxidative and other complex modifications of bases in nucleic acids". Cell Cycle. **8** *(11): 1698–1710. doi:10.4161/cc.8.11.8580. PMC 2995806*. PMID 19411852.

31. van Luenen HG, Farris C, Jan S, Genest PA, Tripathi P, Velds A, Kerkhoven RM, Nieuwland M, Haydock A, Ramasamy G, Vainio S, Heidebrecht T, Perrakis A, Pagie L, van Steensel B, Myler PJ, Borst P (2012). "Leishmania". Cell. **150** *(5): 909–921. doi:10.1016/j.cell.2012.07.030. PMC 3684241*. PMID 22939620.

32. Hazelbaker DZ, Buratowski S (2012). "Transcription: base J blocks the way". Curr Biol. **22** *(22): R960–2. doi:10.1016/j.cub.2012.10.010. PMC 3648658*. PMID 23174300.

33. Wing R, Drew H, Takano T, Broka C, Tanaka S, Itakura K, Dickerson RE (1980). "Crystal structure analysis of a complete turn of B-DNA". Nature. **287** *(5784): 755–8. Bibcode*:1980Natur.287..755W. doi:10.1038/287755a0. PMID 7432492.

34. to:[a] [b] Pabo CO, Sauer RT (1984). "Protein-DNA recognition". Annu Rev Biochem. **53**: 293–321. doi:10.1146/annurev.bi.53.070184.001453. PMID 6236744.

35. Clausen-Schaumann H, Rief M, Tolksdorf C, Gaub HE (2000). "Mechanical stability of single DNA molecules". Biophys J. **78** *(4): 1997–2007. Bibcode*:2000BpJ....78.1997C. doi:10.1016/S0006-3495(00)76747-6. PMC 1300792. PMID 10733978.

36. Chalikian TV, Völker J, Plum GE, Breslauer KJ (1999). "A more unified picture for the thermodynamics of nucleic acid duplex melting: A characterization by calorimetric and volumetric techniques". Proc Natl Acad Sci USA. **96** *(14): 7853–8. Bibcode*:1999PNAS...96.7853C. doi:10.1073/pnas.96.14.7853. PMC 22151. PMID 10393911.

37. deHaseth PL, Helmann JD (1995). "Open complex formation by Escherichia coli RNA polymerase: the mechanism of polymerase-induced strand separation of double helical DNA". Mol Microbiol. **16** *(5): 817–24. doi:10.1111/j.1365-2958.1995.tb02309.x. PMID 7476180.*

38. Isaksson J, Acharya S, Barman J, Cheruku P, Chattopadhyaya J (2004). "Single-stranded adenine-rich DNA and RNA retain structural characteristics of their respective double-stranded conformations and show directional differences in stacking pattern". Biochemistry. **43** *(51): 15996–6010.* doi:10.1021/bi048221v. PMID 15609994.

39. Designation of the two strands of DNA JCBN/NC-IUB Newsletter 1989. Retrieved 7 May 2008

40. Hüttenhofer A, Schattner P, Polacek N (2005). "Non-coding RNAs: hope or hype?". *Trends Genet.* **21** *(5): 289–97. doi:10.1016/j.tig.2005.03.007. PMID 15851066.*

41. Munroe SH (2004). "Diversity of antisense regulation in eukaryotes: multiple mechanisms, emerging patterns". J Cell Biochem. **93** *(4): 664–71. doi:10.1002/jcb.20252. PMID 15389973.*

42. Makalowska I, Lin CF, Makalowski W (2005). "Overlapping genes in vertebrate genomes". Comput Biol Chem. **29** *(1): 1–12. doi:10.1016/j.compbiolchem.2004.12.006.*

PMID 15680581.

43. Johnson ZI, Chisholm SW (2004). “Properties of overlapping genes are conserved across microbial genomes”. Genome Res. **14** *(11): 2268–72. doi:10.1101/gr.2433104. PMC 525685*. PMID 15520290.

44. Lamb RA, Horvath CM (1991). “Diversity of coding strategies in influenza viruses”. Trends Genet. **7** *(8): 261–6.* doi:10.1016/0168-9525(91)90326-L. PMID 1771674.

45. Benham CJ, Mielke SP (2005). “DNA mechanics”. Annu Rev Biomed Eng. **7**: 21–53. doi:10.1146/annurev.bioeng.6.062403.132016. PMID 16004565.

46. Champoux JJ (2001). “DNA topoisomerases: structure, function, and mechanism”. Annu Rev Biochem. **70**: 369–413. doi:10.1146/annurev.biochem.70.1.369. PMID 11395412.

47. Wang JC (2002). “Cellular roles of DNA topoisomerases: a molecular perspective”. Nature Reviews Molecular Cell Biology. **3** *(6): 430–40.* doi:10.1038/nrm831. PMID 12042765.

48. Basu HS, Feuerstein BG, Zarling DA, Shafer RH, Marton LJ (1988). “Recognition of Z-RNA and Z-DNA determinants by polyamines in solution: experimental and theoretical studies”. J Biomol Struct Dyn. **6** *(2): 299–309. doi:10.1080/07391102.1988.10507714 . PMID 2482766.*

49. Franklin RE, Gosling RG (6 March 1953). “The Structure of Sodium Thymonucleate Fibres I. The Influence of Water Content” (PDF). Acta Crystallogr. **6** *(8–9): 673–7.* doi:10.1107/S0365110X53001939. Franklin RE, Gosling RG (1953). “The structure of sodium thymonucleate fibres. II. The cylindrically symmetrical Patterson function”. Acta Crystallogr. **6** *(8–9): 678–85.* doi:10.1107/S0365110X53001940.

50. Franklin RE, Gosling RG (1953). “Molecular Configuration in Sodium Thymonucleate. Franklin R. and Gosling R.G” (PDF). Nature. **171** *(4356): 740–1. Bibcode*:1953Natur.171..740F. doi:10.1038/171740a0. PMID 13054694.

51. Wilkins MH, Stokes AR, Wilson HR (1953). “Molecular Structure of Deoxypentose Nucleic Acids” (PDF). Nature. **171** *(4356): 738–740. Bibcode*:1953Natur.171..738W. doi:10.1038/171738a0. PMID 13054693.

52. Leslie AG, Arnott S, Chandrasekaran R, Ratliff RL (1980). “Polymorphism of DNA double helices”. J. Mol. Biol. **143** *(1): 49–72.* doi:10.1016/0022-2836(80)90124-2. PMID 7441761.

53. Baianu, I.C. (1980). “Structural Order and Partial Disorder in Biological systems”. Bull. Math. Biol. **42** *(4): 137–141.* doi:10.1007/BF02462372. http://cogprints.org/3822/

54. Hosemann R., Bagchi R.N., Direct analysis of diffraction by matter, North-Holland Publs., Amsterdam – New York, 1962.

55. Baianu, I.C. (1978). “X-ray scattering by partially disordered membrane systems”. Acta Crystallogr A. **34** *(5): 751–753. Bibcode*:1978AcCrA..34..751B. doi:10.1107/S0567739478001540.

56. Wahl MC, Sundaralingam M (1997). “Crystal structures of A-DNA duplexes”. Biopolymers. **44** *(1): 45–63.* doi:10.1002/(SICI)1097-0282(1997)44:1<45::AID-BIP4>3.0.CO;2-#. PMID 9097733.

57. Lu XJ, Shakked Z, Olson WK (2000). “A-form conformational motifs in ligand-bound DNA structures”. J. Mol. Biol. **300** *(4): 819–40. doi:10.1006/jmbi.2000.3690. PMID 10891271.*

58. Rothenburg S, Koch-Nolte F, Haag F (2001). “DNA methylation and Z-DNA formation as mediators of quantitative differences in the expression of alleles”. Immunol Rev. **184**:

286–98. doi:10.1034/j.1600-065x.2001.1840125.x. PMID 12086319.

59. Oh DB, Kim YG, Rich A (2002). "Z-DNA-binding proteins can act as potent effectors of gene expression in vivo". Proc. Natl. Acad. Sci. U.S.A. **99** *(26): 16666–71. Bibcode*:2002PNAS...9916666O. doi:10.1073/pnas.262672699. PMC 139201. PMID 12486233.
60. Palmer, Jason (2 December 2010). "Arsenic-loving bacteria may help in hunt for alien life". BBC News. Retrieved 2 December 2010.
61. Bortman, Henry (2 December 2010). "Arsenic-Eating Bacteria Opens New Possibilities for Alien Life". Space.com. Retrieved 2 December 2010.
62. Katsnelson, Alla (2 December 2010). "Arsenic-eating microbe may redefine chemistry of life". Nature News. doi:10.1038/news.2010.645.
63. Cressey, Daniel (3 October 2012). "'Arsenic-life' Bacterium Prefers Phosphorus after all". Nature News. doi:10.1038/nature.2012.11520.
64. Greider CW, Blackburn EH (1985). "Identification of a specific telomere terminal transferase activity in Tetrahymena extracts". Cell. **43** *(2 Pt 1): 405–13.* doi:10.1016/0092-8674(85)90170-9. PMID 3907856.
65. Nugent CI, Lundblad V (1998). "The telomerase reverse transcriptase: components and regulation". Genes Dev. **12** *(8): 1073–85. doi:10.1101/gad.12.8.1073. PMID 9553037.*
66. Wright WE, Tesmer VM, Huffman KE, Levene SD, Shay JW (1997). "Normal human chromosomes have long G-rich telomeric overhangs at one end". Genes Dev. **11** *(21): 2801–9. doi:10.1101/gad.11.21.2801. PMC 316649.* PMID 9353250.
67. Created from
68. Burge S, Parkinson GN, Hazel P, Todd AK, Neidle S (2006). "Quadruplex DNA: sequence, topology and structure". Nucleic Acids Res. **34** *(19): 5402–15. doi:10.1093/* nar/gkl655. PMC 1636468. PMID 17012276.
69. Parkinson GN, Lee MP, Neidle S (2002). "Crystal structure of parallel quadruplexes from human telomeric DNA". Nature. **417** *(6891): 876–80. Bibcode*:2002Natur.417..876P. doi:10.1038/nature755. PMID 12050675.
70. Griffith JD, Comeau L, Rosenfield S, Stansel RM, Bianchi A, Moss H, de Lange T (1999). "Mammalian telomeres end in a large duplex loop". Cell. **97** *(4): 503–14.* doi:10.1016/S0092-8674(00)80760-6. PMID 10338214.
71. Seeman NC (2005). "DNA enables nanoscale control of the structure of matter". Q. Rev. Biophys. **38** *(4): 363–71.* doi:10.1017/S0033583505004087. PMC 3478329. PMID 16515737.
72. Hu Q, Rosenfeld MG (2012). "Epigenetic regulation of human embryonic stem cells". Frontiers in Genetics. **3**: 238. doi:10.3389/fgene.2012.00238. PMC 3488762. PMID 23133442.
73. Klose RJ, Bird AP (2006). "Genomic DNA methylation: the mark and its mediators". Trends Biochem Sci. **31** *(2): 89–97. doi:10.1016/j.tibs.2005.12.008. PMID 16403636.*
74. Bird A (2002). "DNA methylation patterns and epigenetic memory". Genes Dev. **16** *(1): 6–21. doi:10.1101/gad.947102. PMID 11782440.*
75. Walsh CP, Xu GL (2006). "Cytosine methylation and DNA repair". Curr Top Microbiol Immunol. Current Topics in Microbiology and Immunology. **301**: 283–315. doi:10.1007/3-540-31390-7_11. ISBN 3-540-29114-8. PMID 16570853.
76. Kriaucionis S, Heintz N (2009). "The nuclear DNA base 5-hydroxymethylcytosine

is present in Purkinje neurons and the brain". Science. **324** *(5929): 929–30. Bibcode*:2009Sci...324..929K. doi:10.1126/science.1169786. PMC 3263819. PMID 19372393.

77. Ratel D, Ravanat JL, Berger F, Wion D (2006). "N6-methyladenine: the other methylated base of DNA". BioEssays. **28** *(3): 309–15. doi:10.1002/bies.20342. PMC 2754416.* PMID 16479578.

78. Gommers-Ampt JH, Van Leeuwen F, de Beer AL, Vliegenthart JF, Dizdaroglu M, Kowalak JA, Crain PF, Borst P (1993). "beta-D-glucosyl-hydroxymethyluracil: a novel modified base present in the DNA of the parasitic protozoan T. brucei". Cell. **75** *(6): 1129–36.* doi:10.1016/0092-8674(93)90322-H. PMID 8261512.

79. Created from PDB 1JDG

80. Douki T, Reynaud-Angelin A, Cadet J, Sage E (2003). "Bipyrimidine photoproducts rather than oxidative lesions are the main type of DNA damage involved in the genotoxic effect of solar UVA radiation". Biochemistry. **42** *(30): 9221–6.* doi:10.1021/bi034593c. PMID 12885257.

81. Cadet J, Delatour T, Douki T, Gasparutto D, Pouget JP, Ravanat JL, Sauvaigo S (1999). "Hydroxyl radicals and DNA base damage". Mutat Res. **424** *(1–2): 9–21.* doi:10.1016/S0027-5107(99)00004-4. PMID 10064846.

82. Beckman KB, Ames BN (1997). "Oxidative decay of DNA". J. Biol. Chem. **272** *(32): 19633–6. doi:10.1074/jbc.272.32.19633. PMID 9289489.*

83. Valerie K, Povirk LF (2003). "Regulation and mechanisms of mammalian double-strand break repair". Oncogene. **22** *(37): 5792–812. doi:10.1038/sj.onc.1206679. PMID 12947387.*

84. Johnson, George (28 December 2010). "Unearthing Prehistoric Tumors, and Debate". The New York Times. If we lived long enough, sooner or later we all would get cancer.

85. Alberts, B; Johnson A, Lewis J; et al. (2002). "The Preventable Causes of Cancer". Molecular biology of the cell (4th ed.). New York: Garland Science. ISBN 0-8153-4072-9. A certain irreducible background incidence of cancer is to be expected regardless of circumstances: mutations can never be absolutely avoided, because they are an inescapable consequence of fundamental limitations on the accuracy of DNA replication, as discussed in Chapter 5. If a human could live long enough, it is inevitable that at least one of his or her cells would eventually accumulate a set of mutations sufficient for cancer to develop.

86. Bernstein H, Payne CM, Bernstein C, Garewal H, Dvorak K (2008). Cancer and aging as consequences of un-repaired DNA damage. In: New Research on DNA Damages (Editors: Honoka Kimura and Aoi Suzuki) Nova Science Publishers, Inc., New York, Chapter 1, pp. 1–47. open access, but read only https://www.novapublishers.com/catalog/product_info.php?products_id=43247 ISBN 978-1604565812

87. Hoeijmakers JH (October 2009). "DNA damage, aging, and cancer". N. Engl. J. Med. **361** *(15): 1475–85.* doi:10.1056/NEJMra0804615. PMID 19812404.

88. Freitas AA, de Magalhães JP (2011). "A review and appraisal of the DNA damage theory of ageing". Mutat. Res. **728** *(1–2): 12–22. doi:10.1016/j.mrrev.2011.05.001. PMID 21600302.*

89. Ferguson LR, Denny WA (1991). "The genetic toxicology of acridines". Mutat Res. **258**(2): 123–60. doi:10.1016/0165-1110(91)90006-H. PMID 1881402.

90. Stephens TD, Bunde CJ, Fillmore BJ (2000). "Mechanism of action in thalidomide teratogenesis". Biochem Pharmacol. **59** *(12): 1489–99.* doi:10.1016/S0006-2952(99)00388-3. PMID 10799645.

91. Jeffrey AM (1985). "DNA modification by chemical carcinogens". Pharmacol Ther. **28** *(2): 237–72.* doi:10.1016/0163-7258(85)90013-0. PMID 3936066.

92. Braña MF, Cacho M, Gradillas A, de Pascual-Teresa B, Ramos A (2001). "Intercalators as anticancer drugs". Curr Pharm Des. **7** *(17): 1745–80.* doi:10.2174/1381612013397113. PMID 11562309.

93. Venter JC, Adams MD, Myers EW, Li PW, Mural RJ, Sutton GG, et al. (2001). "The sequence of the human genome". Science. **291** *(5507): 1304–51. Bibcode*:2001Sci...291.1304V. doi:10.1126/science.1058040. PMID 11181995.

94. Thanbichler M, Wang SC, Shapiro L (2005). "The bacterial nucleoid: a highly organized and dynamic structure". J Cell Biochem. **96** *(3): 506–21. doi:10.1002/jcb.20519. PMID 15988757.*

95. Wolfsberg TG, McEntyre J, Schuler GD (2001). "Guide to the draft human genome". Nature. **409** *(6822): 824–6. Bibcode*:2001Natur.409..824W. doi:10.1038/35057000. PMID 11236998.

96. Gregory TR (2005). "The C-value enigma in plants and animals: a review of parallels and an appeal for partnership". Annals of Botany. **95** *(1): 133–46. doi:10.1093/*aob/mci009. PMID 15596463.

97. Birney E, Stamatoyannopoulos JA, Dutta A, Guigó R, Gingeras TR, Margulies EH, et al. (2007). "Identification and analysis of functional elements in 1% of the human genome by the ENCODE pilot project". Nature. **447** *(7146): 799–816. Bibcode*:2007Natur.447..799B. doi:10.1038/nature05874. PMC 2212820. PMID 17571346.

98. Created from PDB 1MSW

99. Pidoux AL, Allshire RC (2005). "The role of heterochromatin in centromere function". Philosophical Transactions of the Royal Society B. **360** *(1455): 569–79. doi:10.1098/rstb.2004.1611. PMC 1569473*. PMID 15905142.

100. Harrison PM, Hegyi H, Balasubramanian S, Luscombe NM, Bertone P, Echols N, Johnson T, Gerstein M (2002). "Molecular Fossils in the Human Genome: Identification and Analysis of the Pseudogenes in Chromosomes 21 and 22". Genome Res. **12** *(2): 272–80. doi:10.1101/gr.207102. PMC 155275*. PMID 11827946.

101. Harrison PM, Gerstein M (2002). "Studying genomes through the aeons: protein families, pseudogenes and proteome evolution". J Mol Biol. **318** *(5): 1155–74.* doi:10.1016/S0022-2836(02)00109-2. PMID 12083509.

102. Albà M (2001). "Replicative DNA polymerases". Genome Biol. **2** *(1): reviews3002.1–reviews3002.4.* doi:10.1186/gb-2001-2-1-reviews3002. PMC 150442. PMID 11178285.

103. Tani, Katsuji; Nasu, Masao (2010). "Roles of Extracellular DNA in Bacterial Ecosystems". In Kikuchi, Yo; Rykova, Elena Y. Extracellular Nucleic Acids. Springer. pp. 25–38. ISBN 978-3-642-12616-1.

104. Vlassov, V. V.; Laktionov, P. P.; Rykova, E. Y. (2007). "Extracellular nucleic acids". BioEssays. **29**: 654–667. doi:10.1002/bies.20604.

105. Finkel, S. E.; Kolter, R. (2001). "DNA as a nutrient: novel role for bacterial competence gene homologs". J. Bacteriol. **183**: 6288–6293. doi:10.1128/JB.183.21.6288-6293.2001.

106. Mulcahy, H.; Charron-Mazenod, L.; Lewenza, S. (2008). "Extracellular DNA chelates cations and induces antibiotic resistance in Pseudomonas aeruginosa biofilms". PLoS

Pathog. **4**: e1000213. doi:10.1371/journal.ppat.1000213.

107. Berne, C.; Kysela, D. T.; Brun, Y. V. (2010). "A bacterial extracellular DNA inhibits settling of motile progeny cells within a biofilm". Mol. Microbiol. **77**: 815–829. doi:10.1111/j.1365-2958.2010.07267.x.
108. Whitchurch, C. B.; Tolker-Nielsen, T.; Ragas, P. C.; Mattick, J. S. (2002). "Extracellular DNA required for bacterial biofilm formation" (PDF). Science. **295**: 1487. doi:10.1126/science.295.5559.1487.
109. Hu, W.; Li, L.; Sharma, S.; Wang, J.; McHardy, I.; Lux, R.; Yang, Z.; He, X.; Gimzewski, J. K.; Li, Y.; Shi, W. (2012). "DNA Builds and Strengthens the Extracellular Matrix in Myxococcus xanthus Biofilms by Interacting with Exopolysaccharides". PLoS ONE. **7** *(12): e51905. Bibcode*:2012PLoSO...751905H. doi:10.1371/journal.pone.0051905.
110. Sandman K, Pereira SL, Reeve JN (1998). "Diversity of prokaryotic chromosomal proteins and the origin of the nucleosome". Cell Mol Life Sci. **54** *(12): 1350–64.* doi:10.1007/s000180050259. PMID 9893710.
111. Dame RT (2005). "The role of nucleoid-associated proteins in the organization and compaction of bacterial chromatin". Mol. Microbiol. **56** *(4): 858–70. doi:10.1111/j.1365-2958.2005.04598.x. PMID 15853876.*
112. Luger K, Mäder AW, Richmond RK, Sargent DF, Richmond TJ (1997). "Crystal structure of the nucleosome core particle at 2.8 A resolution". Nature. **389** *(6648): 251–60. Bibcode*:1997Natur.389..251L. doi:10.1038/38444. PMID 9305837.
113. Jenuwein T, Allis CD (2001). "Translating the histone code". Science. **293** *(5532): 1074–80. doi:10.1126/science.1063127. PMID 11498575.*
114. Ito T (2003). "Nucleosome assembly and remodelling". Curr Top Microbiol Immunol. Current Topics in Microbiology and Immunology. **274**: 1–22. doi:10.1007/978-3-642-55747-7_1. ISBN 978-3-540-44208-0. PMID 12596902.
115. Thomas JO (2001). "HMG1 and 2: architectural DNA-binding proteins". Biochem Soc Trans. **29** *(*Pt 4): 395–401. doi:10.1042/BST0290395. PMID 11497996.
116. Grosschedl R, Giese K, Pagel J (1994). "HMG domain proteins: architectural elements in the assembly of nucleoprotein structures". Trends Genet. **10** *(3): 94–100.* doi:10.1016/0168-9525(94)90232-1. PMID 8178371.
117. Iftode C, Daniely Y, Borowiec JA (1999). "Replication protein A (RPA): the eukaryotic SSB". Crit Rev Biochem Mol Biol. **34** *(3): 141–80.* doi:10.1080/10409239991209255. PMID 10473346.
118. Created from PDB 1LMB
119. Myers LC, Kornberg RD (2000). "Mediator of transcriptional regulation". Annu Rev Biochem. **69**: 729–49. doi:10.1146/annurev.biochem.69.1.729. PMID 10966474.
120. Spiegelman BM, Heinrich R (2004). "Biological control through regulated transcriptional coactivators". Cell. **119** *(2): 157–67. doi:10.1016/j.cell.2004.09.037. PMID 15479634.*
121. Li Z, Van Calcar S, Qu C, Cavenee WK, Zhang MQ, Ren B (2003). "A global transcriptional regulatory role for c-Myc in Burkitt's lymphoma cells". Proc Natl Acad Sci USA. **100** *(14): 8164–9. Bibcode*:2003PNAS..100.8164L. doi:10.1073/pnas.1332764100. PMC 166200. PMID 12808131.
122. Created from PDB 1RVA
123. Bickle TA, Krüger DH (1993). "Biology of DNA restriction". Microbiol Rev. **57** *(2): 434–50. PMC 372918.* PMID 8336674.

124. Doherty AJ, Suh SW (2000). "Structural and mechanistic conservation in DNA ligases". Nucleic Acids Res. **28** *(21): 4051–8. doi:10.1093/*nar/28.21.4051. PMC 113121. PMID 11058099.

125. Schoeffler AJ, Berger JM (2005). "Recent advances in understanding structure-function relationships in the type II topoisomerase mechanism". Biochem Soc Trans. **33** *(*Pt 6): 1465–70. doi:10.1042/BST20051465. PMID 16246147.

126. Tuteja N, Tuteja R (2004). "Unraveling DNA helicases. Motif, structure, mechanism and function". Eur J Biochem. **271** *(10): 1849–63. doi:10.1111/j.1432-1033.2004.04094.x. PMID 15128295.*

127. Joyce CM, Steitz TA (1995). "Polymerase structures and function: variations on a theme?". *J* Bacteriol. **177** *(22): 6321–9. PMC 177480.* PMID 7592405.

128. Hubscher U, Maga G, Spadari S (2002). "Eukaryotic DNA polymerases". Annu Rev Biochem. **71**: 133–63. doi:10.1146/annurev.biochem.71.090501.150041. PMID 12045093.

129. Johnson A, O'Donnell M (2005). "Cellular DNA replicases: components and dynamics at the replication fork". Annu Rev Biochem. **74**: 283–315. doi:10.1146/annurev.biochem.73.011303.073859. PMID 15952889.

130. Tarrago-Litvak L, Andréola ML, Nevinsky GA, Sarih-Cottin L, Litvak S (1 May 1994). "The reverse transcriptase of HIV-1: from enzymology to therapeutic intervention". FASEB J. **8**(8): 497–503. PMID 7514143.

131. Martinez E (2002). "Multi-protein complexes in eukaryotic gene transcription". Plant Mol Biol. **50** *(6): 925–47.* doi:10.1023/A:1021258713850. PMID 12516863.

132. Created from PDB 1M6G

133. Cremer T, Cremer C (2001). "Chromosome territories, nuclear architecture and gene regulation in mammalian cells". Nature Reviews Genetics. **2** *(4): 292–301.* doi:10.1038/35066075. PMID 11283701.

134. Pál C, Papp B, Lercher MJ (2006). "An integrated view of protein evolution". Nature Reviews Genetics. **7** *(5): 337–48.* doi:10.1038/nrg1838. PMID 16619049.

135. O'Driscoll M, Jeggo PA (2006). "The role of double-strand break repair – insights from human genetics". Nature Reviews Genetics. **7** *(1): 45–54.* doi:10.1038/nrg1746. PMID 16369571.

136. Vispé S, Defais M (1997). "Mammalian Rad51 protein: a RecA homologue with pleiotropic functions". Biochimie. **79** *(9–10): 587–92.* doi:10.1016/S0300-9084(97)82007-X. PMID 9466696.

137. Neale MJ, Keeney S (2006). "Clarifying the mechanics of DNA strand exchange in meiotic recombination". Nature. **442** *(7099): 153–8. Bibcode*:2006Natur.442..153N. doi:10.1038/nature04885. PMID 16838012.

138. Dickman MJ, Ingleston SM, Sedelnikova SE, Rafferty JB, Lloyd RG, Grasby JA, Hornby DP (2002). "The RuvABC resolvasome". Eur J Biochem. **269** *(22): 5492–501. doi:10.1046/j.1432-1033.2002.03250.x. PMID 12423347.*

139. Joyce GF (2002). "The antiquity of RNA-based evolution". Nature. **418** *(6894): 214–21. Bibcode*:2002Natur.418..214J. doi:10.1038/418214a. PMID 12110897.

140. Orgel LE (2004). "Prebiotic chemistry and the origin of the RNA world". Crit Rev Biochem Mol Biol. **39** *(2): 99–123.* doi:10.1080/10409230490460765. PMID 15217990.

141. Davenport RJ (2001). "Ribozymes. Making copies in the RNA world". Science. **292**(5520): 1278. doi:10.1126/science.292.5520.1278a. PMID 11360970.

142. Szathmáry E (1992). "What is the optimum size for the genetic alphabet?". Proc Natl Acad Sci USA. **89** *(7): 2614–8. Bibcode*:1992PNAS...89.2614S. doi:10.1073/pnas.89.7.2614. PMC 48712. PMID 1372984.

143. Lindahl T (1993). "Instability and decay of the primary structure of DNA". Nature. **362**(6422): 709–15. Bibcode:1993Natur.362..709L. doi:10.1038/362709a0. PMID 8469282.

144. Vreeland RH, Rosenzweig WD, Powers DW (2000). "Isolation of a 250 million-year-old halotolerant bacterium from a primary salt crystal". Nature. **407** *(6806): 897–900.* doi:10.1038/35038060. PMID 11057666.

145. Hebsgaard MB, Phillips MJ, Willerslev E (2005). "Geologically ancient DNA: fact or artefact?". *Trends* Microbiol. **13** *(5): 212–20. doi:10.1016/j.tim.2005.03.010. PMID 15866038.*

146. Nickle DC, Learn GH, Rain MW, Mullins JI, Mittler JE (2002). "Curiously modern DNA for a "250 million-year-old" bacterium". J Mol Evol. **54** *(1): 134–7.* doi:10.1007/s00239-001-0025-x. PMID 11734907.

147. Callahan MP, Smith KE, Cleaves HJ, Ruzicka J, Stern JC, Glavin DP, House CH, Dworkin JP (August 2011). "Carbonaceous meteorites contain a wide range of extraterrestrial nucleobases". Proc. Natl. Acad. Sci. U.S.A. **108** *(34): 13995–8. Bibcode*:2011PNAS..10813995C. doi:10.1073/pnas.1106493108. PMC 3161613. PMID 21836052.

148. Steigerwald, John (8 August 2011). "NASA Researchers: DNA Building Blocks Can Be Made in Space". NASA. Retrieved 10 August 2011.

149. ScienceDaily Staff (9 August 2011). "DNA Building Blocks Can Be Made in Space, NASA Evidence Suggests". ScienceDaily. Retrieved 9 August 2011.

150. Marlaire, Ruth (3 March 2015). "NASA Ames Reproduces the Building Blocks of Life in Laboratory". NASA. Retrieved 5 March 2015.

151. Goff SP, Berg P (1976). "Construction of hybrid viruses containing SV40 and lambda phage DNA segments and their propagation in cultured monkey cells". Cell. **9** *(4 PT 2): 695–705.* doi:10.1016/0092-8674(76)90133-1. PMID 189942.

152. Houdebine LM (2007). "Transgenic animal models in biomedical research". Methods Mol Biol. **360**: 163–202. doi:10.1385/1-59745-165-7:163. ISBN 1-59745-165-7. PMID 17172731.

153. Daniell H, Dhingra A (2002). "Multigene engineering: dawn of an exciting new era in biotechnology". Current Opinion in Biotechnology. **13** *(2): 136–41.* doi:10.1016/S0958-1669(02)00297-5. PMC 3481857. PMID 11950565.

154. Job D (2002). "Plant biotechnology in agriculture". Biochimie. **84** *(11): 1105–10.* doi:10.1016/S0300-9084(02)00013-5. PMID 12595138.

155. Collins A, Morton NE (1994). "Likelihood ratios for DNA identification". Proc Natl Acad Sci USA. **91** *(13): 6007–11. Bibcode*:1994PNAS...91.6007C. doi:10.1073/pnas.91.13.6007. PMC 44126. PMID 8016106.

156. Weir BS, Triggs CM, Starling L, Stowell LI, Walsh KA, Buckleton J (1997). "Interpreting DNA mixtures". J Forensic Sci. **42** *(2): 213–22. PMID 9068179.*

157. Jeffreys AJ, Wilson V, Thein SL (1985). "Individual-specific ‹fingerprints' of human DNA". Nature. **316** *(6023): 76–9. Bibcode*:1985Natur.316...76J. doi:10.1038/316076a0. PMID 2989708.

158. Colin Pitchfork — first murder conviction on DNA evidence also clears the prime suspect Forensic Science Service Accessed 23 December 2006

159. "DNA Identification in Mass Fatality Incidents". National Institute of Justice. September 2006.

160. "Paternity Blood Tests That Work Early in a Pregnancy" New York Times June 20, 2012

161. Breaker, Ronald R.; Joyce, Gerald F. (1994-01-12). "A DNA enzyme that cleaves RNA". Chemistry & Biology. **1** *(4): 223–229.* doi:10.1016/1074-5521(94)90014-0. ISSN 1074-5521. PMID 9383394.

162. Chandra, Madhavaiah; Sachdeva, Amit; Silverman, Scott K. "DNA-catalyzed sequence-specific hydrolysis of DNA". Nature Chemical Biology. **5** *(10): 718–720. doi:10.1038/nchembio.201. PMC 2746877.* PMID 19684594.

163. Carmi, Nir; Shultz, Lisa A.; Breaker, Ronald R. (1996-01-12). "In vitro selection of self-cleaving DNAs". Chemistry & Biology. **3** *(12): 1039–1046.* doi:10.1016/S1074-5521(96)90170-2. ISSN 1074-5521. PMID 9000012.

164. Torabi, Seyed-Fakhreddin; Wu, Peiwen; McGhee, Claire E.; Chen, Lu; Hwang, Kevin; Zheng, Nan; Cheng, Jianjun; Lu, Yi (2015-05-12). "In vitro selection of a sodium-specific DNAzyme and its application in intracellular sensing". Proceedings of the National Academy of Sciences. **112** *(19): 5903–5908. doi:10.1073/pnas.1420361112. ISSN 0027-8424. PMC 4434688.* PMID 25918425.

165. Baldi, Pierre; Brunak, Soren (2001). Bioinformatics: The Machine Learning Approach. MIT Press. ISBN 978-0-262-02506-5. OCLC 45951728.

166. Gusfield, Dan. Algorithms on Strings, Trees, and Sequences: Computer Science and Computational Biology. Cambridge University Press, 15 January 1997. ISBN 978-0-521-58519-4.

167. Sjölander K (2004). "Phylogenomic inference of protein molecular function: advances and challenges". Bioinformatics. **20** *(2): 170–9. doi:10.1093/bioinformatics/bth021. PMID 14734307.*

168. Mount DM (2004). Bioinformatics: Sequence and Genome Analysis (2 ed.). Cold Spring Harbor, NY: Cold Spring Harbor Laboratory Press. ISBN 0-87969-712-1. OCLC 55106399.

169. Rothemund PW (2006). "Folding DNA to create nanoscale shapes and patterns". Nature. **440** *(7082): 297–302. Bibcode*:2006Natur.440..297R. doi:10.1038/nature04586. PMID 16541064.

170. Andersen ES, Dong M, Nielsen MM, Jahn K, Subramani R, Mamdouh W, Golas MM, Sander B, Stark H, Oliveira CL, Pedersen JS, Birkedal V, Besenbacher F, Gothelf KV, Kjems J (2009). "Self-assembly of a nanoscale DNA box with a controllable lid". Nature. **459** *(7243): 73–6. Bibcode*:2009Natur.459...73A. doi:10.1038/nature07971. PMID 19424153.

171. Ishitsuka Y, Ha T (2009). "DNA nanotechnology: a nanomachine goes live". Nat Nanotechnol. **4** *(5): 281–2. Bibcode*:2009NatNa...4..281I. doi:10.1038/nnano.2009.101. PMID 19421208.

172. Aldaye FA, Palmer AL, Sleiman HF (2008). "Assembling materials with DNA as the guide". Science. **321** *(5897): 1795–9. Bibcode*:2008Sci...321.1795A. doi:10.1126/science.1154533. PMID 18818351.

173. Wray GA (2002). "Dating branches on the Tree of Life using DNA". Genome Biol. **3**(1): reviews0001.1–reviews0001.7. doi:10.1046/j.1525-142X.1999.99010.x. PMC 150454. PMID 11806830.

174. Lost Tribes of Israel, Nova, PBS airdate: 22 February 2000. Transcript available from PBS.org. Retrieved 4 March 2006.

175. Kleiman, Yaakov. "The Cohanim/DNA Connection: The fascinating story of how DNA studies confirm an ancient biblical tradition". aish.com (13 January 2000). Retrieved 4 March 2006.

176. Goldman N, Bertone P, Chen S, Dessimoz C, LeProust EM, Sipos B, Birney E (23 January 2013). "Towards practical, high-capacity, low-maintenance information storage in synthesized DNA". Nature. **494** *(7435): 77–80. Bibcode*:2013Natur.494...77G. doi:10.1038/nature11875. PMC 3672958. PMID 23354052.

177. Naik, Gautam (24 January 2013). "Storing Digital Data in DNA". Wall Street Journal. Retrieved 24 January 2013.

178. Comment by Dandekar, T., Lopez, D., Schaack, D. (2013) http://www.nature.com/nature/journal/v494/n7435/abs/nature11875.html#comment-57415

179. Emerging Technology Final, Dandekar T., Lopez, D., Programmable bacterial membranes with active DNA storage; presentation for the University of Würzburg for the Royal Society for Chemistry, London, 2016-06-29

180. Patent "Molecular highly integrated data storage via active control DNA", DE102013004584 A1, https://www.google.com/patents/DE102013004584A1"

181. Miescher, Friedrich (1871) "Ueber die chemische Zusammensetzung der Eiterzellen"(On the chemical composition of pus cells), Medicinisch-chemische Untersuchungen, **4** : 441–460. From p. 456: "Ich habe mich daher später mit meinen Versuchen an die ganzen Kerne gehalten, die Trennung der Körper, die ich einstweilen ohne weiteres Präjudiz als lösliches und unlösliches Nuclein bezeichnen will, einem günstigeren Material *überlassend*." (Therefore, in my experiments I subsequently limited myself to the whole nucleus, leaving to a more favorable material the separation of the substances, that for the present, without further prejudice, I will designate as soluble and insoluble nuclear material ("Nuclein").)

182. Dahm R (2008). "Discovering DNA: Friedrich Miescher and the early years of nucleic acid research". Hum. Genet. **122** *(6): 565–81.* doi:10.1007/s00439-007-0433-0. PMID 17901982.

183. Albrect Kossel (1879) "Ueber Nucleïn der Hefe" (On nuclein in yeast) Zeitschrift für physiologische Chemie, **3** : 284-291.

184. Albrect Kossel (1880) "Ueber Nucleïn der Hefe II" (On nuclein in yeast, Part 2) Zeitschrift für physiologische Chemie, **4** : 290-295.

185. Albrect Kossel (1881) "Ueber die Verbreitung des Hypoxanthins im Thier- und Pflanzenreich" (On the distribution of hypoxanthins in the animal and plant kingdoms) Zeitschrift für physiologische Chemie, **5** : 267-271.

186. Albrect Kossel, Untersuchungen über die Nucleine und ihre Spaltungsprodukte[Inves tigations into nuclein and its cleavage products] (Strassburg, Germany: K.J. Trübne, 1881), 19 pages.

187. Albrect Kossel (1882) "Ueber Xanthin und Hypoxanthin" (On xanthin and hypoxanthin), Zeitschrift für physiologische Chemie, **6** : 422-431.

188. Albrect Kossel (1883) "Zur Chemie des Zellkerns" (On the chemistry of the cell nucleus), Zeitschrift für physiologische Chemie, **7** : 7-22.

189. Albrect Kossel (1886) "Weitere Beiträge zur Chemie des Zellkerns" (Further contributions to the chemistry of the cell nucleus), Zeitschrift für Physiologische Chemie, **10** : 248-264. Available on-line at: Max Planck Institute for the History of Science, Berlin, Germany. On p. 264, Kossel remarked presciently: "Der Erforschung der quantitativen Verhältnisse

der vier stickstoffreichen Basen, der Abhängigkeit ihrer Menge von den physiologischen Zuständen der Zelle, verspricht wichtige Aufschlüsse über die elementaren physiologisch-chemischen Vorgänge." (The study of the quantitative relations of the four nitrogenous bases — [and] of the dependence of their quantity on the physiological states of the cell — promises important insights into the fundamental physiological-chemical processes.)

190. Jones ME (September 1953). "Albrecht Kossel, A Biographical Sketch". Yale Journal of Biology and Medicine. National Center for Biotechnology Information. **26** *(1): 80–97. PMC 2599350.* PMID 13103145.

191. Levene P (1 December 1919). "The structure of yeast nucleic acid". J Biol Chem. **40** *(2): 415–24.*

192. W. T. Astbury and Florence O. Bell (1938) "Some recent developments in the X-ray study of proteins and related structures," Cold Spring Harbor Symposia on Quantitative Biology, **6** : 109-121. Available on-line at: University of Leeds.

193. Astbury, W. T., (1947) "X-ray studies of nucleic acids," Symposia of the Society for Experimental Biology, **1** : 66-76. Available on-line at: Oregon State University.

194. Koltsov proposed that a cell's genetic information was encoded in a long chain of amino acids. See:

195. Н. К. Кольцов, "Физико-химические основы морфологии» (The physical-chemical basis of morphology) -- speech given at the 3rd All-Union Meeting of Zoologist, Anatomists, and Histologists at Leningrad, U.S.S.R., December 12, 1927.

196. Reprinted in German as: Nikolaj K. Koltzoff (1928) "Physikalisch-chemische Grundlagen der Morphologie" (The physical-chemical basis of morphology), Biologisches Zentralblatt, **48** (6) : 345-369.

197. In 1934, Koltsov contended that the proteins that contain a cell's genetic information replicate. See: N. K. Koltzoff (October 5, 1934) "The structure of the chromosomes in the salivary glands of Drosophila," Science, **80** (2075) : 312-313. From page 313: "I think that the size of the chromosomes in the salivary glands [of Drosophila] is determined through the multiplication of genonemes. By this term I designate the axial thread of the chromosome, in which the geneticists locate the linear combination of genes; ... In the normal chromosome there is usually only one genoneme; before cell-division this genoneme has become divided into two strands."

198. Soyfer VN (2001). "The consequences of political dictatorship for Russian science". Nature Reviews Genetics. **2** *(9): 723–729.* doi:10.1038/35088598. PMID 11533721.

199. Griffith F (January 1928). "The significance of pneumococcal types". The Journal of Hygiene (London). **27** *(2): 113–59.* doi:10.1017/S0022172400031879. PMC 2167760. PMID 20474956.

200. Lorenz MG, Wackernagel W (1994). "Bacterial gene transfer by natural genetic transformation in the environment". Microbiol. Rev. **58** *(3): 563–602. PMC 372978.* PMID 7968924.

201. Avery OT, Macleod CM, McCarty M (1944). "Studies on the Chemical Nature of the Substance Inducing Transformation of Pneumococcal Types: Induction of Transformation by a Desoxyribonucleic Acid Fraction Isolated from Pneumococcus Type Iii". J Exp Med. **79** *(2): 137–158. doi:10.1084/jem.79.2.137. PMC 2135445.* PMID 19871359.

202. Hershey AD, Chase M (1952). "Independent Functions of Viral Protein and Nucleic Acid in Growth of Bacteriophage". J Gen Physiol. **36** *(1): 39–56. doi:10.1085/jgp.36.1.39. PMC 2147348.* PMID 12981234.

203. The B-DNA X-ray pattern on the right of this linked image was obtained by Rosalind Franklin and Raymond Gosling in May 1952 at high hydration levels of DNA and it has been labeled as "Photo 51"

204. Nature Archives Double Helix of DNA: 50 Years

205. "Original X-ray diffraction image". Osulibrary.oregonstate.edu. Retrieved 6 February2011.

206. The Nobel Prize in Physiology or Medicine 1962 Nobelprize .org Accessed 22 December 06

207. Maddox B (23 January 2003). "The double helix and the 'wronged heroine'" (PDF). Nature. **421** *(6921): 407–408. Bibcode*:2003Natur.421..407M. doi:10.1038/nature01399. PMID 12540909.

208. Crick, F.H.C. On degenerate templates and the adaptor hypothesis (PDF).genome. wellcome.ac.uk (Lecture, 1955). Retrieved 22 December 2006.

209. Meselson M, Stahl FW (1958). "The replication of DNA in Escherichia coli". Proc Natl Acad Sci USA. **44** *(7): 671–82. Bibcode*:1958PNAS...44..671M. doi:10.1073/pnas.44.7.671. PMC 528642. PMID 16590258.

210. The Nobel Prize in Physiology or Medicine 1968 Nobelprize.org Accessed 22 December 06

CHAPTER 6

RNA Molecule

Chapter Contents

Ribonucleic acid, or RNA is one of the three major biological macromolecules that are essential for all known forms of life (along with DNA and proteins). A central tenet of molecular biology states that the flow of genetic information in a cell is from DNA through RNA to proteins: "DNA makes RNA makes protein". Proteins are the workhorses of the cell; they play leading roles in the cell as enzymes, as structural components, and in cell signaling, to name just a few.

INTRODUCTION

Ribonucleic acid (RNA) is a polymeric molecule essential in various biological roles in coding, decoding, regulation, and expression of genes. RNA and DNA are nucleic acids, and, along with proteins and carbohydrates, constitute the four major macromolecules essential for all known forms of life. Like DNA, RNA is assembled as a chain of nucleotides, but unlike DNA it is more often found in nature as a single-strand folded onto itself, rather than a paired double-strand. Cellular organisms use messenger RNA (*mRNA*) to convey genetic information (using the letters G, U, A, and C to denote the nitrogenous bases guanine, uracil, adenine, and cytosine) that directs synthesis of specific proteins. Many viruses encode their genetic information using an RNA genome.

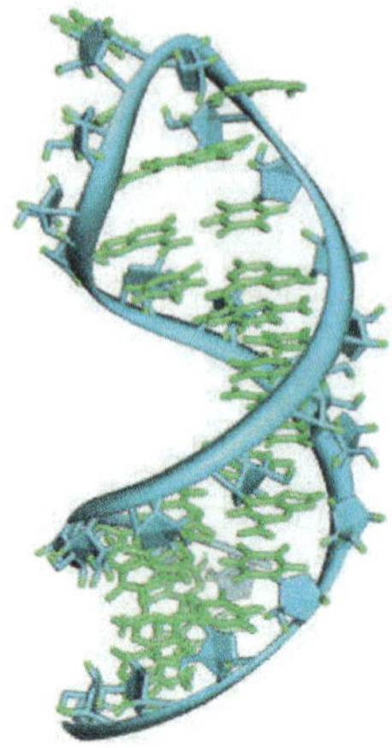

A hairpin loop from a pre-mRNA. Highlighted are the nucleobases(green) and the ribose-phosphate backbone (blue). Note that this is a single strand of RNA that folds back upon itself.

Some RNA molecules play an active role within cells by catalyzing biological reactions, controlling gene expression, or sensing and communicating responses to cellular signals. One of these active processes is protein synthesis, a universal function where RNA molecules direct the assembly of proteins on ribosomes. This process uses transfer RNA (***tRNA***) molecules to deliver amino acids to the ribosome, where ribosomal RNA (***rRNA***) then links amino acids together to form proteins.

COMPARISON WITH DNA

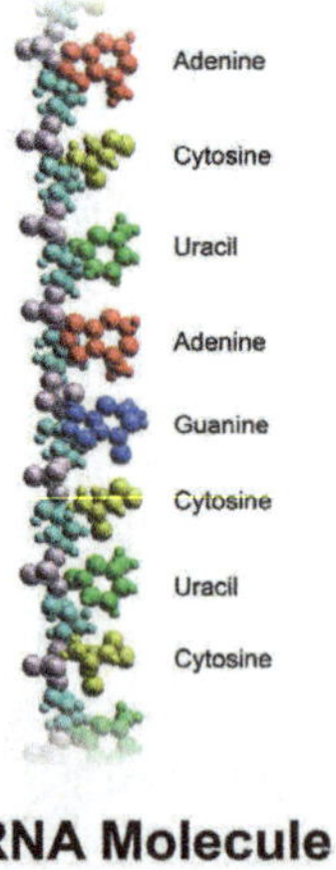

Bases in an RNA molecule.

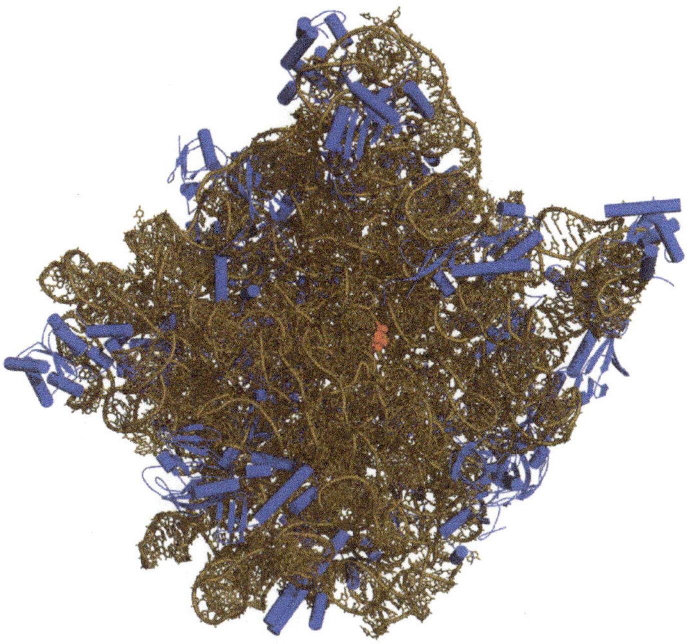

Three-dimensional representation of the 50S ribosomal subunit. Ribosomal RNA is in ochre, proteins in blue. The active site is a small segment of rRNA, indicated in red.

The chemical structure of RNA is very similar to that of DNA, but differs in three main ways:

- Unlike double-stranded DNA, RNA is a single-stranded molecule in many of its biological roles and has a much shorter chain of nucleotides. However, RNA can, by complementary base pairing, form intrastrand (i.e., single-strand) double helixes, as in tRNA.
- While DNA contains *deoxyribose*, RNA contains *ribose* (in deoxyribose there is no hydroxyl group attached to the pentose ring in the 2' position). These hydroxyl groups make RNA less stable than DNA because it is more prone to hydrolysis.
- The complementary base to adenine in DNA is thymine, whereas in RNA, it is uracil, which is an unmethylated form of thymine.

Like DNA, most biologically active RNAs, including mRNA, tRNA, rRNA, snRNAs, and other non-coding RNAs, contain self-complementary sequences that allow parts of the RNA to fold and pair with itself to form double helices. Analysis of these RNAs has revealed that they are highly structured. Unlike DNA, their structures do not consist of long double helices, but rather collections of short helices packed together into structures akin to proteins. In this fashion, RNAs can achieve chemical catalysis (like enzymes). For instance, determination of the structure of the ribosome—an enzyme that catalyzes peptide bond formation—revealed that its active site is composed entirely of RNA.

STRUCTURE

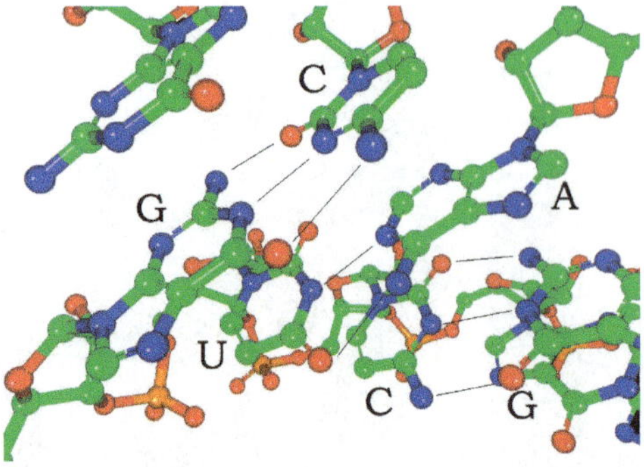

Watson-Crick base pairs in a siRNA(hydrogen atoms are not shown)

Each nucleotide in RNA contains a ribose sugar, with carbons numbered 1' through 5'. A base is attached to the 1' position, in general, adenine (A), cytosine (C), guanine (G), or uracil (U). Adenine and guanine are purines, cytosine and uracil are pyrimidines. A phosphate group is attached to the 3' position of one ribose and the 5' position of the next. The phosphate groups have a negative charge each, making RNA a charged molecule (polyanion). The bases form hydrogen bonds between cytosine and guanine, between adenine and uracil and between guanine and uracil. However, other interactions are possible, such as a group of adenine bases binding to each other in a bulge, or the GNRA tetraloop that has a guanine–adenine base-pair.

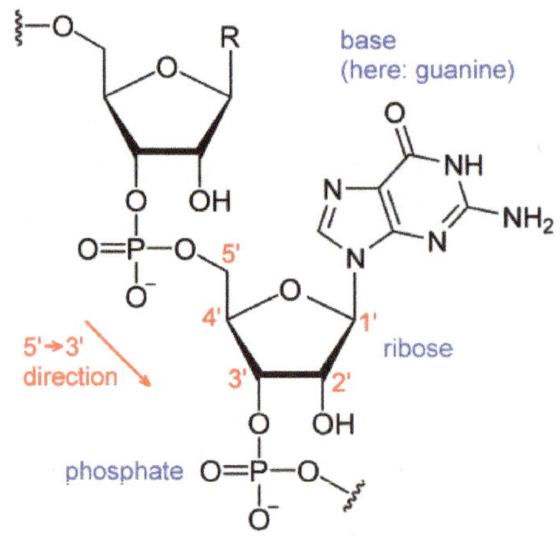

Chemical structure of RNA

An important structural feature of RNA that distinguishes it from DNA is the presence of a hydroxyl group at the 2' position of the ribose sugar. The presence of this functional group causes the helix to mostly adopt the A-form geometry, although in single strand dinucleotide contexts, RNA can rarely also adopt the B-form most commonly observed in

DNA. The A-form geometry results in a very deep and narrow major groove and a shallow and wide minor groove. A second consequence of the presence of the 2'-hydroxyl group is that in conformationally flexible regions of an RNA molecule (that is, not involved in formation of a double helix), it can chemically attack the adjacent phosphodiester bond to cleave the backbone.

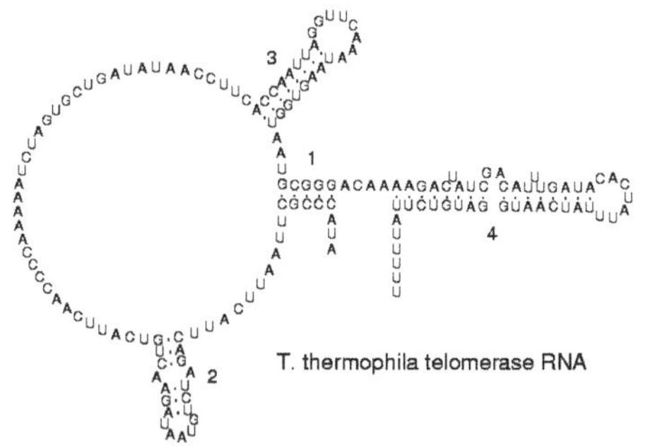

Secondary structure of a telomerase RNA.

RNA is transcribed with only four bases (adenine, cytosine, guanine and uracil), but these bases and attached sugars can be modified in numerous ways as the RNAs mature. Pseudouridine (Ψ), in which the linkage between uracil and ribose is changed from a C–N bond to a C–C bond, and ribothymidine (T) are found in various places (the most notable ones being in the TΨC loop of tRNA). Another notable modified base is hypoxanthine, a deaminated adenine base whose nucleoside is called inosine (I). Inosine plays a key role in the wobble hypothesis of the genetic code.

There are more than 100 other naturally occurring modified nucleosides, The greatest structural diversity of modifications can be found in tRNA, while pseudouridine and nucleosides with 2'-O-methylribose often present in rRNA are the most common. The specific roles of many of these modifications in RNA are not fully understood. However, it is notable that, in ribosomal RNA, many of the post-transcriptional modifications occur in highly functional regions, such as the peptidyl transferase center and the subunit interface, implying that they are important for normal function.

The functional form of single-stranded RNA molecules, just like proteins, frequently requires a specific tertiary structure. The scaffold for this structure is provided by secondary structural elements that are hydrogen bonds within the molecule. This leads to several recognizable "domains" of secondary structure like hairpin loops, bulges, and internal loops. Since RNA is charged, metal ions such as Mg^{2+} are needed to stabilise many secondary and tertiary structures.

The naturally occurring enantiomer of RNA is D-RNA composed of D-ribonucleotides. All chirality centers are located in the D-ribose. By the use of L-ribose or rather L-ribonucleotides, L-RNA can be synthesized. L-RNA is much more stable against degradation by RNase.

Like other structured biopolymers such as proteins, one can define topology of a folded RNA molecule. This is often done based on arrangement of intra-chain contacts within a folded RNA, termed as circuit topology.

SYNTHESIS

Synthesis of RNA is usually catalyzed by an enzyme—RNA polymerase—using DNA as a template, a process known as transcription. Initiation of transcription begins with the binding of the enzyme to a promoter sequence in the DNA (usually found "upstream" of a gene). The DNA double helix is unwound by the helicase activity of the enzyme. The enzyme then progresses along the template strand in the 3' to 5' direction, synthesizing a complementary RNA molecule with elongation occurring in the 5' to 3' direction. The DNA sequence also dictates where termination of RNA synthesis will occur.

Primary transcript RNAs are often modified by enzymes after transcription. For example, a poly(A) tail and a 5' cap are added to eukaryotic pre-mRNA and introns are removed by the spliceosome.

There are also a number of RNA-dependent RNA polymerases that use RNA as their template for synthesis of a new strand of RNA. For instance, a number of RNA viruses (such as poliovirus) use this type of enzyme to replicate their genetic material. Also, RNA-dependent RNA polymerase is part of the RNA interference pathway in many organisms.

TYPES OF RNA

Overview

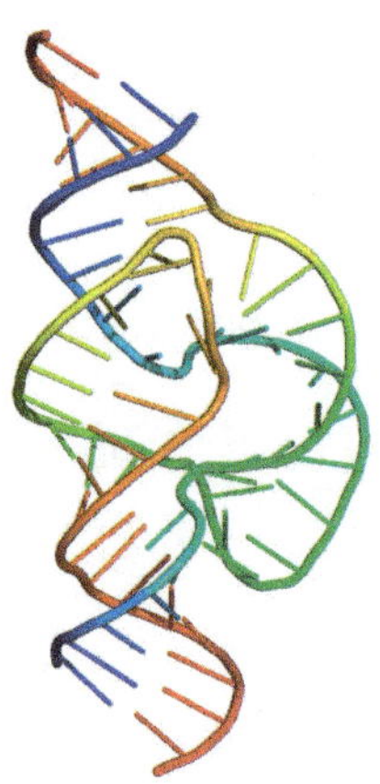

Structure of a hammerhead ribozyme, a ribozyme that cuts RNA

Messenger RNA (mRNA) is the RNA that carries information from DNA to the ribosome, the sites of protein synthesis (translation) in the cell. The coding sequence of the mRNA determines the amino acid sequence in the protein that is produced. However, many RNAs do not code for protein (about 97% of the transcriptional output is non-protein-coding in eukaryotes).

These so-called non-coding RNAs ("ncRNA") can be encoded by their own genes (RNA genes), but can also derive from mRNA introns. The most prominent examples of non-coding RNAs are transfer RNA (tRNA) and ribosomal RNA (rRNA), both of which are involved in the process of translation. There are also non-coding RNAs involved in gene regulation, RNA processing and other roles. Certain RNAs are able to catalyse chemical reactions such as cutting and ligating other RNA molecules, and the catalysis of peptide bond formation in the ribosome; these are known as ribozymes.

In Length

According to the length of RNA chain, RNA includes small RNA and long RNA. Usually, small RNAs are <200 nt in length, and long RNA are >200 nt. Long RNAs, also called large RNAs, mainly include long non-coding RNA (lncRNA) and mRNA. Small RNAs mainly include 5.8S ribosomal RNA (rRNA), 5S rRNA, transfer RNA (tRNA), microRNA (miRNA), small interfering RNA (siRNA), small nucleolar RNA (snoRNAs), Piwi-interacting RNA (piRNA), tRNA-derived small RNA (tsRNA) and small rDNA-derived RNA (srRNA).

In Translation

Messenger RNA (mRNA) carries information about a protein sequence to the ribosomes, the protein synthesis factories in the cell. It is coded so that every three nucleotides (a codon) corresponds to one amino acid. In eukaryotic cells, once precursor mRNA (pre-mRNA) has been transcribed from DNA, it is processed to mature mRNA. This removes its introns—non-coding sections of the pre-mRNA. The mRNA is then exported from the nucleus to the cytoplasm, where it is bound to ribosomes and translated into its corresponding protein form with the help of tRNA. In prokaryotic cells, which do not have nucleus and cytoplasm compartments, mRNA can bind to ribosomes while it is being transcribed from DNA. After a certain amount of time the message degrades into its component nucleotides with the assistance of ribonucleases.

Transfer RNA (tRNA) is a small RNA chain of about 80 nucleotides that transfers a specific amino acid to a growing polypeptide chain at the ribosomal site of protein synthesis during translation. It has sites for amino acid attachment and an anticodon region for codon recognition that binds to a specific sequence on the messenger RNA chain through hydrogen bonding.

Ribosomal RNA (rRNA) is the catalytic component of the ribosomes. Eukaryotic ribosomes contain four different rRNA molecules: 18S, 5.8S, 28S and 5S rRNA. Three of the rRNA molecules are synthesized in the nucleolus, and one is synthesized elsewhere. In the cytoplasm, ribosomal RNA and protein combine to form a nucleoprotein called a ribosome. The ribosome binds mRNA and carries out protein synthesis. Several ribosomes may be attached to a single mRNA at any time. Nearly all the RNA found in a typical eukaryotic cell is rRNA.

Transfer-messenger RNA (tmRNA) is found in many bacteria and plastids. It tags proteins encoded by mRNAs that lack stop codons for degradation and prevents the ribosome from stalling.

Regulatory RNAs

Several types of RNA can downregulate gene expression by being complementary to a part of an mRNA or a gene's DNA. MicroRNAs (miRNA; 21-22 nt) are found in eukaryotes and act through RNA interference (RNAi), where an effector complex of miRNA and enzymes can cleave complementary mRNA, block the mRNA from being translated, or accelerate its degradation.

While small interfering RNAs (siRNA; 20-25 nt) are often produced by breakdown of viral RNA, there are also endogenous sources of siRNAs. siRNAs act through RNA interference in a fashion similar to miRNAs. Some miRNAs and siRNAs can cause genes they target to be methylated, thereby decreasing or increasing transcription of those genes. Animals have Piwi-interacting RNAs (piRNA; 29-30 nt) that are active in germline cells and are thought to be a defense against transposons and play a role in gametogenesis.

Many prokaryotes have CRISPR RNAs, a regulatory system similar to RNA interference. Antisense RNAs are widespread; most downregulate a gene, but a few are activators of transcription. One way antisense RNA can act is by binding to an mRNA, forming double-stranded RNA that is enzymatically degraded. There are many long noncoding RNAs that regulate genes in eukaryotes, one such RNA is Xist, which coats one X chromosome in female mammals and inactivates it.

An mRNA may contain regulatory elements itself, such as riboswitches, in the 5' untranslated region or 3' untranslated region; these cis-regulatory elements regulate the activity of that mRNA. The untranslated regions can also contain elements that regulate other genes.

In RNA Processing

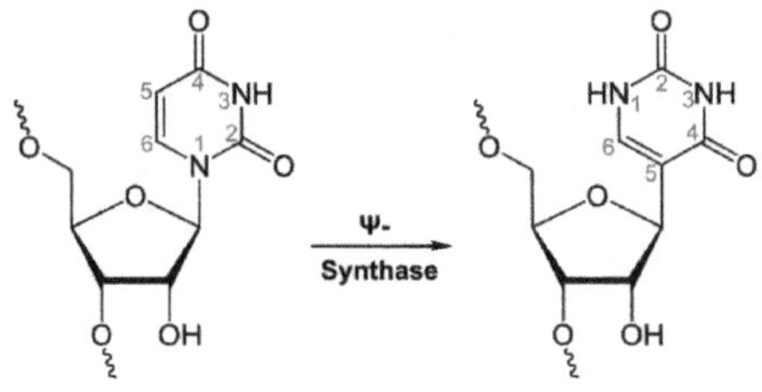

Uridine to pseudouridine is a common RNA modification.

Many RNAs are involved in modifying other RNAs. Introns are spliced out of pre-mRNA by spliceosomes, which contain several small nuclear RNAs (snRNA), or the introns can be ribozymes that are spliced by themselves. RNA can also be altered by having its nucleotides modified to nucleotides other than A, C, G and U. In eukaryotes, modifications of RNA nucleotides are in general directed by small nucleolar RNAs (snoRNA; 60-300 nt), found in the nucleolus and cajal bodies. snoRNAs associate with enzymes and guide them to a spot on an RNA by basepairing to that RNA. These enzymes then perform the nucleotide modification. rRNAs and tRNAs are extensively modified, but snRNAs and mRNAs can also be the target of base modification. RNA can also be methylated.

RNA Genomes

Like DNA, RNA can carry genetic information. RNA viruses have genomes composed of RNA that encodes a number of proteins. The viral genome is replicated by some of those proteins, while other proteins protect the genome as the virus particle moves to a new host cell. Viroids are another group of pathogens, but they consist only of RNA, do not encode any protein and are replicated by a host plant cell's polymerase.

In Reverse Transcription

Reverse transcribing viruses replicate their genomes by reverse transcribing DNA copies from their RNA; these DNA copies are then transcribed to new RNA. Retrotransposonsalso spread by copying DNA and RNA from one another, and telomerase contains an RNA that is used as template for building the ends of eukaryotic chromosomes.

Double-stranded RNA

Double-stranded RNA (dsRNA) is RNA with two complementary strands, similar to the DNA found in all cells. dsRNA forms the genetic material of some viruses (double-stranded RNA viruses). Double-stranded RNA such as viral RNA or siRNA can trigger RNA interference in eukaryotes, as well as interferon response in vertebrates.

Circular RNA

Recently, it was shown that there is a single stranded covalently closed, *i.e.* circular form of RNA expressed throughout the animal and plant kingdom. circRNAs are thought to arise via a "back-splice" reaction where the spliceosome joins a downstream donor to an upstream acceptor splice site. So far the function of circRNAs is largely unknown, although for few examples a microRNA sponging activity has been demonstrated.

KEY DISCOVERIES IN RNA BIOLOGY

Robert W. Holley, left, poses with his research team.

Research on RNA has led to many important biological discoveries and numerous Nobel Prizes. Nucleic acids were discovered in 1868 by Friedrich Miescher, who called the material ‹nuclein' since it was found in the nucleus. It was later discovered that prokaryotic cells, which do not have a nucleus, also contain nucleic acids. The role of RNA in protein synthesis was suspected already in 1939. Severo Ochoa won the 1959 Nobel Prize in Medicine (shared with Arthur Kornberg) after he discovered an enzyme that can synthesize RNA in the laboratory. However, the enzyme discovered by Ochoa (polynucleotide phosphorylase) was later shown to be responsible for RNA degradation, not RNA synthesis. In 1956 Alex Rich and David Davies hybridized two separate strands of RNA to form the first crystal of RNA whose structure could be determined by X-ray crystallography.

The sequence of the 77 nucleotides of a yeast tRNA was found by Robert W. Holley in 1965, winning Holley the 1968 Nobel Prize in Medicine (shared with Har Gobind Khorana and Marshall Nirenberg). In 1967, Carl Woese hypothesized that RNA might be catalytic and suggested that the earliest forms of life (self-replicating molecules) could have relied on RNA both to carry genetic information and to catalyze biochemical reactions—an RNA world.

During the early 1970s, retroviruses and reverse transcriptase were discovered, showing for the first time that enzymes could copy RNA into DNA (the opposite of the usual route for transmission of genetic information). For this work, David Baltimore, Renato Dulbecco and Howard Temin were awarded a Nobel Prize in 1975. In 1976, Walter Fiersand his team determined the first complete nucleotide sequence of an RNA virus genome, that of bacteriophage MS2.

In 1977, introns and RNA splicing were discovered in both mammalian viruses and in cellular genes, resulting in a 1993 Nobel to Philip Sharp and Richard Roberts. Catalytic RNA molecules (ribozymes) were discovered in the early 1980s, leading to a 1989 Nobel award to Thomas Cech and Sidney Altman. In 1990, it was found in *Petunia* that introduced genes can silence similar genes of the plant's own, now known to be a result of RNA interference.

At about the same time, 22 nt long RNAs, now called microRNAs, were found to have a role in the development of *C. elegans*. Studies on RNA interference gleaned a Nobel Prize for Andrew Fire and Craig Mello in 2006, and another Nobel was awarded for studies on the transcription of RNA to Roger Kornberg in the same year. The discovery of gene regulatory RNAs has led to attempts to develop drugs made of RNA, such as siRNA, to silence genes.

EVOLUTION

In March 2015, complex DNA and RNA organic compounds of life, including uracil, cytosine and thymine, were reportedly formed in the laboratory under outer space conditions, using starter chemicals, such as pyrimidine, found in meteorites. Pyrimidine, like polycyclic aromatic hydrocarbons (PAHs), the most carbon-rich chemical found in the Universe, may have been formed in red giants or in interstellar dust and gas clouds, according to the scientists.

REFERENCES

1. "RNA: The Versatile Molecule". University of Utah. 2015.
2. "Nucleotides and Nucleic Acids" (PDF). University of California, Los Angeles.
3. R.N. Shukla. Analysis of Chromosomes. ISBN 9789384568177.
4. Berg JM; Tymoczko JL; Stryer L (2002). Biochemistry (5th ed.). WH Freeman and Company. pp. 118–19, 781–808. ISBN 0-7167-4684-0. OCLC 179705944.
5. I. Tinoco & C. Bustamante (1999). "How RNA folds". J. Mol. Biol. **293** *(2): 271–281. doi:10.1006/jmbi.1999.3001. PMID 10550208.*
6. Higgs PG (2000). "RNA secondary structure: physical and computational aspects". Quarterly Reviews of Biophysics. **33** *(3): 199–253.* doi:10.1017/S0033583500003620. PMID 11191843.
7. Nissen P; Hansen J; Ban N; Moore PB; Steitz TA (2000). "The structural basis of ribosome activity in peptide bond synthesis". Science. **289** *(5481): 920–30. Bibcode:*2000Sci...289..920N. doi:10.1126/science.289.5481.920. PMID 10937990.
8. Lee JC; Gutell RR (2004). "Diversity of base-pair conformations and their occurrence in rRNA structure and RNA structural motifs". J. Mol. Biol. **344** *(5): 1225–49. doi:10.1016/j.jmb.2004.09.072. PMID 15561141.*
9. Barciszewski J; Frederic B; Clark C (1999). RNA biochemistry and biotechnology. Springer. pp. 73–87. ISBN 0-7923-5862-7. OCLC 52403776.
10. Salazar M; Fedoroff OY; Miller JM; Ribeiro NS; Reid BR (1992). "The DNA strand in DNAoRNA hybrid duplexes is neither B-form nor A-form in solution". Biochemistry. **32** *(16): 4207–15.* doi:10.1021/bi00067a007. PMID 7682844.
11. Sedova A; Banavali NK (2016). "RNA approaches the B-form in stacked single strand dinucleotide contexts". Biopolymers. **105** *(2): 65–82. doi:10.1002/bip.22750. PMID 26443416.*
12. Hermann T; Patel DJ (2000). "RNA bulges as architectural and recognition motifs". Structure. **8** *(3): R47–R54.* doi:10.1016/S0969-2126(00)00110-6. PMID 10745015.
13. Mikkola S; Stenman E; Nurmi K; Yousefi-Salakdeh E; Strömberg R; Lönnberg H (1999). "The mechanism of the metal ion promoted cleavage of RNA phosphodiester bonds involves a general acid catalysis by the metal aquo ion on the departure of the leaving group". Perkin transactions 2 (8): 1619–26. doi:10.1039/a903691a.
14. Jankowski JAZ; Polak JM (1996). Clinical gene analysis and manipulation: Tools, techniques and troubleshooting. Cambridge University Press. p. 14. ISBN 0-521-47896-0. OCLC 33838261.
15. Yu Q; Morrow CD (2001). "Identification of critical elements in the tRNA acceptor stem and TΨC loop necessary for human immunodeficiency virus type 1 infectivity". J Virol. **75** *(10): 4902–6. doi:10.1128/JVI.75.10.4902-4906.2001. PMC 114245.* PMID 11312362.

16. Elliott MS; Trewyn RW (1983). "Inosine biosynthesis in transfer RNA by an enzymatic insertion of hypoxanthine". J. Biol. Chem. **259** *(4): 2407–10. PMID 6365911.*

17. Cantara, WA; Crain, PF; Rozenski, J; McCloskey, JA; Harris, KA; Zhang, X; Vendeix, FA; Fabris, D; Agris, PF (January 2011). "The RNA Modification Database, RNAMDB: 2011 update". Nucleic Acids Research. **39** *(Database issue): D195–201. doi:10.1093/*nar/gkq1028. PMC 3013656. PMID 21071406.

18. Söll D; RajBhandary U (1995). TRNA: Structure, biosynthesis, and function. ASM Press. p. 165. ISBN 1-55581-073-X. OCLC 183036381.

19. Kiss T (2001). "Small nucleolar RNA-guided post-transcriptional modification of cellular RNAs". The EMBO Journal. **20** *(14): 3617–22. doi:10.1093/* emboj/20.14.3617. PMC 125535. PMID 11447102.

20. King TH; Liu B; McCully RR; Fournier MJ (2002). "Ribosome structure and activity are altered in cells lacking snoRNPs that form pseudouridines in the peptidyl transferase center". Molecular Cell. **11** *(2): 425–35.* doi:10.1016/S1097-2765(03)00040-6. PMID 12620230.

21. Mathews DH; Disney MD; Childs JL; Schroeder SJ; Zuker M; Turner DH (2004). "Incorporating chemical modification constraints into a dynamic programming algorithm for prediction of RNA secondary structure". Proc. Natl. Acad. Sci. USA. **101** *(19): 7287–92. Bibcode:*2004PNAS..101.7287M. doi:10.1073/pnas.0401799101. PMC 409911. PMID 15123812.

22. Tan ZJ; Chen SJ (2008). "Salt dependence of nucleic acid hairpin stability". Biophys. J. **95** *(2): 738–52. Bibcode:*2008BpJ....95..738T. doi:10.1529/biophysj.108.131524. PMC 2440479. PMID 18424500.

23. Vater A; Klussmann S (January 2015). "Turning mirror-image oligonucleotides into drugs: the evolution of Spiegelmer therapeutics". Drug Discovery Today. **20** *(1): 147–155. doi:10.1016/j.drudis.2014.09.004. PMID 25236655.*

24. Nudler E; Gottesman ME (2002). "Transcription termination and anti-termination in E. coli". Genes to Cells. **7** *(8): 755–68. doi:10.1046/j.1365-2443.2002.00563.x. PMID 12167155.*

25. Jeffrey L Hansen; Alexander M Long; Steve C Schultz (1997). "Structure of the RNA-dependent RNA polymerase of poliovirus". Structure. **5** *(8): 1109–22.* doi:10.1016/S0969-2126(97)00261-X. PMID 9309225.

26. Ahlquist P (2002). "RNA-Dependent RNA Polymerases, Viruses, and RNA Silencing". Science. **296** *(5571): 1270–73. Bibcode:*2002Sci...296.1270A. doi:10.1126/science.1069132. PMID 12016304.

27. Cooper GC; Hausman RE (2004). The Cell: A Molecular Approach (3rd ed.). Sinauer. pp. 261–76, 297, 339–44. ISBN 0-87893-214-3. OCLC 174924833.

28. Mattick JS; Gagen MJ (1 September 2001). "The evolution of controlled multitasked gene networks: the role of introns and other noncoding RNAs in the development of complex organisms". Mol. Biol. Evol. **18** *(9): 1611–30. doi:10.1093/oxfordjournals.molbev.a003951. PMID 11504843.*

29. Mattick, JS (2001). "Noncoding RNAs: the architects of eukaryotic complexity". EMBO Reports. **2** *(11): 986–91. doi:10.1093/*embo-reports/kve230. PMC 1084129. PMID 11713189.

30. Mattick JS (October 2003). "Challenging the dogma: the hidden layer of non-protein-coding RNAs in complex organisms" (PDF). BioEssays. **25** *(10): 930–9. doi:10.1002/bies.10332. PMID 14505360.*

31. Mattick JS (October 2004). "The hidden genetic program of complex organisms". Scientific American. **291** *(4): 60–7.* doi:10.1038/scientificamerican1004-60. PMID 15487671.

32. Wirta W (2006). Mining the transcriptome – methods and applications. Stockholm: School of Biotechnology, Royal Institute of Technology. ISBN 91-7178-436-5. OCLC 185406288.

33. Rossi JJ (2004). "Ribozyme diagnostics comes of age". Chemistry & Biology. **11** *(7): 894–95. doi:10.1016/j.chembiol.2004.07.002. PMID 15271347.*

34. Storz, G (17 May 2002). "An expanding universe of noncoding RNAs.". *Science.* **296**(5571): 1260–3. doi:10.1126/science.1072249. PMID 12016301.

35. Fatica, A; Bozzoni, I (January 2014). "Long non-coding RNAs: new players in cell differentiation and development.". *Nature Reviews Genetics.* **15** *(1): 7–21.* doi:10.1038/nrg3606. PMID 24296535.

36. Chen, Q; Yan, M; Cao, Z; Li, X; Zhang, Y; Shi, J; Feng, GH; Peng, H; Zhang, X; Zhang, Y; Qian, J; Duan, E; Zhai, Q; Zhou, Q (22 January 2016). "Sperm tsRNAs contribute to intergenerational inheritance of an acquired metabolic disorder.". *Science.* **351** *(6271): 397–400. doi:10.1126/science.aad7977. PMID 26721680.*

37. Wei, H; Zhou, B; Zhang, F; Tu, Y; Hu, Y; Zhang, B; Zhai, Q (2013). "Profiling and identification of small rDNA-derived RNAs and their potential biological functions.". *PLOS ONE.* **8** *(2): e56842. doi:10.1371/journal.pone.0056842. PMC 3572043.* PMID 23418607.

38. Gueneau de Novoa P, Williams KP; Williams (2004). "The tmRNA website: reductive evolution of tmRNA in plastids and other endosymbionts". Nucleic Acids Res. **32**(Database issue): D104–8. doi:10.1093/nar/gkh102. PMC 308836. PMID 14681369.

39. Carthew, RW; Sontheimer, EJ. (February 2009). "Origins and Mechanisms of miRNAs and siRNAs.". *Cell.* **136**: 642–55. doi:10.1016/j.cell.2009.01.035. PMC 2675692. PMID 19239886.

40. Liang, Kung-Hao; Yeh, Chau-Ting (2013). "A gene expression restriction network mediated by sense and antisense Alu sequences located on protein-coding messenger RNAs.". *BMC Genomics.* 14: 325. doi:10.1186/1471-2164-14-325. PMC 3655826. PMID 23663499. Retrieved 2013-05-11.

41. Wu L; Belasco JG (January 2008). "Let me count the ways: mechanisms of gene regulation by miRNAs and siRNAs". Mol. Cell. 29 *(1): 1–7. doi:10.1016/j.molcel.2007.12.010. PMID 18206964.*

42. Matzke MA; Matzke AJM (2004). "Planting the seeds of a new paradigm". PLoS Biology. 2 *(5): e133. doi:10.1371/journal.pbio.0020133. PMC 406394.* PMID 15138502.

43. Vazquez F; Vaucheret H; Rajagopalan R; Lepers C; Gasciolli V; Mallory AC; Hilbert J; Bartel DP; Crété P (2004). "Endogenous trans-acting siRNAs regulate the accumulation of Arabidopsis mRNAs". Molecular Cell. 16 *(1): 69–79. doi:10.1016/j.molcel.2004.09.028. PMID 15469823.*

44. Watanabe T, Totoki Y, Toyoda A, et al. (May 2008). "Endogenous siRNAs from naturally formed dsRNAs regulate transcripts in mouse oocytes". Nature. 453 *(7194): 539–43. Bibcode:*2008Natur.453..539W. doi:10.1038/nature06908. PMID 18404146.

45. Sontheimer EJ; Carthew RW (July 2005). "Silence from within: endogenous siRNAs and miRNAs". Cell. 122 *(1): 9–12. doi:10.1016/j.cell.2005.06.030. PMID 16009127.*

46. Doran G (2007). "RNAi – Is one suffix sufficient?". *Journal of RNAi and Gene Silencing.* 3 *(1): 217–19.*

47. Pushparaj PN; Aarthi JJ; Kumar SD; Manikandan J (2008). "RNAi and RNAa — The Yin and Yang of RNAome". Bioinformation. 2 *(6): 235–7.* doi:10.6026/97320630002235. PMC 2258431. PMID 18317570.

48. Horwich MD; Li C; Matranga C; Vagin V; Farley G; Wang P; Zamore PD (2007). "The Drosophila RNA methyltransferase, DmHen1, modifies germline piRNAs and single-stranded siRNAs in RISC". Current Biology. 17 *(14): 1265–72. doi:10.1016/j.cub.2007.06.030. PMID 17604629.*

49. Girard A; Sachidanandam R; Hannon GJ; Carmell MA (2006). "A germline-specific class of small RNAs binds mammalian Piwi proteins". Nature. 442 *(7099): 199–202. Bibcode:*2006Natur.442..199G. doi:10.1038/nature04917. PMID 16751776.

50. Horvath P; Barrangou R (2010). "CRISPR/Cas, the Immune System of Bacteria and Archaea". Science. 327 *(5962): 167–70. Bibcode:*2010Sci...327..167H. doi:10.1126/science.1179555. PMID 20056882.

51. Wagner EG; Altuvia S; Romby P (2002). "Antisense RNAs in bacteria and their genetic elements". Adv Genet. Advances in Genetics. 46: 361–98. doi:10.1016/S0065-2660(02)46013-0. ISBN 9780120176465. PMID 11931231.

52. Gilbert SF (2003). Developmental Biology (7th ed.). Sinauer. pp. 101–3. ISBN 0-87893-258-5. OCLC 154656422.

53. Amaral PP; Mattick JS (October 2008). "Noncoding RNA in development". Mammalian genome : official journal of the International Mammalian Genome Society. 19 *(7–8): 454–92.* doi:10.1007/s00335-008-9136-7. PMID 18839252.

54. Heard E; Mongelard F; Arnaud D; Chureau C; Vourc'h C; Avner P (1999). "Human XISTyeast artificial chromosome transgenes show partial X inactivation center function in mouse embryonic stem cells". Proc. Natl. Acad. Sci. USA. 96 *(12): 6841–46. Bibcode:*1999PNAS...96.6841H. doi:10.1073/pnas.96.12.6841. PMC 22003. PMID 10359800.

55. Batey RT (2006). "Structures of regulatory elements in mRNAs". Curr. Opin. Struct. Biol. 16 *(3): 299–306. doi:10.1016/j.sbi.2006.05.001. PMID 16707260.*

56. Scotto L; Assoian RK (June 1993). "A GC-rich domain with bifunctional effects on mRNA and protein levels: implications for control of transforming growth factor beta 1 expression". Mol. Cell. Biol. 13 *(6): 3588–97. PMC 359828.* PMID 8497272.

57. Steitz TA; Steitz JA (1993). "A general two-metal-ion mechanism for catalytic RNA". Proc. Natl. Acad. Sci. U.S.A. 90 *(14): 6498–502. Bibcode:*1993PNAS...90.6498S. doi:10.1073/pnas.90.14.6498. PMC 46959. PMID 8341661.

58. Xie J; Zhang M; Zhou T; Hua X; Tang L; Wu W (2007). "Sno/scaRNAbase: a curated database for small nucleolar RNAs and cajal body-specific RNAs". Nucleic Acids Res. 35(Database issue): D183–7. doi:10.1093/nar/gkl873. PMC 1669756. PMID 17099227.

59. Omer AD; Ziesche S; Decatur WA; Fournier MJ; Dennis PP (2003). "RNA-modifying machines in archaea". Molecular Microbiology. 48 *(3): 617–29. doi:10.1046/j.1365-2958.2003.03483.x. PMID 12694609.*

60. Cavaillé J; Nicoloso M; Bachellerie JP (1996). "Targeted ribose methylation of RNA in vivo directed by tailored antisense RNA guides". Nature. 383 *(6602): 732–5. Bibcode:*1996Natur.383..732C. doi:10.1038/383732a0. PMID 8878486.

61. Kiss-László Z; Henry Y; Bachellerie JP; Caizergues-Ferrer M; Kiss T (1996). "Site-specific ribose methylation of preribosomal RNA: a novel function for small nucleolar RNAs". Cell. 85 *(7): 1077–88.* doi:10.1016/S0092-8674(00)81308-2. PMID 8674114.

62. Daròs JA; Elena SF; Flores R (2006). "Viroids: an Ariadne's thread into the RNA labyrinth". EMBO Rep. 7 *(6): 593–8. doi:10.1038/sj.embor.7400706. PMC 1479586.* PMID 16741503.

63. Kalendar R; Vicient CM; Peleg O; Anamthawat-Jonsson K; Bolshoy A; Schulman AH (2004). "Large retrotransposon derivatives: abundant, conserved but nonautonomous retroelements of barley and related genomes". Genetics. 166 *(3): 1437–50. doi:10.1534/genetics.166.3.1437. PMC 1470764.* PMID 15082561.

64. Podlevsky JD; Bley CJ; Omana RV; Qi X; Chen JJ (2008). "The telomerase database". Nucleic Acids Res. 36 *(Database issue): D339–43. doi:10.1093/*nar/gkm700. PMC 2238860. PMID 18073191.

65. Blevins, T.; Rajeswaran, R.; Shivaprasad, PV.; Beknazariants, D.; Si-Ammour, A.; Park, HS.; Vazquez, F.; Robertson, D.; Meins, F. (2006). "Four plant Dicers mediate viral small RNA biogenesis and DNA virus induced silencing". Nucleic Acids Res. 34 *(21): 6233–46. doi:10.1093/*nar/gkl886. PMC 1669714. PMID 17090584.

66. Jana S; Chakraborty C; Nandi S; Deb JK (2004). "RNA interference: potential therapeutic targets". Appl. Microbiol. Biotechnol. 65 *(6): 649–57.* doi:10.1007/s00253-004-1732-1. PMID 15372214.

67. Schultz U; Kaspers B; Staeheli P (2004). "The interferon system of non-mammalian vertebrates". Dev. Comp. Immunol. 28 *(5): 499–508. doi:10.1016/j.*

dci.2003.09.009. PMID 15062646.

68. Whitehead, K. A.; Dahlman, J. E.; Langer, R. S.; Anderson, D. G. (2011). "Silencing or Stimulation? SiRNA Delivery and the Immune System". Annual Review of Chemical and Biomolecular Engineering. 2: 77–96. doi:10.1146/annurev-chembioeng-061010-114133. PMID 22432611.
69. Dahm R (2005). "Friedrich Miescher and the discovery of DNA". Developmental Biology. 278 *(2): 274–88. doi:10.1016/j.ydbio.2004.11.028. PMID 15680349.*
70. Caspersson T; Schultz J (1939). "Pentose nucleotides in the cytoplasm of growing tissues". Nature. 143 *(3623): 602–3. Bibcode*:1939Natur.143..602C. doi:10.1038/143602c0.
71. Ochoa S (1959). "Enzymatic synthesis of ribonucleic acid" (PDF). Nobel Lecture.
72. Rich A; Davies, D (1956). "A New Two-Stranded Helical Structure: Polyadenylic Acid and Polyuridylic Acid". Journal of the American Chemical Society. 78 *(14): 3548–3549.* doi:10.1021/ja01595a086.
73. Holley RW, et al. (1965). "Structure of a ribonucleic acid". Science. 147 *(3664): 1462–65. Bibcode*:1965Sci...147.1462H. doi:10.1126/science.147.3664.1462. PMID 14263761.
74. Siebert S (2006). "Common sequence structure properties and stable regions in RNA secondary structures" (PDF). Dissertation, Albert-Ludwigs-Universität, Freiburg im Breisgau. p. 1. Archived from the original (PDF) on March 9, 2012.
75. Szathmáry E (1999). "The origin of the genetic code: amino acids as cofactors in an RNA world". Trends Genet. 15 *(6): 223–9.* doi:10.1016/S0168-9525(99)01730-8. PMID 10354582.
76. Fiers W, et al. (1976). "Complete nucleotide-sequence of bacteriophage MS2-RNA: primary and secondary structure of replicase gene". Nature. 260 *(5551): 500–7. Bibcode*:1976Natur.260..500F. doi:10.1038/260500a0. PMID 1264203.
77. Napoli C; Lemieux C; Jorgensen R (1990). "Introduction of a chimeric chalcone synthase gene into petunia results in reversible co-suppression of homologous genes in trans". Plant Cell. 2 *(4): 279–89. doi:10.1105/tpc.2.4.279. PMC 159885.* PMID 12354959.
78. Dafny-Yelin M; Chung SM; Frankman EL; Tzfira T (December 2007). "pSAT RNA interference vectors: a modular series for multiple gene down-regulation in plants". Plant Physiol. 145 *(4): 1272–81. doi:10.1104/pp.107.106062. PMC 2151715.* PMID 17766396.
79. Ruvkun G (2001). "Glimpses of a tiny RNA world". Science. 294 *(5543): 797–99. doi:10.1126/science.1066315. PMID 11679654.*
80. Fichou Y; Férec C (2006). "The potential of oligonucleotides for therapeutic applications". Trends in Biotechnology. 24 *(12): 563–70. doi:10.1016/j.tibtech.2006.10.003. PMID 17045686.*
81. Marlaire, Ruth (3 March 2015). "NASA Ames Reproduces the Building Blocks of Life in Laboratory". NASA. Retrieved 5 March 2015.

CHAPTER 7

Fermentation

Chapter Contents

Fermentation, chemical process by which molecules such as glucose are broken down anaerobically. More broadly, fermentation is the foaming that occurs during the manufacture of wine and beer, a process at least 10,000 years old. The frothing results from the evolution of carbon dioxide gas, though this was not recognized until the 17th century. French chemist and microbiologist Louis Pasteur in the 19th century used the term fermentation in a narrow sense to describe the changes brought about by yeasts and other microorganisms growing in the absence of air (anaerobically); he also recognized that ethyl alcohol and carbon dioxide are not the only products of fermentation.

INTRODUCTION

Fermentation is a metabolic process that converts sugar to acids, gases, or alcohol. It occurs in yeast and bacteria, and also in oxygen-starved muscle cells, as in the case of lactic acid fermentation. Fermentation is also used more broadly to refer to the bulk growth of microorganisms on a growth medium, often with the goal of producing a specific chemical product. French microbiologist Louis Pasteur is often remembered for his insights into fermentation and its microbial causes. The science of fermentation is known as zymology.

Fin progress: Bubbles of CO_2 form a froth on top of the fermentation mixture.

Fermentation takes place when the electron transport chain is unusable (often due to lack of a final electron receptor, such as oxygen). In this case it becomes the cell's primary means of ATP (energy) production. It turns NADH and pyruvate produced in glycolysis into $NAD^+$and an organic molecule (which varies depending on the type of fermentation; see examples below). In the presence of O_2, NADH and pyruvate are used to generate ATP in respiration. This is called oxidative phosphorylation, and it generates much more ATP than glycolysis alone. For that reason, cells generally benefit from avoiding fermentation when oxygen is available, the exception being obligate anaerobes which cannot tolerate oxygen.

The first step, glycolysis, is common to all fermentation pathways:

$$C_6H_{12}O_6 + 2\ NAD^+ + 2\ ADP + 2\ P_i \rightarrow 2\ CH_3COCOO^- + 2\ NADH + 2\ ATP + 2\ H_2O + 2H^+$$

Pyruvate is CH_3COCOO^-. P_i is inorganic phosphate. Two ADP molecules and two P_i are converted to two ATP and two water molecules via substrate-level phosphorylation. Two molecules of NAD^+ are also reduced to NADH.

In oxidative phosphorylation the energy for ATP formation is derived from an electrochemical proton gradient generated across the inner mitochondrial membrane (or, in the case of bacteria, the plasma membrane) via the electron transport chain. Glycolysis has substrate-level phosphorylation (ATP generated directly at the point of reaction).

Humans have used fermentation to produce drinks and beverages since the Neolithic age. For example, fermentation is used for preservation in a process that produces lactic acid as found in such sour foods as pickled cucumbers, kimchi and yogurt (see fermentation in food processing), as well as for producing alcoholic beverages such as wine (see fermentation in winemaking) and beer. Fermentation can even occur within the stomachs of animals, such as humans.

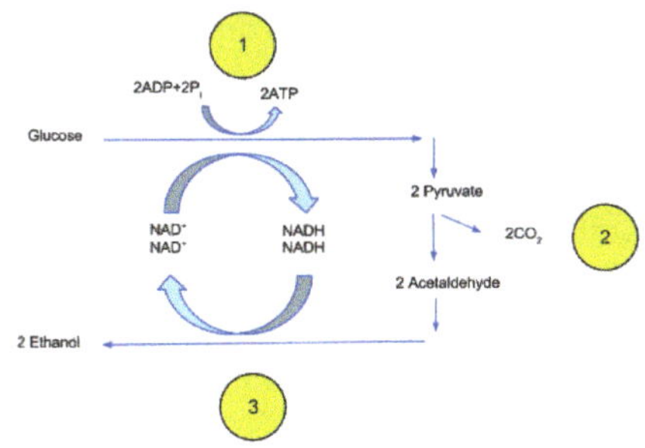

Overview of ethanol fermentation. One glucose molecule breaks down into two pyruvate molecules (1). The energy from this exothermic reaction is used to bind inorganic phosphates to ATP and convert NAD+ to NADH. The two pyruvates are then broken down into two acetaldehyde molecules and give off two CO2 molecules as a waste product (2). The acetaldehyde is then reduced into ethanol using the energy and hydrogen from NADH; in this process the NADH is oxidized into NAD+ so that the cycle may repeat (3).

DEFINITIONS

To many people, fermentation simply means the production of alcohol: grains and fruits are fermented to produce beer and wine. If a food soured, one might say it was 'off' or fermented. Here are some definitions of fermentation. They range from informal, general usage to more scientific definitions.

- Preservation methods for food via microorganisms (general use).
- Any process that produces alcoholic beverages or acidic dairy products (general use).
- Any large-scale microbial process occurring with or without air (common definition used in industry).
- Any energy-releasing metabolic process that takes place only under anaerobic conditions (becoming more scientific).
- Any metabolic process that releases energy from a sugar or other organic molecules, does not require oxygen or an electron transport system, and uses an organic molecule as the final electron acceptor (most scientific).

EXAMPLES

Fermentation does not necessarily have to be carried out in an anaerobic environment. For example, even in the presence of abundant oxygen, yeast cells greatly prefer fermentation to aerobic respiration, as long as sugars are readily available for consumption (a phenomenon known as the Crabtree effect). The antibiotic activity of hops also inhibits aerobic metabolism in yeast.

Fermentation reacts NADH with an endogenous, organic electron acceptor. Usually this is pyruvate formed from the sugar during the glycolysis step. During fermentation, pyruvate is metabolized to various compounds through several processes:

- ethanol fermentation, aka alcoholic fermentation, is the production of ethanol and carbon dioxide
- lactic acid fermentation refers to two means of producing lactic acid:
    - homolactic fermentation is the production of lactic acid exclusively
    - heterolactic fermentation is the production of lactic acid as well as other acids and alcohols.

Sugars are the most common substrate of fermentation, and typical examples of fermentation products are ethanol, lactic acid, carbon dioxide, and hydrogen gas (H_2). However, more exotic compounds can be produced by fermentation, such as butyric acid and acetone. Yeast carries out fermentation in the production of ethanol in beers, wines, and other alcoholic drinks, along with the production of large quantities of carbon dioxide. Fermentation occurs in mammalian muscle during periods of intense exercise where oxygen supply becomes limited, resulting in the creation of lactic acid.

CHEMISTRY

Fermentation products contain chemical energy (they are not fully oxidized), but are considered waste products, since they cannot be metabolized further without the use of oxygen.

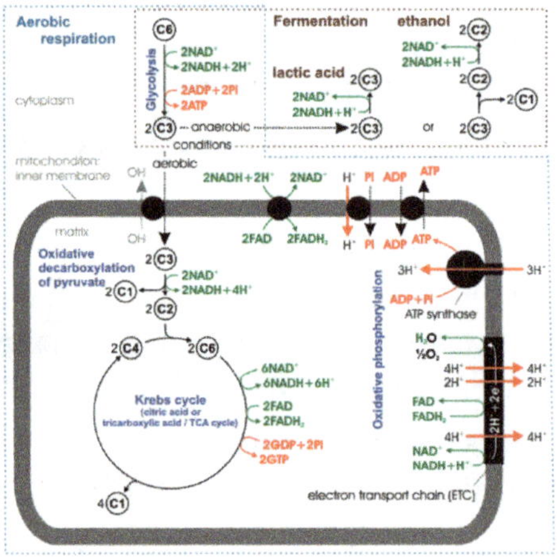

Comparison of aerobic respirationand most known fermentation types in eucaryotic cell. Numbers in circles indicate counts of carbon atoms in molecules, C6 is glucose $C_6H_{12}O_6$, C1 carbon dioxide CO_2. Mitochondrialouter membrane is omitted.

Ethanol Fermentation

The chemical equation below shows the alcoholic fermentation of glucose, whose chemical formula is $C_6H_{12}O_6$. One glucose molecule is converted into two ethanol molecules and two carbon dioxide molecules:

$C_6H_{12}O_6 \rightarrow 2\ C_2H_5OH + 2\ CO_2$

C_2H_5OH is the chemical formula for ethanol.

Before fermentation takes place, one glucose molecule is broken down into two pyruvate molecules. This is known as glycolysis.

Lactic acid Fermentation

Homolactic fermentation (producing only lactic acid) is the simplest type of fermentation. The pyruvate from glycolysis undergoes a simple redox reaction, forming lactic acid. It is unique because it is one of the only respiration processes to not produce a gas as a byproduct. Overall, one molecule of glucose (or any six-carbon sugar) is converted to two molecules of lactic acid: $C_6H_{12}O_6 \rightarrow 2\ CH_3CHOHCOOH$ It occurs in the muscles of animals when they need energy faster than the blood can supply oxygen. It also occurs in some kinds of bacteria (such as lactobacilli) and some fungi. It is this type of bacteria that converts lactose into lactic acid in yogurt, giving it its sour taste. These lactic acid bacteria can carry out either homolactic fermentation, where the end-product is mostly lactic acid, or

Heterolactic fermentation, where some lactate is further metabolized and results in ethanol and carbon dioxide (via the phosphoketolase pathway), acetate, or other metabolic products, e.g.: $C_6H_{12}O_6 \rightarrow CH_3CHOHCOOH + C_2H_5OH + CO_2$

If lactose is fermented (as in yogurts and cheeses), it is first converted into glucose and galactose (both six-carbon sugars with the same atomic formula): $C_{12}H_{22}O_{11} + H_2O \rightarrow 2\ C_6H_{12}O_6$ Heterolactic fermentation is in a sense intermediate between lactic acid fermentation, and other types, e.g. alcoholic fermentation (see below). The reasons to go further and convert lactic acid into anything else are:

- The acidity of lactic acid impedes biological processes; this can be beneficial to the fermenting organism as it drives out competitors that are unadapted to the acidity; as a result, the food will have a longer shelf life (part of the reason foods are purposely fermented in the first place); however, beyond a certain point, the acidity starts affecting the organism that produces it.
- The high concentration of lactic acid (the final product of fermentation) drives the equilibrium backwards (Le Chatelier's principle), decreasing the rate at which fermentation can occur, and slowing down growth.
- Ethanol, into which lactic acid can be easily converted, is volatile and will readily escape, allowing the reaction to proceed easily. CO_2 is also produced, but it is only weakly acidic, and even more volatile than ethanol.
- Acetic acid (another conversion product) is acidic, and not as volatile as ethanol; however, in the presence of limited oxygen, its creation from lactic acid releases additional energy. It is a lighter molecule than lactic acid, that forms fewer hydrogen bonds with its surroundings (due to having fewer groups that can form such bonds), thus is more volatile and will also allow the reaction to move forward more quickly.

- If propionic acid, butyric acid, and longer monocarboxylic acids are produced (see mixed acid fermentation), the amount of acidity produced per glucose consumed will decrease, as with ethanol, allowing faster growth.

Aerobic Respiration

In aerobic respiration, the pyruvate produced by glycolysis is oxidized completely, generating additional ATP and NADH in the citric acid cycle and by oxidative phosphorylation. However, this can occur only in the presence of oxygen. Oxygen is toxic to organisms that are obligate anaerobes, and is not required by facultative anaerobic organisms. In the absence of oxygen, one of the fermentation pathways occurs in order to regenerate NAD^+; lactic acid fermentation is one of these pathways.

HYDROGEN GAS PRODUCTION IN FERMENTATION

Hydrogen gas is produced in many types of fermentation (mixed acid fermentation, butyric acid fermentation, caproate fermentation, butanol fermentation, glyoxylatefermentation), as a way to regenerate NAD^+ from NADH. Electrons are transferred to ferredoxin, which in turn is oxidized by hydrogenase, producing H_2. Hydrogen gas is a substrate for methanogens and sulfate reducers, which keep the concentration of hydrogen low and favor the production of such an energy-rich compound, but hydrogen gas at a fairly high concentration can nevertheless be formed, as in flatus.

As an example of mixed acid fermentation, bacteria such as Clostridium pasteurianum ferment glucose producing butyrate, acetate, carbon dioxide and hydrogen gas: The reaction leading to acetate is:

$$C_6H_{12}O_6 + 4\ H_2O \rightarrow 2\ CH_3COO^- + 2\ HCO_3^- + 4\ H^+ + 4\ H_2$$

Glucose could theoretically be converted into just CO_2 and H_2, but the global reaction releases little energy.

METHANE GAS PRODUCTION IN FERMENTATION

Acetic acid can also undergo a dismutation reaction to produce methane and carbon dioxide:

$$CH_3COO^- + H^+ \rightarrow CH_4 + CO_2 \qquad \Delta G° = -36\ \text{kJ/reaction}$$

This disproportionation reaction is catalysed by methanogen archaea in their fermentative metabolism. One electron is transferred from the carbonyl function (e^- donor) of the carboxylic group to the methyl group (e^- acceptor) of acetic acid to respectively produce CO_2 and methane gas.

HISTORY OF HUMAN USE

The use of fermentation, particularly for beverages, has existed since the Neolithic and has been documented dating from 7000–6600 BCE in Jiahu, China, 5000 BCE in India, Ayurveda mentions many Medicated Wines, 6000 BCE in Georgia, 3150 BCE in ancient Egypt, 3000 BCE in Babylon, 2000 BCE in pre-Hispanic Mexico, and 1500 BC in Sudan.

Fermented foods have a religious significance in Judaism and Christianity. The Baltic god Rugutis was worshiped as the agent of fermentation.

The first solid evidence of the living nature of yeast appeared between 1837 and 1838 when three publications appeared by C. Cagniard de la Tour, T. Swann, and F. Kuetzing, each of whom independently concluded as a result of microscopic investigations that yeast is a living organism that reproduces by budding. It is perhaps because wine, beer, and bread were each basic foods in Europe that most of the early studies on fermentation were done on yeasts, with which they were made. Soon, bacteria were also discovered; the term was first used in English in the late 1840s, but it did not come into general use until the 1870s, and then largely in connection with the new germ theory of disease.

Louis Pasteur in his laboratory

Louis Pasteur (1822–1895), during the 1850s and 1860s, showed that fermentation is initiated by living organisms in a series of investigations. In 1857, Pasteur showed that lactic acid fermentation is caused by living organisms. In 1860, he demonstrated that bacteria cause souring in milk, a process formerly thought to be merely a chemical change, and his work in identifying the role of microorganisms in food spoilage led to the process of pasteurization. In 1877, working to improve the French brewing industry, Pasteur published his famous paper on fermentation, "Etudes sur la Bière", which was translated into English in 1879 as "Studies on fermentation". He defined fermentation (incorrectly) as "Life without air", but correctly showed that specific types of microorganisms cause specific types of fermentations and specific end-products.

Although showing fermentation to be the result of the action of living microorganisms was a breakthrough, it did not explain the basic nature of the fermentation process, or prove that it is caused by the microorganisms that appear to be always present. Many scientists, including Pasteur, had unsuccessfully attempted to extract the fermentation enzyme from yeast. Success came in 1897 when the German chemist Eduard Buechner ground up yeast, extracted a juice from them, then found to his amazement that this "dead" liquid would ferment a sugar solution, forming carbon dioxide and alcohol much like living yeasts. Buechner's results are considered to mark the birth of biochemistry. The "unorganized ferments" behaved just like the organized ones. From that time on, the term enzyme came to be applied to all ferments. It was then understood that fermentation is caused by enzymes that are produced by microorganisms. In 1907, Buechner won the Nobel Prize in chemistry for his work.

Advances in microbiology and fermentation technology have continued steadily up until the present. For example, in the late 1970s, it was discovered that microorganisms could be mutated with physical and chemical treatments to be higher-yielding, faster-growing, tolerant of less oxygen, and able to use a more concentrated medium. Strain selection and hybridization developed as well, affecting most modern food fermentations. Other approaches to advancing the fermentation industry has been done by companies such as BioTork, a biotechnology company that naturally evolves microorganisms to improve fermentation processes. This approach differs from the more popular genetic modification, which has become the current industry standard.

ETYMOLOGY

The word "ferment" is derived from the Latin verb fervere, which means to boil. It is thought to have been first used in the late 14th century in alchemy, but only in a broad sense. It was not used in the modern scientific sense until around 1600.

ACETONE–BUTANOL–ETHANOL FERMENTATION

Acetone–butanol–ethanol (ABE) fermentation is a process that uses bacterial fermentation to produce acetone, n-Butanol, and ethanol from carbohydrates such as starchand glucose. It was developed by the chemist Chaim Weizmann and was thc primary process used to make acetone during World War I, such as to produce cordite, a substance essential for the British war industry.

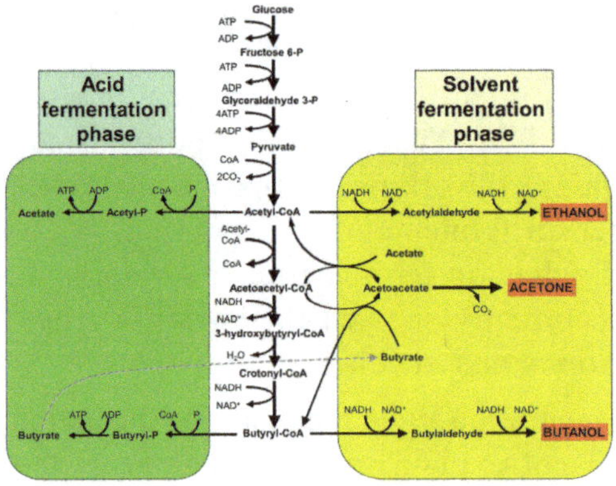

This figure shows pathway of acetone–butanol–ethanol fermentation by clostridia.

The Process

The process may be likened to how yeast ferments sugars to produce ethanol for wine, beer, or fuel, but the organisms that carry out the ABE fermentation are strictly anaerobic (obligate anaerobes). The ABE fermentation produces solvents in a ratio of 3 parts acetone, 6 parts butanol to 1 part ethanol. It usually uses a strain of bacteria from the Class Clostridia (Family Clostridiaceae). Clostridium acetobutylicum is the most well-studied and widely used species, although Clostridium beijerinckii has also been used with good results.

For gas stripping, the most common gases used are the off-gases from the fermentation itself, a mixture of carbon dioxide and hydrogen gas.

The production of butanol by biological means was first performed by Louis Pasteur in 1861.

In 1905, Austrian biochemist Franz Schardinger found that acetone could similarly be produced.

In 1910 Auguste Fernbach (1860-1939) developed a bacterial fermentation process using potato starch as a feedstock in the production of butanol.

Industrial exploitation of ABE fermentation started in 1916, during World War I, with Chaim Weizmann's isolation of Clostridium acetobutylicum, as described in U.S. patent 1315585.

The Weizmann process was operated by Commercial Solvents Corporation from about 1920 to 1964 with plants in the US (Terre Haute, IN, and Peoria, IL), and Liverpool, England. The Peoria plant was the largest of the three; it used molasses as feedstock and had 96 50,000-gallon fermenters.

After World War II, ABE fermentation became generally non-profitable, compared to the production of the same three solvents (acetone, butanol, ethanol) from petroleum.

During the 1950s and 1960s, ABE fermentation was replaced by petroleum chemical plants. Due to different raw material costs, ABE fermentation was viable in South Africa until the early 1980s, with the last plant closing in 1983.

Currently there are no operating ABE plants.

Improvement Attempts

In order to make ABE fermentation profitable, many in-situ product recovery systems have been developed. These include gas stripping, pervaporation, membrane extraction, adsorption, and reverse osmosis. However, at this time none of them have been implemented at an industrial scale.

Current Perspectives

ABE fermentation is attracting renewed interest with an eye on butanol as a renewable biofuel.

FERMENTATION LOCK

The fermentation lock or airlock is a device used in beer brewing and wine making that allows carbon dioxide released during fermentation to escape the fermenter, while not allowing air to enter the fermenter, thus avoiding oxidation.

There are two main designs for the fermentation lock, or airlock. These designs work when half filled with water. When the pressure of the gasinside the fermentation vessel exceeds the prevailing atmospheric pressure the gas will push its way through the water as individual bubbles into the outside air. A sanitizing solution or vodka is sometimes placed in the fermentation lock to prevent contamination of the beer in case the water is inadvertently drawn into the fermenter.

This device may take the form of a tube connected to the headspace of the fermenting vessel into a tub of sanitized liquid or a simpler device mounted directly on top of the fermentation vessel.

Currently, a popular fermentation lock that mounts on top of the fermentation vessel is the three-piece fermentation lock. Other models contain three bulbous chambers allowing for a broader range of pressure equalization. These bulbous fermentation locks were generally made of hand blown glass and are nowadays often made of clear plastic.

The use of perforated rubber balloons offers an easy and inexpensive alternative to conventional airlocks: as used primarily in homebrewing, the balloon is stretched over the orifice of the fermentation vessel and, if necessary, tightened with rubber bands. The balloon is then perforated with a needle. These punctures, while not completely airtight, sufficiently protect the vessel's contents from contamination and allow the gases produced by fermentation to evacuate from the vessel as the pressure rises and the balloon inflates.

DARK FERMENTATION

Dark fermentation is the fermentative conversion of organic substrate to biohydrogen. It is a complex process manifested by diverse groups of bacteria, involving a series of biochemical reactions using three steps similar to anaerobic conversion. Dark fermentation differs from photofermentation in that it proceeds without the presence of light.

Fermentative/hydrolytic microorganisms hydrolyze complex organic polymers to monomers which are further converted to a mixture of lower-molecular-weight organic acids and alcohols by obligatory producing acidogenic bacteria.

Utilization of wastewater as a potential substrate for biohydrogen production has been drawing considerable interest in recent years especially in the dark fermentation process. Industrial wastewater as a fermentative substrate for H_2 production addresses most of the criteria required for substrate selection viz., availability, cost and biodegradability(Angenent, et al., 2004; Kapdan and Kargi, 2006). Chemical wastewater (Venkata Mohan, et al., 2007a,b), cattle wastewater (Tang, et al., 2008), dairy process wastewater (Venkata Mohan, et al. 2007c, Rai et al. 2012), starch hydrolysate wastewater (Chen, et al., 2008) and designed synthetic wastewater (Venkata Mohan, et al., 2007a, 2008b) have been reported to produce biohydrogen apart from wastewater treatment from dark fermentation processes using selectively enriched mixed cultures under acidophilic conditions. Various wastewaters viz., paper mill wastewater (Idania, et al., 2005), starch effluent (Zhang, et al., 2003), food processing wastewater (Shin et al., 2004, van Ginkel, et al., 2005), domestic wastewater (Shin, et al., 2004, 2008e), rice winery wastewater (Yu et al., 2002), distillery and molasses based wastewater (Ren, et al., 2007, Venkata Mohan, et al., 2008a), wheat straw wastes (Fan, et al., 2006) and palm oil mill wastewater (Vijayaraghavan and Ahmed, 2006) have been studied as fermentable substrates for H_2 production along with wastewater treatment.

Using wastewater as a fermentable substrate facilitates both wastewater treatment apart from H_2 production. The efficiency of the dark fermentative H_2 production process was found to depend on pre-treatment of the mixed consortia used as a biocatalyst, operating pH, and organic loading rate apart from wastewater characteristics (Venkata Mohan, et al., 2007d, 2008c, d, Vijaya Bhaskar, et al., 2008d).

In spite of its advantages, the main challenge observed with fermentative H_2 production processes is the relatively low energy conversion efficiency from the organic source. Typical H_2 yields range from 1 to 2 mol of H_2/mol of glucose, which results in 80-90% of the initial COD remaining in the wastewater in the form of various volatile organic acids (VFAs) and solvents, such as acetic acid, propionic acid, butyric acid, and ethanol. Even under optimal conditions about 60-70% of the original organic matter remains in solution. Bioaugmentation with selectively enriched acidogenic consortia to enhance H_2 production was also reported (Venkata Mohan, et al., 2007b). Generation and accumulation of soluble acid metabolites causes a sharp drop in the system pH and inhibits the H_2 production process. Usage of unutilized carbon sources present in acidogenic process for additional biogas production sustains the practical applicability of the process. One way to utilize/recover the remaining organic matter in a usable form is to produce additional H_2 by terminal integration of photo-fermentative processes of H_2 production (Venkata Mohan, et al. 2008e, Rai et al. 2012) and methane by integrating acidogenic processes to terminal methanogenic processes.

ETHANOL FERMENTATION

Ethanol fermentation, also called alcoholic fermentation, is a biological process which converts sugars such as glucose, fructose, and sucrose into cellular energy, producing ethanol and carbon dioxide as a side-effect. Because yeasts perform this conversion in the absence of oxygen, alcoholic fermentation is considered an anaerobic process.

Ethanol fermentation has many uses, including the production of alcoholic beverages, the production of ethanol fuel, and bread baking.

In ethanol fermentation, (1) one glucose molecule breaks down into two pyruvates. The energy from this exothermic reaction is used to bind inorganic phosphates to ADP and convert NAD+ to NADH. (2) The two pyruvates are then broken down into two acetaldehydes and give off two CO2 as a waste product. (3) The two acetaldehydes are then converted to two ethanol by using the H- ions from NADH, converting NADH back into NAD+.

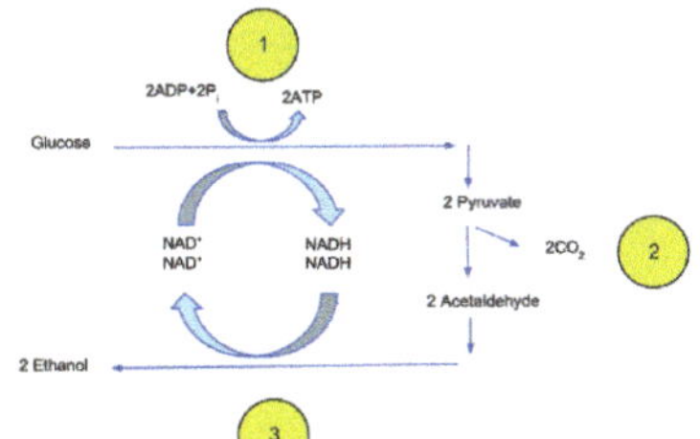

Biochemical Process of Fermentation of Sucrose

A laboratory vessel being used for the fermentation of straw.

The chemical equations below summarize the fermentation of sucrose ($C_{12}H_{22}O_{11}$) into ethanol (C_2H_5OH). Alcoholic fermentation converts one mole of glucose into two moles of ethanol and two moles of carbon dioxide, producing two moles of ATP in the process.

The overall chemical formula for alcoholic fermentation is:

$$C_6H_{12}O_6 \rightarrow 2\ C_2H_5OH + 2\ CO_2$$

Sucrose is a dimer of glucose and fructose molecules. In the first step of alcoholic fermentation, the enzyme invertase cleaves the glycosidic linkage between the glucose and fructose molecules.

$$C_{12}H_{22}O_{11} + H_2O + \text{invertase} \rightarrow 2\ C_6H_{12}O_6$$

Next, each glucose molecule is broken down into two pyruvate molecules in a process known as glycolysis. Glycolysis is summarized by the equation:

$$C_6H_{12}O_6 + 2\ ADP + 2\ P_i + 2\ NAD^+ \rightarrow 2\ CH_3COCOO^- + 2\ ATP + 2\ NADH + 2\ H_2O + 2\ H^+$$

The chemical formula of pyruvate is CH_3COCOO^-. P_i stands for the inorganic phosphate.

Finally, pyruvate is converted to ethanol and CO_2 in two steps, regenerating oxidized NAD+ needed for glycolysis:

1. $CH_3COCOO^- + H^+ \rightarrow CH_3CHO + CO_2$

catalyzed by pyruvate decarboxylase

2. $CH_3CHO + NADH{+}H^+ \rightarrow C_2H_5OH + NAD^+$

This reaction is catalyzed by alcohol dehydrogenase (ADH1 in baker's yeast).

As shown by the reaction equation, glycolysis causes the reduction of two molecules of NAD^+ to NADH. Two ADP molecules are also converted to two ATP and two water molecules via substrate-level phosphorylation.

Related Processes

Fermentation of sugar to ethanol and CO_2 can also be done by Zymomonas mobilis, however the path is slightly different since formation of pyruvate does not happen by glycolysis but instead by the Entner–Doudoroff pathway. Other microorganisms can produce ethanol from sugars by fermentation but often only as a side product. Examples are

- Heterolactic acid fermentation in which Leuconostoc bacterias produce Lactate + Ethanol + CO_2
- Mixed acid fermentation where Escherichia produce Ethanol mixed with Lactate, Acetate, Succinate, Formate, CO_2 and H_2
- 2,3-butanediol fermentation by Enterobacter producing Ethanol, Butanediol, Lactate, Formate, CO_2 and H_2

Gallery

Grapes fermenting during wine production.

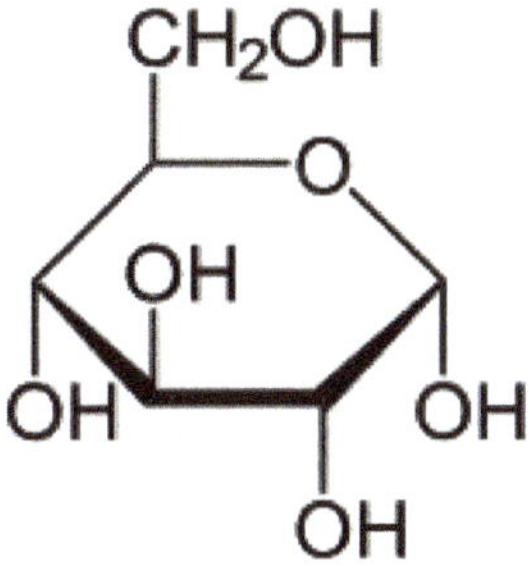

Glucose depicted in Haworth projection

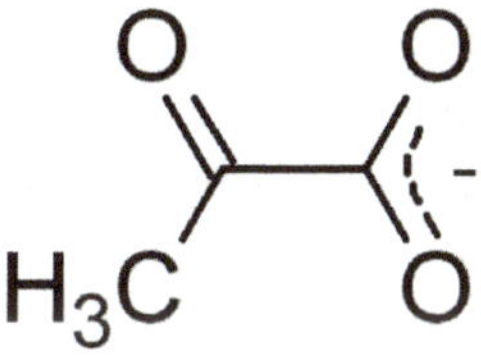

Pyruvate

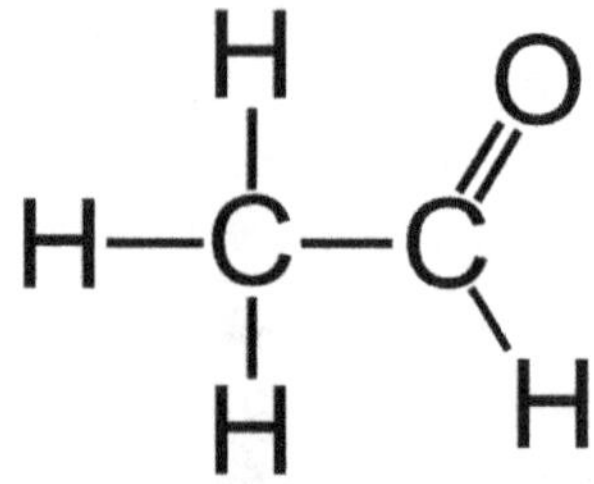

Acetaldehyde

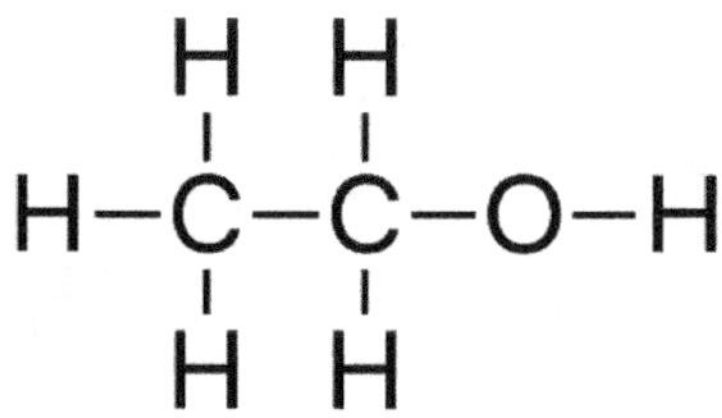

Ethanol

Effect of Oxygen

Fermentation does not require oxygen. If oxygen is present, some species of yeast (e.g., Kluyveromyces lactis or Kluyveromyces lipolytica) will oxidize pyruvate completely to carbon dioxide and water in a process called cellular respiration, hence these species of yeast will produce ethanol only in an anaerobic environment (not cellular respiration).

However, many yeasts such as the commonly used baker's yeast Saccharomyces cerevisiae, or fission yeast Schizosaccharomyces pombe, prefer fermentation to respiration. These yeasts will produce ethanol even under aerobic conditions, if they are provided with the right kind of nutrition. During batch fermentation, the rate of ethanol production per milligram of cell protein is maximal for a brief period early in this process and declines progressively as ethanol accumulates in the surrounding broth. Studies demonstrate that the removal of this accumulated ethanol does not immediately restore fermentative activity, and they provide evidence that the decline in metabolic rate is due to physiological changes (including possible ethanol damage) rather than to the presence of ethanol. Several potential causes for the decline in fermentative activity have been investigated. Viability remained at or above 90%, internal pH remained near neutrality, and the specific activities of the glycolytic and alcohologenic enzymes (measured in vitro) remained high throughout batch fermentation. None of these factors appears to be causally related to the fall in fermentative activity during batch fermentation.

Bread Baking

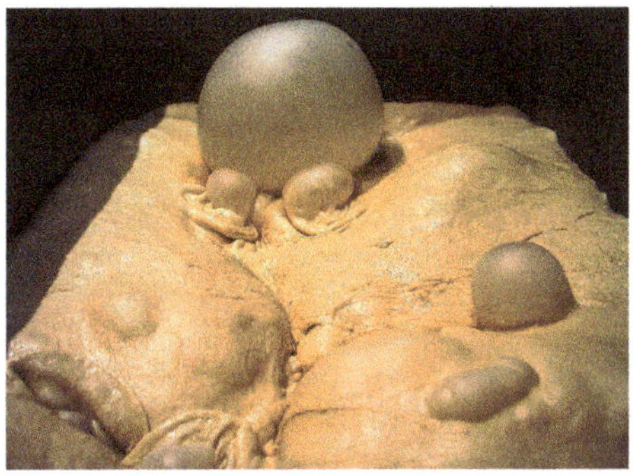

The formation of carbon dioxide — a byproduct of ethanol fermentation — causes bread to rise.

Ethanol fermentation causes bread dough to rise. Yeast organisms consume sugars in the dough and produce ethanol and carbon dioxide as waste products. The carbon dioxide forms bubbles in the dough, expanding it into something of a foam. Less than 2% ethanol remains after baking.

Alcoholic Beverages

Primary fermentation cellar, Budweiser Brewery, Fort Collins, Colorado

All ethanol contained in alcoholic beverages (including ethanol produced by carbonic maceration) is produced by means of fermentation induced by yeast.

- Wine is produced by fermentation of the natural sugars present in grapes; cider and perry are produced by similar fermentation of natural sugar in apples and pears, respectively; and other fruit wines are produced from the fermentation of the sugars in any other kinds of fruit. Brandy and eaux de vie (e.g. slivovitz) are produced by distillation of these fruit-fermented beverages.

- Mead is produced by fermentation of the natural sugars present in honey.
- Beer, whiskey, and vodka are produced by fermentation of grain starches that have been converted to sugar by the enzyme amylase, which is present in grain kernels that have been malted (i.e. germinated). Other sources of starch (e.g. potatoes and unmalted grain) may be added to the mixture, as the amylase will act on those starches as well. Whiskey and vodka are also distilled; gin and related beverages are produced by the addition of flavoring agents to a vodka-like feedstock during distillation.
- Rice wines (including sake) are produced by the fermentation of grain starches converted to sugar by the mold Aspergillus oryzae. Baijiu, soju, and shōchū are distilled from the product of such fermentation.
- Rum and some other beverages are produced by fermentation and distillation of sugarcane. Rum is usually produced from the sugarcane product molasses.

In all cases, fermentation must take place in a vessel that allows carbon dioxide to escape but prevents outside air from coming in. This is because exposure to oxygen would prevent the formation of ethanol, while a buildup of carbon dioxide creates a risk the vessel will rupture or fail catastrophically, causing injury and property damage.

Feedstocks for Fuel Production

Yeast fermentation of various carbohydrate products is also used to produce the ethanol that is added to gasoline.

The dominant ethanol feedstock in warmer regions is sugarcane. In temperate regions, corn or sugar beets are used.

In the United States, the main feedstock for the production of ethanol is currently corn. Approximately 2.8 gallons of ethanol are produced from one bushel of corn (0.42 liter per kilogram). While much of the corn turns into ethanol, some of the corn also yields by-products such as DDGS (distillers dried grains with solubles) that can be used as feed for livestock. A bushel of corn produces about 18 pounds of DDGS (320 kilograms of DDGS per metric ton of maize). Although most of the fermentation plants have been built in corn-producing regions, sorghum is also an important feedstock for ethanol production in the Plains states. Pearl millet is showing promise as an ethanol feedstock for the southeastern U.S. and the potential of duckweed is being studied.

In some parts of Europe, particularly France and Italy, grapes have become a de facto feedstock for fuel ethanol by the distillation of surplus wine. In Japan, it has been proposed to use rice normally made into sake as an ethanol source.

Cassava as Ethanol Feedstock

Ethanol can be made from mineral oil or from sugars or starches. Starches are cheapest. The starchy crop with highest energy content per acre is cassava, which grows in tropical countries.

Thailand already had a large cassava industry in the 1990s, for use as cattle feed and as a cheap admixture to wheat flour. Nigeria and Ghana are already establishing cassava-to-ethanol plants. Production of ethanol from cassava is currently economically feasible when crude oil prices are above US$120 per barrel.

New varieties of cassava are being developed, so the future situation remains uncertain. Currently, cassava can yield between 25-40 tonnes per hectare (with irrigation and fertilizer), and from a tonne of cassava roots, circa 200 liters of ethanol can be produced (assuming cassava with 22% starch content). A liter of ethanol contains circa 21.46 MJ of energy. The overall energy efficiency of cassava-root to ethanol conversion is circa 32%.

The yeast used for processing cassava is Endomycopsis fibuligera, sometimes used together with bacterium Zymomonas mobilis.

Byproducts of Fermentation

Ethanol fermentation produces unharvested byproducts such as heat, carbon dioxide, food for livestock, water, methanol, fuels, fertilizer and alcohols

REFERENCES

1. Klein, Donald W.; Lansing M.; Harley, John (2006). Microbiology (6th ed.). New York: McGraw-Hill. ISBN 978-0-07-255678-0.
2. AP Biology. Anestis, Mark. 2nd Edition. McGraw-Hill Professional. 2006. ISBN 978-0-07-147630-0. p. 61
3. Tortora, Gerard J.; Funke, Berdell R.; Case, Christine L. (2010). "5". Microbiology An Introduction (10 ed.). San Francisco, CA 94111, USA: Pearson Benjamin Cummings. p. 135. ISBN 978-0-321-58202-7.
4. Dickinson, J. R. (1999). "Carbon metabolism". In J. R. Dickinson; M. Schweizer. The metabolism and molecular physiology of Saccharomyces cerevisiae. Philadelphia, PA: Taylor & Francis. ISBN 978-0-7484-0731-6.
5. Voet, Donald; Voet, Judith G. (1995). Biochemistry (2nd ed.). New York, NY: John Wiley & Sons. ISBN 978-0-471-58651-7.
6. Stryer, Lubert (1995). Biochemistry (fourth ed.). New York - Basingstoke: W. H. Freeman and Company. ISBN 978-0716720096.
7. Life, the science of biology. Purves, William Kirkwood. Sadava, David. Orians, Gordon H. 7th Edition. Macmillan Publishers. 2004. ISBN 978-0-7167-9856-9. pp. 139–140
8. Stryer, Lubert (1975). Biochemistry. W. H. Freeman and Company. ISBN 0-7167-0174-X.
9. Introductory Botany: plants, people, and the Environment. Berg, Linda R. Cengage Learning, 2007. ISBN 978-0-534-46669-5. p. 86
10. A dictionary of applied chemistry, Volume 3. Thorpe, Sir Thomas Edward. Longmans, Green and Co., 1922. p.159
11. Madigan, Michael T.; Martinko, John M.; Parker, Jack (1996). Brock biology of microorganisms (8th ed.). Prentice Hall. ISBN 978-0-13-520875-5.

12. Thauer, R.K.; Jungermann, K.; Decker, K. (1977). "Energy conservation in chemotrophic anaerobic bacteria". Bacteriological Reviews. 41 (1): 100–80. ISSN 0005-3678. PMC 413997. PMID 860983.

13. Ferry, J.G. (1992). "Methane from acetate". Journal of Bacteriology. 174 (17): 5489–5495. PMC 206491. PMID 1512186. Retrieved 2011-11-05.

14. Vogels, G.D.; Keltjens J.T.; Van Der Drift C. (1988). "Biochemistry of methane production". In Zehnder A.J.B. Biology of anaerobic microorganisms. New York: Wiley. pp. 707–770.

15. McGovern, P. E.; Zhang, J.; Tang, J.; Zhang, Z.; Hall, G. R.; Moreau, R. A.; Nunez, A.; Butrym, E. D.; Richards, M. P.; Wang, C. -S.; Cheng, G.; Zhao, Z.; Wang, C. (2004). "Fermented beverages of pre- and proto-historic China". Proceedings of the National Academy of Sciences. 101 (51): 17593–17598. doi:10.1073/pnas.0407921102. PMC 539767. PMID 15590771.

16. Vouillamoz, J. F.; McGovern, P. E.; Ergul, A.; Söylemezoğlu, G. K.; Tevzadze, G.; Meredith, C. P.; Grando, M. S. (2006). "Genetic characterization and relationships of traditional grape cultivars from Transcaucasia and Anatolia". Plant Genetic Resources: characterization and utilization. 4 (2): 144. doi:10.1079/PGR2006114.

17. Cavalieri, D; McGovern P.E.; Hartl D.L.; Mortimer R.; Polsinelli M. (2003). "Evidence for S. cerevisiae fermentation in ancient wine" (PDF). Journal of Molecular Evolution. 57 Suppl 1: S226–32. doi:10.1007/s00239-003-0031-2. PMID 15008419. 15008419. Archived from the original (PDF) on December 9, 2006. Retrieved 2007-01-28.

18. Fermented fruits and vegetables. A global perspective". FAO Agricultural Services Bulletins - 134. Archived from the original on January 19, 2007. Retrieved 2007-01-28.

19. Dirar, H., (1993), The Indigenous Fermented Foods of the Sudan: A Study in African Food and Nutrition, CAB International, UK

20. «Gintaras Beresneviius. M. Strijkovskio Kronikos" lietuvi diev sraas". spauda.lt.

21. Rūgutis. Mitologijos enciklopedija, 2 tomas. Vilnius. Vaga. 1999. 293 p.

22. A brief history of fermentation, East and West. Soyinfocenter.com. Retrieved on 2011-01-04.

23. Accomplishments of Louis Pasteur. Fjcollazo.com (2005-12-30). Retrieved on 2011-01-04.

24. HowStuffWorks "Louis Pasteur". Science.howstuffworks.com (2009-07-01). Retrieved on 2011-01-04.

25. Louis Pasteur (1879) Studies on fermentation: The diseases of beer, their causes, and the means of preventing them. Macmillan Publishers.

26. Modern History Sourcebook: Louis Pasteur (1822–1895): Physiological theory of fermentation, 1879. Translated by F. Faulkner, D.C. Robb.

27. New beer in an old bottle: Eduard Buchner and the Growth of Biochemical Knowledge. Cornish-Bowden, Athel. Universitat de Valencia. 1997. ISBN 978-84-370-3328-0. p. 25.

28. The enigma of ferment: from the philosopher's stone to the first biochemical Nobel prize. Lagerkvist, Ulf. World Scientific Publishers. 2005. ISBN 978-981-256-421-4. p. 7.

29. A treasury of world science, Volume 1962, Part 1. Runes, Dagobert David. Philosophical Library Publishers. 1962. p. 109.

30. Wang, H. L.; Swain, E. W.; Hesseltine, C. W. (1980). "Phytase of molds used in oriental food fermentation". Journal of Food Science. 45 (5): 1262. doi:10.1111/j.1365-2621.1980.tb06534.x.

31. Mark R. Wilkins and Hasan Atiye (2012). "Fermentation". In Nurhan Turgut Dunford. Food and Industrial Bioproducts and Bioprocessing. Wiley. p. 195. ISBN 9781119946052.

32. Qureshi N, Blaschek, HP. 2001. Recent advances in ABE fermentation: hyper-butanol producing Clostridium beijerinckii BA101. J Ind Microbiol Biotechnol 27(5):287-291.

33. http://www.pasteur.fr/infosci/archives/frb0.html

34. GB application 191504845, Charles Weizmann, "Improvements in the Bacterial Fermentation of Carbohydrates and in Bacterial Cultures for the same", published 1919-03-06, assigned to Charles Weizmann

35. Fred C. Kelly (1936). One Thing Leads to Another: The Growth of an Industry, Houghton Mifflin

36. http://news.nationalgeographic.com/news/energy/2013/06/130614-scotland-whisky-waste-to-biofuel/

37. Angenent, L.T., Karim, K., Al-Dahhan, M.H., Wrenn, B.A., Domíguez-Espinosa, R., 2004. "Production of bioenergy and biochemicals from industrial and agricultural wastewater". Trends in Biotechnology 22, 477-85

38. Chen, S.-D., Lee, K.-S., Lo, Y.-C., Chen, W.-M., Wu, J.-F., Lin, C.-Y., Chang, J.-S.,2008, "Batch and continuous biohydrogen production from starch hydrolysate by Clostridium species". Int J Hydrogen Energy 33, 1803-12

39. Dabrock, B., Bahl, H., Gottschalk, G., 1992. "Parameters affecting solvent production by Clostridium pasteurianum", Appl Environ Microbiol, 58, 1233-9

40. Das, D., Veziroglu, T.N., 2001. "Hydrogen production by biological process: a survey of literature". Int J Hydrogen Energy 26, 13-28

41. Das, D., 2008, "International workshop on biohydrogen production technology" (IWBT 2008),7–9 February 2008, IIT Kharapgur. Int J Hydrogen Energy 33, 2627-8

42. Fan, Y.T, Zhang, Y.H., Zhang, S.F., Hou, H-W., Ren, B-Z., 2006. "Efficient conversion of wheat straw wastes into biohydrogen gas by cow dung compost". Biores Technol 97, 500-5

43. Ferchichi, M., Crabbe, E., Gwang-Hoon, G., Hintz, W., Almadidy, A., 2005. "Influence of initial pH on hydrogen production from cheese whey". J Biotechnol 120, 402-9

44. Idania, V.V., Richard, S., Derek, R., Noemi, R.S., Hector, M.P.V., 2005. "Hydrogen generation via anaerobic fermentation of paper mill wastes". Biores Technol 96, 1907-13

45. Kapdan, I. K., Kargi, F., 2006. "Bio-hydrogen production from waste materials", Enzyme Microb Technol 38, 569–82

46. Kim, J., Park, C., Kim, T-H., Lee, M., Kim, S., Kim, S., Seung-Wook., Lee, J., 2003. "Effects of various pretreatments for enhanced anaerobic digestion with waste activated sludge". J. Biosci. Bioeng 95, 271-5

47. Kraemer, J.T., Bagley, D.M., 2007. "Improving the yield from fermentative hydrogen production". Biotechnol Let 29, 685–95

48. Logan, B.E., 2004. Feature article: "Biologically extracting energy from wastewater: Biohydrogen production and microbial fuel cells". Environ Sci Technol 38, 160A-167A

49. Logan, B.E., Oh, S.E., van Ginkel, S., Kim, I.S., 2002. "Biological hydrogen production measured in batch anaerobic respirometers". Environ Sci Technol 36, 2530-5

50. Rai, Pankaj K, Singh, S.P & Asthana, R.K . "Biohydrogen production from cheese whey wastewater in a two-step anaerobic process". Applied Biochemistry and Biotechnology2012, 167 (6) 1540-9

51. Ren, N.Q., Chua, H., Chan, S.Y., Tsang, Y.F., Wang, Y.J., Sin, N., 2007. "Assessing optimal fermentation type for bio-hydrogen production in continuous flow acidogenic reactors", Biores Technol 98, 1774-80

52. Roy Chowdhury, S., Cox, D., Levandowsky, M., 1988. "Production of hydrogen by microbial fermentation". Int J Hydrogen Energy 13, 407-10

53. Shin, H.S., Youn, J.H., Kim, S.H., 2004. "Hydrogen production from food waste in anaerobic mesophilic and thermophilic acidogenesis". Int J Hydrogen Energy 29, 1355-63

54. Sparling, R., Risbey, D., Poggi-Varaldo, H.M., 1997. "Hydrogen production from inhibited anaerobic composters". Int J Hydrogen Energy 22, 563–6

55. Tang, G., Huang, J., Sun, Z., Tang, Q., Yan, C., Liu, G., 2008. "Biohydrogen production from cattle wastewater by enriched anaerobic mixed consortia: Influence of fermentation temperature and pH". J Biosci Bioengng., 106, 80-7

56. Valdez-Vazquez, I., Rıos-Leal, E., Munoz-Paez, K.M., Carmona-Martınez, A., Poggi-Varaldo, H.M., 2006. "Effect of inhibition treatment, type of Inocula, and incubation temperature on batch H2 production from organic solid waste". Biotechnol Bioeng 95, 342-9

57. van Ginkel, S.W., Oh, S.E., Logan. B. E., 2005. "Biohydrogen gas production from food processing and domestic wastewaters". Int. J. Hydrogen Energy 30, 1535-42

58. Venkata Mohan, S., Vijaya Bhaskar, Y., Sarm, P.N., 2007a. "Biohydrogen

production from chemical wastewater treatment by selectively enriched anaerobic mixed consortia in biofilm configured reactor operated in periodic discontinuous batch mode". Water Res 41, 2652-64

59. Venkata Mohan, S., Mohanakrishna G., Veer Raghuvulu S., Sarma, P.N., 2007b. "Enhancing biohydrogen production from chemical wastewater treatment in anaerobic sequencing batch biofilm reactor (AnSBBR) by bioaugmenting with selectively enriched kanamycin resistant anaerobic mixed consortia". Int J Hydrogen Energy 32, 3284–92
60. Venkata Mohan, S., Lalit Babu, V., Sarma, P.N., 2007c. "Anaerobic biohydrogen production from dairy wastewater treatment in sequencing batch reactor (AnSBR): Effect of organic loading rate". Enzyme and Microbial Technology 41(4), 506-15
61. Venkata Mohan, S., Bhaskar, Y.B., Krishna, T.M., Chandrasekhara Rao N., Lalit Babu V., Sarma, P.N., 2007d. "Biohydrogen production from chemical wastewater as substrate by selectively enriched anaerobic mixed consortia: Influence of fermentation pH and substrate composition". Int J Hydrogen Energy, 32, 2286–95
62. Venkata Mohan, S., Mohanakrishna, G., Ramanaiah, S.V, Sarma, P.N., 2008a. "Simultaneous biohydrogen production and wastewater treatment in biofilm configured anaerobic periodic discontinuous batch reactor using distillery wastewater". Int J Hydrogen Energy 33(2), 550-8
63. Venkata Mohan, S., Mohanakrishna, G., Ramanaiah, S.V, Sarma, P.N., 2008b. "Integration of acidogenic and methanogenic processes for simultaneous production of biohydrogen and methane from wastewater treatment". Int J Hydrogen Energy 33, 2156–66
64. Venkata Mohan, S., Lalit Babu, V., Sarma, P.N., 2008c. "Effect of various pre-treatment methods on anaerobic mixed microflora to enhance biohydrogen production utilizing dairy wastewater as substrate". Biores Technol 99, 59-67
65. Venkata Mohan, S., Lalit Babu, V., Srikanth, S., Sarma, P.N., 2008d. "Bio-electrochemical behavior of fermentative hydrogen production process with the function of feeding pH". Int J Hydrogen Energy doi:10.1016/j.ijhydene.2008.05.073
66. Venkata Mohan, S., Srikanth, S., Dinakar, P., Sarma, P.N., 2008e. "Photo-biological hydrogen production by the adopted mixed culture: Data enveloping analysis". Int J Hydrogen Energy 33(2), 559-69
67. Venkata Mohan, S., Mohanakrishna,G., Reddy, S.S., Raju, B.D., Rama Rao, K.S., Sarma, P,N., 2008f. "Self-immobilization of acidogenic mixed consortia on mesoporous material (SBA-15) and activated carbon to enhance fermentative hydrogen production". Int J Hydrogen Energy doi:10.1016/j.ijhydene.2008.07.096
68. Vijaya Bhaskar, Y., Venkata Mohan S, Sarma, P.N., 2008. "Effect of substrate loading rate of chemical wastewater on fermentative biohydrogen production in biofilm configured sequencing batch reactor". Biores Technol 99, 6941–8
69. Vijayaraghavan, K., Ahmad, D., "Biohydrogen generation from palm oil mill effluent using anaerobic contact filter". Int J Hydrogen Energy 31, 1284-91

70. Yu, H., Zhu, Z., Hu, W., Zhang, H., 2002. "Hydrogen production from rice winery wastewater in an upflow anaerobic reactor by using mixed anaerobic cultures", Int J Hydrogen Energy 27, 1359-65

71. Zhang, T., Liu, H., Fang, H.H.P., 2003. "Biohydrogen production from starch in wastewater under thermophilic condition". J Environ Manag 69, 149-56

72. Zhu, H., Beland, M., 2006, "Evaluation of alternative methods of preparing hydrogen producing seeds from digested wastewater sludge". Int J Hydrogen Energy 31, 1980-8

73. Stryer, Lubert (1975). Biochemistry. W. H. Freeman and Company. ISBN 0-7167-0174-X.

74. Raj SB, Ramaswamy S, Plapp BV. "Yeast alcohol dehydrogenase structure and catalysis.". Biochemistry. 53: 5791-803. doi:10.1021/bi5006442. PMID 25157460.

75. http://web.oranim.ac.il/courses/microbiology/Bacterial%20Fermentation%20 Nature.pdf

76. Logan, BK; Distefano, S (1997). "Ethanol content of various foods and soft drinks and their potential for interference with a breath-alcohol test.". Journal of analytical toxicology. 22 (3): 181–3. PMID 9602932.

77. «The Alcohol Content of Bread.". Canadian Medical Association journal. 16 (11): 1394–5. November 1926. PMID 20316063.

78. James Jacobs, Ag Economist. "Ethanol from Sugar". United States Department of Agriculture. Retrieved 2007-09-04.

79. «Economic Feasibility of Ethanol Production from Sugar in the United States" (PDF). United States Department of Agriculture. July 2006. Archived from the original (pdf) on 2007-08-15. Retrieved 2007-09-04.

80. «Ethanol Biorefinery Locations". Renewable Fuels Association. Archived from the original on 30 April 2007. Retrieved 21 May 2007.

81. Tiny Super-Plant Can Clean Up Hog Farms and Be Used For Ethanol Production

82. Caroline Wyatt (2006-08-10). "Draining France's ‹wine lake'". BBC News. Retrieved 2007-05-21.

83. Japan Plans Its Own Green Fuel by Steve Inskeep. NPR Morning Edition, May 15, 2007

84. Agro2: Ethanol From Cassava

85. Pimentel, D. (Ed.) (1980). CRC Handbook of energy utilization in agriculture. (Boca Raton: CRC Press)

86. Lynn Ellen Doxon. The Alcohol Fuel Handbook. InfinityPublishing.com. ISBN 0-7414-0646-2.

87. Gil, C.; Gómez-Cordovés, C. (1986). "Tryptophol content of young wines made from Tempranillo, Garnacha, Viura and Airén grapes". Food Chemistry. 22: 59. doi:10.1016/0308-8146(86)90009-9.

88. Tryptophol, tyrosol and phenylethanol—The aromatic higher alcohols in beer. Clara M. Szlavko, Journal of the Institute of Brewing, July–August 1973, Volume 79, Issue 4, pages 283–288, doi:10.1002/j.2050-0416.1973.tb03541.x

89. Ribéreau-Gayon, P.; Sapis, J. C. (1965). "On the presence in wine of tyrosol, tryptophol, phenylethyl alcohol and gamma-butyrolactone, secondary products of alcoholic fermentation". Comptes rendus hebdomadaires des seances de l'Academie des sciences. Serie D: Sciences naturelles. 261 (8): 1915–1916. PMID 4954284. (Article in French)

CHAPTER 8

Gene

Chapter Contents

- **Introduction**
- **Molecular Basis**
- **Structure And Function**
- **Gene Expression**
- **Inheritance**
- **Molecular Evolution**
- **Genome**
- **Genetic Engineering**

A gene is a locus (or region) of DNA which is made up of nucleotides and is the molecular unit of heredity.:Glossary The transmission of genes to an organism's offspring is the basis of the inheritance of phenotypic traits. Most biological traits are under the influence of polygenes (many different genes) as well as gene–environment interactions. Some genetic traits are instantly visible, such as eye colour or number of limbs, and some are not, such as blood type, risk for specific diseases, or the thousands of basic biochemical processes that comprise life.

INTRODUCTION

Genes can acquire mutations in their sequence, leading to different variants, known as alleles, in the population. These alleles encode slightly different versions of a protein, which cause different phenotype traits. Colloquial usage of the term "having a gene" (e.g., "good genes," "hair colour gene") typically refers to having a different allele of the gene. Genes evolve due to natural selection or survival of the fittest of the alleles.

The concept of a gene continues to be refined as new phenomena are discovered. For example, regulatory regions of a gene can be far removed from its coding regions, and coding regions can be split into several exons. Some viruses store their genome in RNA instead of DNA and some gene products are functional non-coding RNAs. Therefore, a broad, modern working definition of a gene is any discrete locus of heritable, genomic sequence which affect an organism's traits by being expressed as a functional product or by regulation of gene expression.

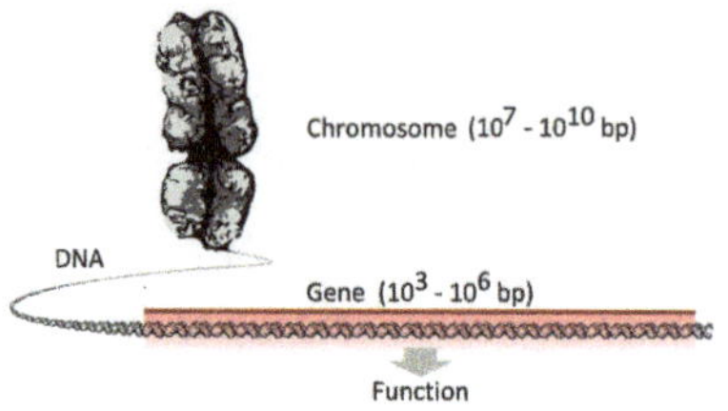

A gene is a region of DNA that encodes function. A chromosome consists of a long strand of DNA containing many genes. A human chromosome can have up to 500 million base pairs of DNA with thousands of genes.

Discovery of Discrete Inherited Units

The existence of discrete inheritable units was first suggested by Gregor Mendel (1822–1884). From 1857 to 1864, he studied inheritance patterns in 8000 common edible pea plants, tracking distinct traits from parent to offspring. He described these mathematically as 2^n combinations where n is the number of differing characteristics in the original peas. Although he did not use the term *gene*, he explained his results in terms of discrete inherited units that give rise to observable physical characteristics. This description prefigured the distinction between genotype (the genetic material of an organism) and phenotype (the visible traits of that organism). Mendel was also the first to demonstrate independent assortment, the distinction between dominant and recessive traits, the distinction between a heterozygote and homozygote, and the phenomenon of discontinuous inheritance.

Gregor Mendel

Prior to Mendel's work, the dominant theory of heredity was one of blending inheritance, which suggested that each parent contributed fluids to the fertilisation process and that the traits of the parents blended and mixed to produce the offspring. Charles Darwin developed a theory of inheritance he termed pangenesis, from Greek pan ("all, whole") and genesis ("birth") / genos ("origin"). Darwin used the term *gemmule* to describe hypothetical particles that would mix during reproduction.

Mendel's work went largely unnoticed after its first publication in 1866, but was rediscovered in the late 19th century by Hugo de Vries, Carl Correns, and Erich von Tschermak, who (claimed to have) reached similar conclusions in their own research. Specifically, in 1889, Hugo de Vries published his book *Intracellular Pangenesis*, in which he postulated that different characters have individual hereditary carriers and that inheritance of specific traits in organisms comes in particles. De Vries called these units "pangenes" (*Pangens* in German), after Darwin's 1868 pangenesis theory.

Sixteen years later, in 1905, the word *genetics* was first used by William Bateson, while Eduard Strasburger, amongst others, still used the term pangene for the fundamental physical and functional unit of heredity. In 1909 the Danish botanist Wilhelm Johannsen shortened the name to "gene".

Discovery of DNA

Advances in understanding genes and inheritance continued throughout the 20th century. Deoxyribonucleic acid (DNA) was shown to be the molecular repository of genetic information by experiments in the 1940s to 1950s. The structure of DNA was studied by Rosalind Franklin and Maurice Wilkins using X-ray crystallography, which led James D. Watson and Francis Crick to publish a model of the double-stranded DNA molecule whose paired nucleotide bases indicated a compelling hypothesis for the mechanism of genetic replication.

In the early 1950s the prevailing view was that the genes in a chromosome acted like discrete entities, indivisible by recombination and arranged like beads on a string. The experiments of Benzer using mutants defective in the rII region of bacteriophage T4 (1955-1959) showed that individual genes have a simple linear structure and are likely to be equivalent to a linear section of DNA.

Collectively, this body of research established the central dogma of molecular biology, which states that proteins are translated from RNA, which is transcribed from DNA. This dogma has since been shown to have exceptions, such as reverse transcription in retroviruses. The modern study of genetics at the level of DNA is known as molecular genetics.

In 1972, Walter Fiers and his team at the University of Ghent were the first to determine the sequence of a gene: the gene for Bacteriophage MS2 coat protein. The subsequent development of chain-termination DNA sequencing in 1977 by Frederick Sanger improved the efficiency of sequencing and turned it into a routine laboratory tool.An automated version of the Sanger method was used in early phases of the Human Genome Project.

Modern Evolutionary Synthesis

The theories developed in the 1930s and 1940s to integrate molecular genetics with Darwinian evolution are called the modern evolutionary synthesis, a term introduced by Julian Huxley. Evolutionary biologists subsequently refined this concept, such as George C. Williams' gene-centric view of evolution. He proposed an evolutionary concept of the gene as a unit of natural selection with the definition: "that which segregates and recombines with appreciable frequency." In this view, the molecular gene *transcribes*as a unit, and the evolutionary gene *inherits* as a unit. Related ideas emphasizing the centrality of genes in evolution were popularized by Richard Dawkins.

MOLECULAR BASIS

DNA

The vast majority of living organisms encode their genes in long strands of DNA (deoxyribonucleic acid). DNA consists of a chain made from four types of nucleotide subunits, each composed of: a five-carbon sugar (2'-deoxyribose), a phosphate group, and one of the four bases adenine, cytosine, guanine, and thymine.

Two chains of DNA twist around each other to form a DNA double helix with the phosphate-sugar backbone spiralling around the outside, and the bases pointing inwards with adenine base pairing to thymine and guanine to cytosine. The specificity of base pairing occurs because adenine and thymine align to form two hydrogen bonds, whereas cytosine and guanine form three hydrogen bonds. The two strands in a double helix must therefore be complementary, with their sequence of bases matching such that the adenines of one strand are paired with the thymines of the other strand, and so on.

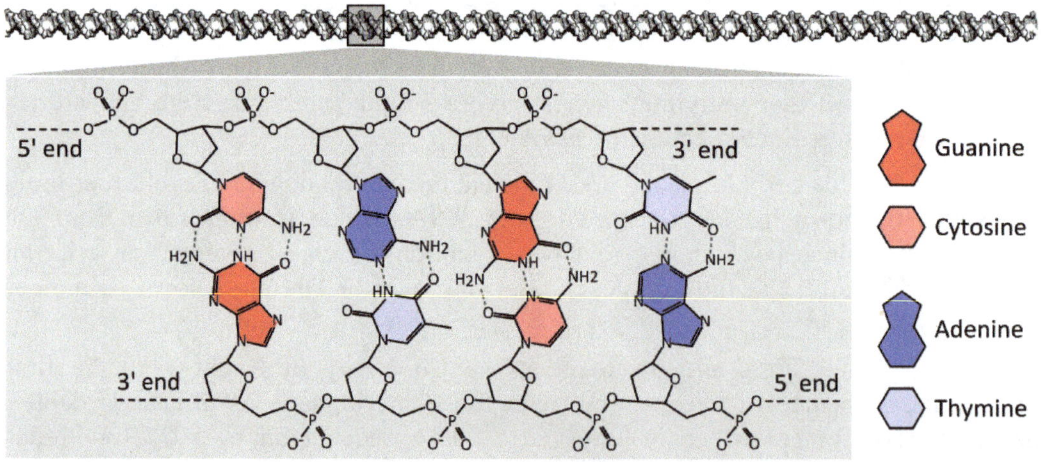

The chemical structure of a four base pair fragment of a DNAdouble helix. The sugar-phosphate backbone chains run in opposite directions with the bases pointing inwards, base-pairing A to T and C to G with hydrogen bonds.

Due to the chemical composition of the pentose residues of the bases, DNA strands have directionality. One end of a DNA polymer contains an exposed hydroxyl group on the deoxyribose; this is known as the 3' end of the molecule. The other end contains an exposed phosphate group; this is the 5' end. The two strands of a double-helix run in opposite directions. Nucleic acid synthesis, including DNA replication and transcription occurs in the 5'→3' direction, because new nucleotides are added via a dehydration reaction that uses the exposed 3' hydroxyl as a nucleophile.

The expression of genes encoded in DNA begins by transcribing the gene into RNA, a second type of nucleic acid that is very similar to DNA, but whose monomers contain the sugar ribose rather than deoxyribose. RNA also contains the base uracil in place of thymine. RNA molecules are less stable than DNA and are typically single-stranded. Genes that encode proteins are composed of a series of three-nucleotide sequences called codons, which serve as the "words" in the genetic "language". The genetic code specifies the correspondence during protein translation between codons and amino acids. The genetic code is nearly the same for all known organisms.

Chromosomes

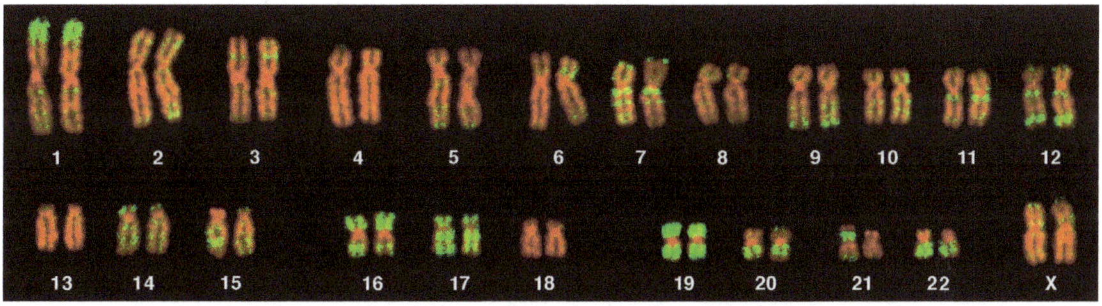

Fluorescent microscopy image of a human female karyotype, showing 23 pairs of chromosomes . The DNA is stained red, with regions rich in housekeeping genes further stained in green. The largest chromosomes are around 10 times the size of the smallest.

The total complement of genes in an organism or cell is known as its genome, which may be stored on one or more chromosomes. A chromosome consists of a single, very long DNA helix on which thousands of genes are encoded. The region of the chromosome at which a particular gene is located is called its locus. Each locus contains one allele of a gene; however, members of a population may have different alleles at the locus, each with a slightly different gene sequence.

The majority of eukaryotic genes are stored on a set of large, linear chromosomes. The chromosomes are packed within the nucleus in complex with storage proteins called histones to form a unit called a nucleosome. DNA packaged and condensed in this way is called chromatin.[:4.2] The manner in which DNA is stored on the histones, as well as chemical modifications of the histone itself, regulate whether a particular region of DNA is accessible for gene expression. In addition to genes, eukaryotic chromosomes contain sequences involved in ensuring that the DNA is copied without degradation of end regions and sorted into daughter cells during cell division: replication origins, telomeres and the centromere. Replication origins are the sequence regions where DNA replication is initiated to make two copies of the chromosome. Telomeres are long stretches of repetitive sequence that cap the ends of the linear chromosomes and prevent degradation of coding and regulatory regions during DNA replication. The length of the telomeres decreases each time the genome is replicated and has been implicated in the aging process. The centromere is required for binding spindle fibres to separate sister chromatids into daughter cells during cell division.

Prokaryotes (bacteria and archaea) typically store their genomes on a single large, circular chromosome. Similarly, some eukaryotic organelles contain a remnant circular chromosome with a small number of genes. Prokaryotes sometimes supplement their chromosome with additional small circles of DNA called plasmids, which usually encode only a few genes and are transferable between individuals. For example, the genes for antibiotic resistance are usually encoded on bacterial plasmids and can be passed between individual cells, even those of different species, via horizontal gene transfer.

Whereas the chromosomes of prokaryotes are relatively gene-dense, those of eukaryotes often contain regions of DNA that serve no obvious function. Simple single-celled eukaryotes have relatively small amounts of such DNA, whereas the genomes of complex multicellular organisms, including humans, contain an absolute majority of DNA without an identified function. This DNA has often been referred to as "junk DNA". However, more recent analyses suggest that, although protein-coding DNA makes up barely 2% of the human genome, about 80% of the bases in the genome may be expressed, so the term "junk DNA" may be a misnomer.

STRUCTURE AND FUNCTION

The structure of a gene consists of many elements of which the actual protein coding sequence is often only a small part. These include DNA regions that are not transcribed as well as untranslated regions of the RNA.

Firstly, flanking the open reading frame, all genes contain a regulatory sequence that is required for their expression. In order to be expressed, genes require a promoter sequence. The promoter is recognized and bound by transcription factors and RNA polymerase to initiate transcription.[:7.1] A gene can have more than one promoter, resulting in messenger RNAs (mRNA) that differ in how far they extend in the 5' end.Promoter regions have a consensus sequence, however highly transcribed genes have "strong" promoter sequences that bind the transcription machinery well, whereas others have "weak" promoters that bind poorly and initiate transcription less frequently.[:7.2] Eukaryotic promoter regions are much more complex and difficult to identify than prokaryotic promoters.

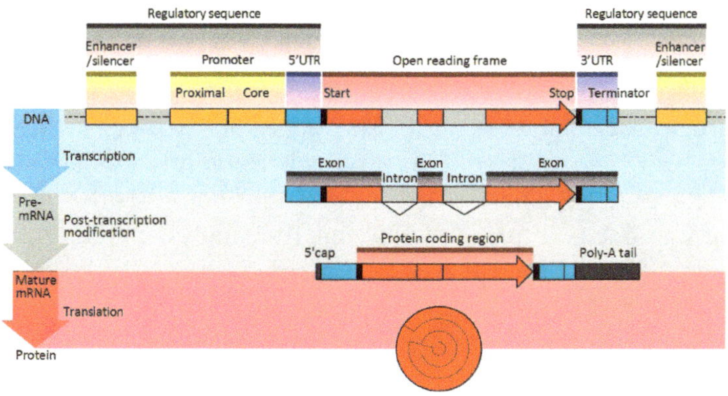

The structure of a eukaryotic protein-coding gene. Regulatory sequence controls when and where expression occurs for the protein coding region (red). Promoter and enhancer regions (yellow) regulate the transcription of the gene into a pre-mRNA which is modified to add a 5' cap and poly-A tail (grey) and remove introns. The mRNA 5' and 3' untranslated regions (blue) regulate translation into the final protein product.

Additionally, genes can have regulatory regions many kilobases upstream or downstream of the open reading frame. These act by binding to transcription factors which then cause the DNA to loop so that the regulatory sequence (and bound transcription factor) become close to the RNA polymerase binding site. For example, enhancers increase transcription by binding an activator protein which then helps to recruit the RNA polymerase to the promoter; conversely silencers bind repressor proteins and make the DNA less available for RNA polymerase.

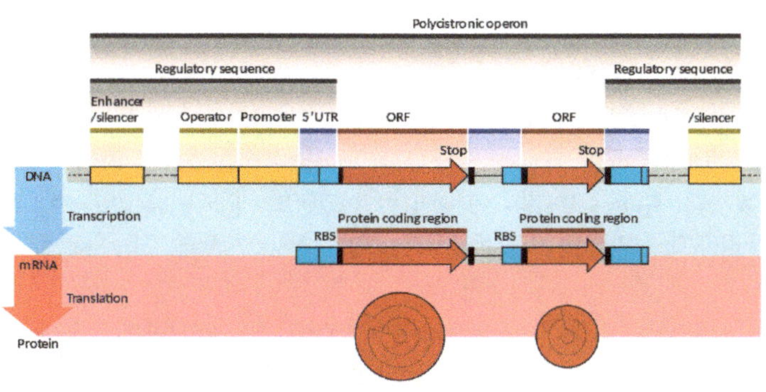

The structure of a prokaryotic operon of protein-coding genes. Regulatory sequence controls when expression occurs for the multiple protein coding regions (red). Promoter, operator and enhancer regions (yellow) regulate the transcription of the gene into an mRNA. The mRNA untranslated regions (blue) regulate translation into the final protein products.

The transcribed pre-mRNA contains untranslated regions at both ends which contain a ribosome binding site, terminator and start and stop codons. In addition, most eukaryotic open reading frames contain untranslated introns which are removed before the exons are translated. The sequences at the ends of the introns, dictate the splice sites to generate the final mature mRNA which encodes the protein or RNA product.

Many prokaryotic genes are organized into operons, with multiple protein-coding sequences that are transcribed as a unit. The genes in an operon are transcribed as a continuous messenger RNA, referred to as a polycistronic mRNA. The term cistron in this context is equivalent to gene. The transcription of an operon's mRNA is often controlled by a repressorthat can occur in an active or inactive state depending on the presence of certain specific metabolites. When active, the repressor binds to a DNA sequence at the beginning of the operon, called the operator region, and represses transcription of the operon; when the repressor is inactive transcription of the operon can occur (see e.g. Lac operon). The products of operon genes typically have related functions and are involved in the same regulatory network.

Functional Definitions

Defining exactly what section of a DNA sequence comprises a gene is difficult. Regulatory regions of a gene such as enhancers do not necessarily have to be close to the coding sequence on the linear molecule because the intervening DNA can be looped out to bring the gene and its regulatory region into proximity. Similarly, a gene's introns can be much larger than its exons. Regulatory regions can even be on entirely different chromosomes and operate *in trans* to allow regulatory regions on one chromosome to come in contact with target genes on another chromosome.

Early work in molecular genetics suggested the concept that one gene makes one protein. This concept (originally called the one gene-one enzyme hypothesis) emerged from an influential 1941 paper by George Beadle and Edward Tatum on experiments with mutants of the fungus Neurospora crassa. Norman Horowitz, an early colleague on the *Neurospora* research, reminisced in 2004 that "these experiments founded the science of what Beadle and Tatum called *biochemical genetics*. In actuality they proved to be the opening gun in what became molecular genetics and all the developments that have followed from that." The one gene-one protein concept has been refined since the discovery of genes that can encode multiple proteins by alternative splicing and coding sequences split in short section across the genome whose mRNAs are concatenated by trans-splicing.

A broad operational definition is sometimes used to encompass the complexity of these diverse phenomena, where a gene is defined as a union of genomic sequences encoding a coherent set of potentially overlapping functional products. This definition categorizes genes by their functional products (proteins or RNA) rather than their specific DNA loci, with regulatory elements classified as *gene-associated* regions.

GENE EXPRESSION

In all organisms, two steps are required to read the information encoded in a gene's DNA and produce the protein it specifies. First, the gene's DNA is *transcribed* to messenger RNA (mRNA).[:6.1] Second, that mRNA is *translated* to protein. RNA-coding genes must still go through the first step, but are not translated into protein. The process of producing a biologically functional molecule of either RNA or protein is called gene expression, and the resulting molecule is called a gene product.

Genetic Code

The nucleotide sequence of a gene's DNA specifies the amino acid sequence of a protein through the genetic code. Sets of three nucleotides, known as codons, each correspond to a specific amino acid. The principle that three sequential bases of DNA code for each amino acid was demonstrated in 1961 using frameshift mutations in the rIIB gene of bacteriophage T4 (see Crick, Brenner et al. experiment).

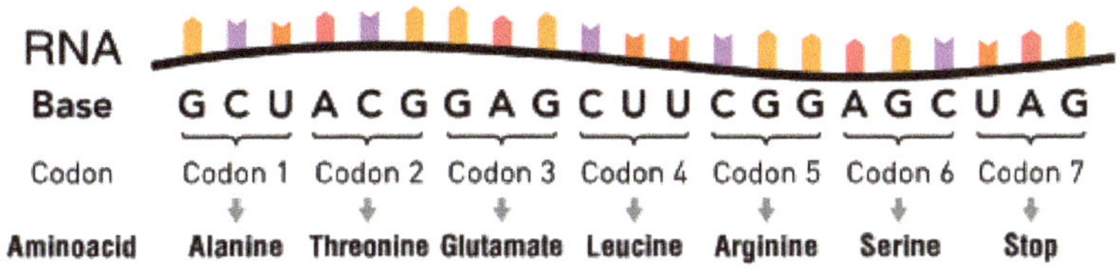

Schematic of a single-stranded RNA molecule illustrating a series of three-base codons. Each three-nucleotide codon corresponds to an amino acid when translated to protein

Additionally, a "start codon", and three "stop codons" indicate the beginning and end of the protein coding region. There are 64 possible codons (four possible nucleotides at each of three positions, hence 4^3 possible codons) and only 20 standard amino acids; hence the code is redundant and multiple codons can specify the same amino acid. The correspondence between codons and amino acids is nearly universal among all known living organisms.

Transcription

Transcription produces a single-stranded RNA molecule known as messenger RNA, whose nucleotide sequence is complementary to the DNA from which it was transcribed.[:6.1] The mRNA acts as an intermediate between the DNA gene and its final protein product. The gene's DNA is used as a template to generate a complementarymRNA. The mRNA matches the sequence of the gene's DNA coding strand because it is synthesised as the complement of the template strand. Transcription is performed by an enzyme called an RNA polymerase, which reads the template strand in the 3' to 5' direction and synthesizes the RNA from 5' to 3'. To initiate transcription, the polymerase first recognizes and binds a promoter region of the gene. Thus, a major mechanism of gene regulation is the blocking or sequestering the promoter region, either by tight binding by repressor molecules that physically block the polymerase, or by organizing the DNA so that the promoter region is not accessible.

In prokaryotes, transcription occurs in the cytoplasm; for very long transcripts, translation may begin at the 5' end of the RNA while the 3' end is still being transcribed. In eukaryotes, transcription occurs in the nucleus, where the cell's DNA is stored. The RNA molecule produced by the polymerase is known as the primary transcript and undergoes post-transcriptional modifications before being exported to the cytoplasm for translation. One of the modifications performed is the splicing of introns which are sequences in the transcribed region that do not encode protein. Alternative splicing mechanisms can result in mature transcripts from the same gene having different sequences and thus coding for different proteins. This is a major form of regulation in eukaryotic cells and also occurs in some prokaryotes.

Translation

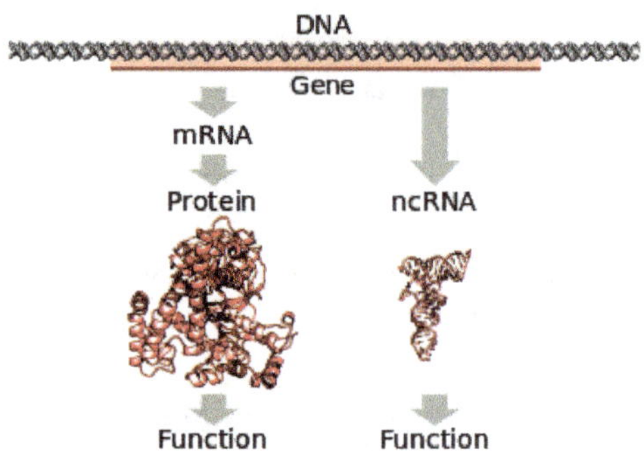

Protein coding genes are transcribed to an mRNAintermediate, then translated to a functional protein. RNA-coding genes are transcribed to a functional non-coding RNA. (PDB: 3BSE, 1OBB, 3TRA)

Translation is the process by which a mature mRNA molecule is used as a template for synthesizing a new protein.[6,2] Translation is carried out by ribosomes, large complexes of RNA and protein responsible for carrying out the chemical reactions to add new amino acids to a growing polypeptide chain by the formation of peptide bonds. The genetic code is read three nucleotides at a time, in units called codons, via interactions with specialized RNA molecules called transfer RNA (tRNA). Each tRNA has three unpaired bases known as the anticodon that are complementary to the codon it reads on the mRNA. The tRNA is also covalently attached to the amino acid specified by the complementary codon. When the tRNA binds to its complementary codon in an mRNA strand, the ribosome attaches its amino acid cargo to the new polypeptide chain, which is synthesized from amino terminus to carboxyl terminus. During and after synthesis, most new proteins must fold to their active three-dimensional structure before they can carry out their cellular functions.[3]

Regulation

Genes are regulated so that they are expressed only when the product is needed, since expression draws on limited resources.[7] A cell regulates its gene expression depending on its external environment (e.g. available nutrients, temperature and other stresses), its internal environment (e.g. cell division cycle, metabolism, infection status), and its specific role if in a multicellular organism. Gene expression can be regulated at any step: from transcriptional initiation, to RNA processing, to post-translational modification of the protein. The regulation of lactose metabolism genes in *E. coli* (*lac* operon) was the first such mechanism to be described in 1961.

RNA Genes

A typical protein-coding gene is first copied into RNA as an intermediate in the manufacture of the final protein product.[:6.1] In other cases, the RNA molecules are the actual functional products, as in the synthesis of ribosomal RNA and transfer RNA. Some RNAs known as ribozymes are capable of enzymatic function, and microRNA has a regulatory role. The DNA sequences from which such RNAs are transcribed are known as non-coding RNA genes.

Some viruses store their entire genomes in the form of RNA, and contain no DNA at all. Because they use RNA to store genes, their cellular hosts may synthesize their proteins as soon as they are infected and without the delay in waiting for transcription. On the other hand, RNA retroviruses, such as HIV, require the reverse transcription of their genome from RNA into DNA before their proteins can be synthesized. RNA-mediated epigenetic inheritance has also been observed in plants and very rarely in animals.

INHERITANCE

Organisms inherit their genes from their parents. Asexual organisms simply inherit a complete copy of their parent's genome. Sexualorganisms have two copies of each chromosome because they inherit one complete set from each parent.[:1]

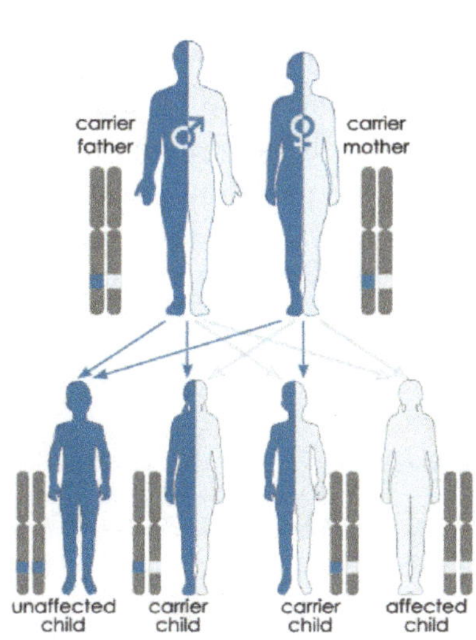

Inheritance of a gene that has two different alleles (blue and white). The gene is located on an autosomal chromosome. The blue allele is recessive to the white allele. The probability of each outcome in the children's generation is one quarter, or 25 percent.

Mendelian Inheritance

According to Mendelian inheritance, variations in an organism's phenotype (observable physical and behavioral characteristics) are due in part to variations in its genotype (particular set of genes). Each gene specifies a particular trait with different sequence of a gene (alleles) giving rise to different phenotypes. Most eukaryotic organisms (such as the pea plants Mendel worked on) have two alleles for each trait, one inherited from each parent.[:20]

Alleles at a locus may be dominant or recessive; dominant alleles give rise to their corresponding phenotypes when paired with any other allele for the same trait, whereas recessive alleles give rise to their corresponding phenotype only when paired with another copy of the same allele. For example, if the allele specifying tall stems in pea plants is dominant over the allele specifying short stems, then pea plants that inherit one tall allele from one parent and one short allele from the other parent will also have tall stems. Mendel's work demonstrated that alleles assort independently in the production of gametes, or germ cells, ensuring variation in the next generation. Although Mendelian inheritance remains a good model for many traits determined by single genes (including a number of well-known genetic disorders) it does not include the physical processes of DNA replication and cell division.

DNA Replication and Cell Division

The growth, development, and reproduction of organisms relies on cell division, or the process by which a single cell divides into two usually identical daughter cells. This requires first making a duplicate copy of every gene in the genome in a process called DNA replication. The copies are made by specialized enzymes known as DNA polymerases, which "read" one strand of the double-helical DNA, known as the template strand, and synthesize a new complementary strand. Because the DNA double helix is held together by base pairing, the sequence of one strand completely specifies the sequence of its complement; hence only one strand needs to be read by the enzyme to produce a faithful copy. The process of DNA replication is semiconservative; that is, the copy of the genome inherited by each daughter cell contains one original and one newly synthesized strand of DNA.[:5.2]

The rate of DNA replication in living cells was first measured as the rate of phage T4 DNA elongation in phage-infected *E. coli* and found to be impressively rapid. During the period of exponential DNA increase at 37 °C, the rate of elongation was 749 nucleotides per second.

After DNA replication is complete, the cell must physically separate the two copies of the genome and divide into two distinct membrane-bound cells.[:18.2] In prokaryotes (bacteria and archaea) this usually occurs via a relatively simple process called binary fission, in which each circular genome attaches to the cell membrane and is separated into the daughter cells as the membrane invaginates to split the cytoplasm into two membrane-bound portions. Binary fission is extremely fast compared to the rates of cell division in eukaryotes. Eukaryotic cell division is a more complex process known as the cell cycle; DNA replication occurs during a phase of this cycle known as S phase, whereas the process of segregating chromosomes and splitting the cytoplasm occurs during M phase.

Molecular Inheritance

The duplication and transmission of genetic material from one generation of cells to the next is the basis for molecular inheritance, and the link between the classical and molecular pictures of genes. Organisms inherit the characteristics of their parents because the cells of the offspring contain copies of the genes in their parents' cells. In asexually reproducing organisms, the offspring will be a genetic copy or clone of the parent organism. In sexually reproducing organisms, a specialized form of cell division called meiosis produces cells called gametes or germ cells that are haploid, or contain only one copy of each gene.[:20.2] The gametes produced by females are called eggs or ova, and those produced by males are called sperm. Two gametes fuse to form a diploid fertilized egg, a single cell that has two sets of genes, with one copy of each gene from the mother and one from the father.

During the process of meiotic cell division, an event called genetic recombination or *crossing-over* can sometimes occur, in which a length of DNA on one chromatid is swapped with a length of DNA on the corresponding homologous non-sister chromatid. This can result in reassortment of otherwise linked alleles.[:5.5] The Mendelian principle of independent assortment asserts that each of a parent's two genes for each trait will sort independently into gametes; which allele an organism inherits for one trait is unrelated to which allele it inherits for another trait. This is in fact only true for genes that do not reside on the same chromosome, or are located very far from one another on the same chromosome. The closer two genes lie on the same chromosome, the more closely they will be associated in gametes and the more often they will appear together; genes that are very close are essentially never separated because it is extremely unlikely that a crossover point will occur between them. This is known as genetic linkage.

MOLECULAR EVOLUTION

Mutation

DNA replication is for the most part extremely accurate, however errors (mutations) do occur. The error rate in eukaryotic cells can be as low as 10^{-8} per nucleotide per replication, whereas for some RNA viruses it can be as high as 10^{-3}. This means that each generation, each human genome accumulates 1–2 new mutations. Small mutations can be caused by DNA replication and the aftermath of DNA damage and include point mutations in which a single base is altered and frameshift mutations in which a single base is inserted or deleted. Either of these mutations can change the gene by missense (change a codon to encode a different amino acid) or nonsense (a premature stop codon). Larger mutations can be caused by errors in recombination to cause chromosomal abnormalities including the duplication, deletion, rearrangement or inversion of large sections of a chromosome. Additionally, DNA repair mechanisms can introduce mutational errors when repairing physical damage to the molecule. The repair, even with mutation, is more important to survival than restoring an exact copy, for example when repairing double-strand breaks.

When multiple different alleles for a gene are present in a species's population it is called polymorphic. Most different alleles are functionally equivalent, however some alleles can give rise to different phenotypic traits. A gene's most common allele is called the wild type, and rare alleles are called mutants. The genetic variation in relative frequencies of different alleles in a population is due to both natural selection and genetic drift. The wild-type allele is not necessarily the ancestor of less common alleles, nor is it necessarily fitter.

Most mutations within genes are neutral, having no effect on the organism's phenotype (silent mutations). Some mutations do not change the amino acid sequence because multiple codons encode the same amino acid (synonymous mutations). Other mutations can be neutral if they lead to amino acid sequence changes, but the protein still functions similarly with the new amino acid (e.g. conservative mutations). Many mutations, however, are deleterious or even lethal, and are removed from populations by natural selection. Genetic disorders are the result of deleterious mutations and can be due to spontaneous mutation in the affected individual, or can be inherited. Finally, a small fraction of mutations are beneficial, improving the organism's fitness and are extremely important for evolution, since their directional selection leads to adaptive evolution.

Sequence Homology

Genes with a most recent common ancestor, and thus a shared evolutionary ancestry, are known as homologs. These genes appear either from gene duplication within an organism's genome, where they are known as paralogous genes, or are the result of divergence of the genes after a speciation event, where they are known as orthologous genes, and often perform the same or similar functions in related organisms. It is often assumed that the functions of orthologous genes are more similar than those of paralogous genes, although the difference is minimal.

The relationship between genes can be measured by comparing the sequence alignment of their DNA.[7.6] The degree of sequence similarity between homologous genes is called conserved sequence. Most changes to a gene's sequence do not affect its function and so genes accumulate mutations over time by neutral molecular evolution. Additionally, any selection on a gene will cause its sequence to diverge at a different rate. Genes under stabilizing selection are constrained and so change more slowly whereas genes under directional selection change sequence more rapidly. The sequence differences between genes can be used for phylogenetic analyses to study how those genes have evolved and how the organisms they come from are related.

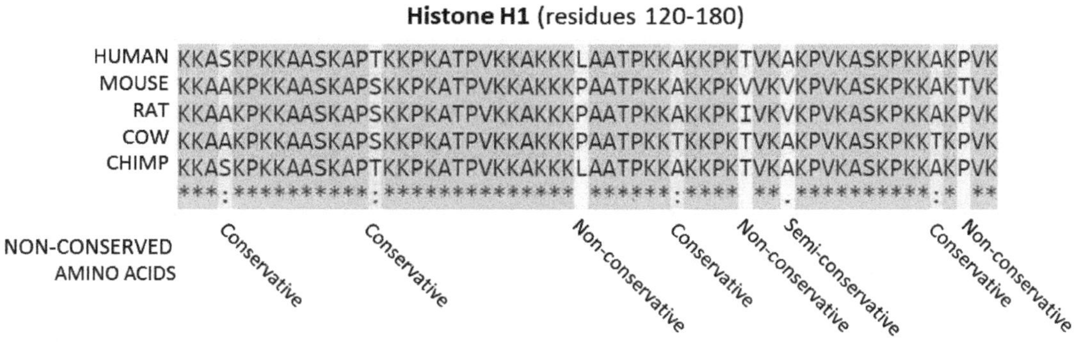

A sequence alignment, produced by ClustalO, of mammalian histoneproteins

Origins of New Genes

The most common source of new genes in eukaryotic lineages is gene duplication, which creates copy number variation of an existing gene in the genome. The resulting genes (paralogs) may then diverge in sequence and in function. Sets of genes formed in this way comprise a gene family. Gene duplications and losses within a family are common and represent a major source of evolutionary biodiversity.Sometimes, gene duplication may result in a nonfunctional copy of a gene, or a functional copy may be subject to mutations that result in loss of function; such nonfunctional genes are called pseudogenes.

"Orphan" genes, whose sequence shows no similarity to existing genes, are less common than gene duplicates. Estimates of the number of genes with no homologs outside humans range from 18 to 60.Two primary sources of orphan protein-coding genes are gene duplication followed by extremely rapid sequence change, such that the original relationship is undetectable by sequence comparisons, and de novo conversion of a previously non-coding sequence into a protein-coding gene. De novo genes are typically shorter and simpler in structure than most eukaryotic genes, with few if any introns. Over long evolutionary time periods, de novo gene birth may be responsible for a significant fraction of taxonomically-restricted gene families.

Horizontal gene transfer refers to the transfer of genetic material through a mechanism other than reproduction. This mechanism is a common source of new genes in prokaryotes, sometimes thought to contribute more to genetic variation than gene duplication. It is a common means of spreading antibiotic resistance, virulence, and adaptive metabolic functions. Although horizontal gene transfer is rare in eukaryotes, likely examples have been identified of protist and alga genomes containing genes of bacterial origin.

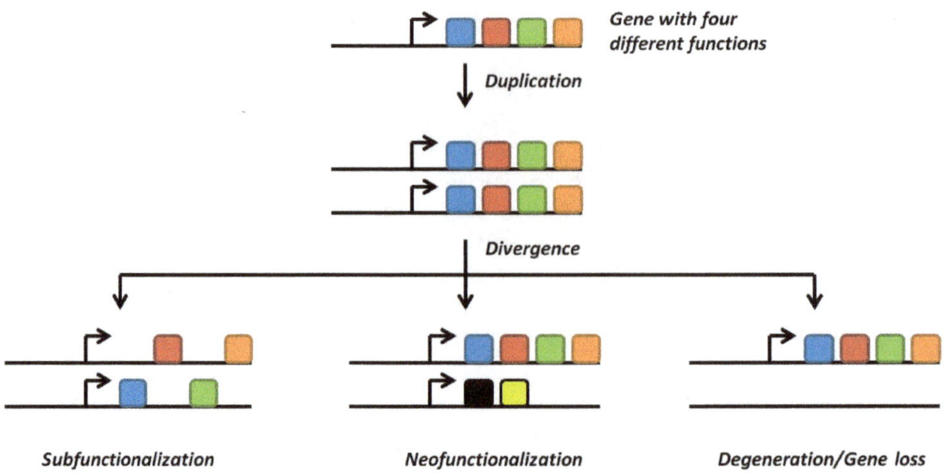

Evolutionary fate of duplicate genes

GENOME

The genome is the total genetic material of an organism and includes both the genes and non-coding sequences.

Number of Genes

The genome size, and the number of genes it encodes varies widely between organisms. The smallest genomes occur in viruses (which can have as few as 2 protein-coding genes), and viroids (which act as a single non-coding RNA gene). Conversely, plants can have extremely large genomes, with rice containing >46,000 protein-coding genes. The total number of protein-coding genes (the Earth's proteome) is estimated to be 5 million sequences.

Although the number of base-pairs of DNA in the human genome has been known since the 1960s, the estimated number of genes has changed over time as definitions of genes, and methods of detecting them have been refined. Initial theoretical predictions of the number of human genes were as high as 2,000,000. Early experimental measures indicated there to be 50,000–100,000 *transcribed* genes (expressed sequence tags). Subsequently, the sequencing in the Human Genome Project indicated that many of these transcripts were alternative variants of the same genes, and the total number of protein-coding genes was revised down to ~20,000 with 13 genes encoded on the mitochondrial genome. Of the human genome, only 1–2% consists of protein-coding genes, with the remainder being ‹noncoding' DNA such as introns, retrotransposons, and noncoding RNAs. Every multicellular organism has all its genes in each cell of its body but not every gene functions in every cell .

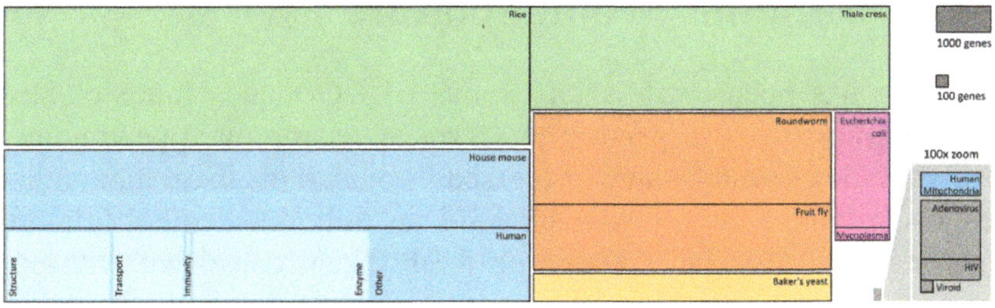

Representative genome sizes for plants (green), vertebrates (blue), invertebrates (red), fungus (yellow), bacteria (purple), and viruses (grey). An inset on the right shows the smaller genomes expanded 100-fold.

Essential Genes

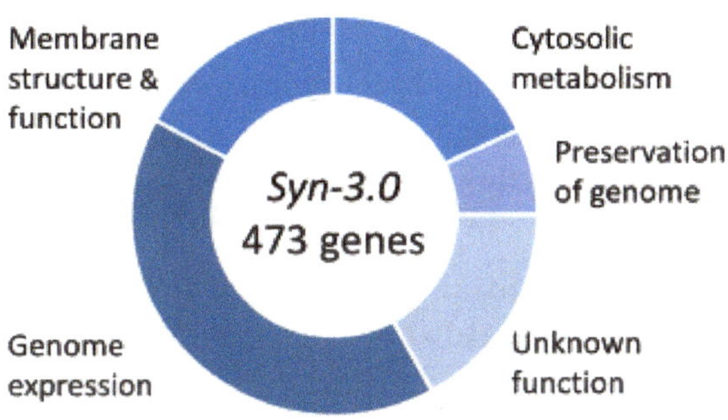

Gene functions in the minimal genome of the synthetic organism, *Syn 3*.

Essential genes are the set of genes thought to be critical for an organism's survival. This definition assumes the abundant availability of all relevant nutrients and the absence of environmental stress. Only a small portion of an organism's genes are essential. In bacteria, an estimated 250–400 genes are essential for *Escherichia coli* and *Bacillus subtilis*, which is less than 10% of their genes. Half of these genes are orthologs in both organisms and are largely involved in protein synthesis. In the budding yeast *Saccharomyces cerevisiae* the number of essential genes is slightly higher, at 1000 genes (~20% of their genes). Although the number is more difficult to measure in higher eukaryotes, mice and humans are estimated to have around 2000 essential genes (~10% of their genes). The synthetic organism, *Syn 3*, has a minimal genome of 473 essential genes and quasi-essential genes (necessary for fast growth), although 149 have unknown function.

Essential genes include Housekeeping genes (critical for basic cell functions) as well as genes that are expressed at different times in the organisms development or life cycle. Housekeeping genes are used as experimental controls when analysing gene expression, since they are constitutively expressed at a relatively constant level.

Genetic and Genomic Nomenclature

Gene nomenclature has been established by the HUGO Gene Nomenclature Committee (HGNC) for each known human gene in the form of an approved gene name and symbol (short-form abbreviation), which can be accessed through a database maintained by HGNC. Symbols are chosen to be unique, and each gene has only one symbol (although approved symbols sometimes change). Symbols are preferably kept consistent with other members of a gene family and with homologs in other species, particularly the mouse due to its role as a common model organism.

GENETIC ENGINEERING

Genetic engineering is the modification of an organism's genome through biotechnology. Since the 1970s, a variety of techniques have been developed to specifically add, remove and edit genes in an organism. Recently developed genome engineering techniques use engineered nuclease enzymes to create targeted DNA repair in a chromosome to either disrupt or edit a gene when the break is repaired. The related term synthetic biology is sometimes used to refer to extensive genetic engineering of an organism.

Genetic engineering is now a routine research tool with model organisms. For example, genes are easily added to bacteria and lineages of knockout mice with a specific gene's function disrupted are used to investigate that gene's function. Many organisms have been genetically modified for applications in agriculture, industrial biotechnology, and medicine.

For multicellular organisms, typically the embryo is engineered which grows into the adult genetically modified organism.However, the genomes of cells in an adult organism can be edited using gene therapy techniques to treat genetic diseases.

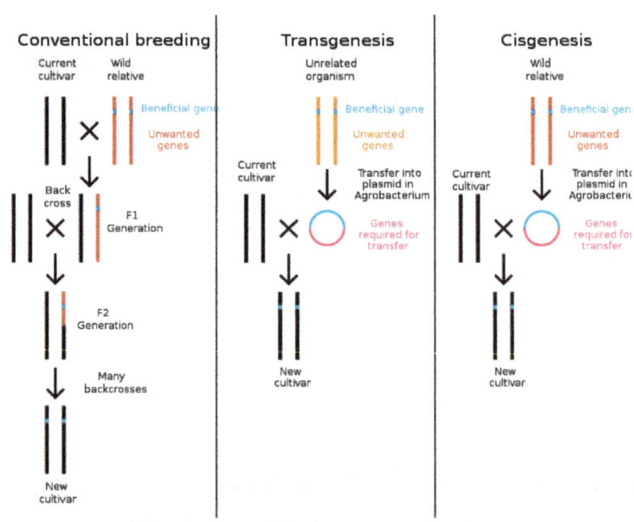

Comparison of conventional plant breeding with transgenic and cisgenic genetic modification.

REFERENCES

1. Slack, J.M.W. Genes-A Very Short Introduction. Oxford University Press 2014
2. Alberts B, Johnson A, Lewis J, Raff M, Roberts K, Walter P (2002). Molecular Biology of the Cell(Fourth ed.). New York: Garland Science. ISBN 978-0-8153-3218-3.
3. Gericke, Niklas Markus; Hagberg, Mariana (5 December 2006). "Definition of historical models of gene function and their relation to students' understanding of genetics". Science & Education. 16 *(7–8): 849–881. Bibcode*:2007Sc&Ed..16..849G. doi:10.1007/s11191-006-9064-4.
4. Pearson H (May 2006). "Genetics: what is a gene?". *Nature.* 441 *(7092): 398–401. Bibcode*:2006Natur.441..398P. doi:10.1038/441398a. PMID 16724031.
5. Pennisi E (June 2007). "Genomics. DNA study forces rethink of what it means to be a gene". Science. 316 *(5831): 1556–1557. doi:10.1126/science.316.5831.1556. PMID 17569836.*
6. Noble D (September 2008). "Genes and causation" (Free full text). Philosophical Transactions. Series A, Mathematical, Physical, and Engineering Sciences. 366 *(1878): 3001–3015. Bibcode*:2008RSPTA.366.3001N. doi:10.1098/rsta.2008.0086. PMID 18559318.
7. "genesis". Oxford English Dictionary (3rd ed.). Oxford University Press. September 2005. (Subscription or UK public library membership required.)
8. Magner, Lois N. (2002). A History of the Life Sciences (Third ed.). Marcel Dekker, CRC Press. p. 371. ISBN 978-0-203-91100-6.
9. Henig, Robin Marantz (2000). The Monk in the Garden: The Lost and Found Genius of Gregor Mendel, the Father of Genetics. Boston: Houghton Mifflin. pp. 1–9. ISBN 978-0395-97765-1.
10. Vries, H. de, Intracellulare Pangenese, Verlag von Gustav Fischer, Jena, 1889. Translated in 1908 from German to English by C. Stuart Gager as Intracellular Pangenesis, Open Court Publishing Co., Chicago, 1910
11. Gerstein MB, Bruce C, Rozowsky JS, Zheng D, Du J, Korbel JO, Emanuelsson O, Zhang ZD, Weissman S, Snyder M (June 2007). "What is a gene, post-ENCODE? History and updated definition". Genome Research. 17 *(6): 669–681. doi:10.1101/gr.6339607. PMID 17567988.*
12. Gager, C.S., Translator's preface to Intracellular Pangenesis, page viii.
13. "The Human Genome Project Timeline". Retrieved 13 September 2006.
14. Avery, OT; MacLeod, CM; McCarty, M (1944). "Studies on the Chemical Nature of the Substance Inducing Transformation of Pneumococcal Types: Induction of Transformation by a Desoxyribonucleic Acid Fraction Isolated from Pneumococcus Type III". The Journal of Experimental Medicine. 79 *(2): 137–58. doi:10.1084/jem.79.2.137. PMC 2135445.* PMID 19871359. Reprint: Avery, OT; MacLeod, CM; McCarty, M (1979). "Studies on the chemical nature of the substance inducing transformation of pneumococcal types. Inductions of

transformation by a desoxyribonucleic acid fraction isolated from pneumococcus type III". The Journal of Experimental Medicine. 149 *(2): 297–326. doi:10.1084/jem.149*.2.297. PMC 2184805. PMID 33226.

15. Hershey, AD; Chase, M (1952). "Independent functions of viral protein and nucleic acid in growth of bacteriophage". The Journal of General Physiology. 36 *(1): 39–56. doi:10.1085/jgp.36.1.39. PMC 2147348*. PMID 12981234.
16. Judson, Horace (1979). The Eighth Day of Creation: Makers of the Revolution in Biology. Cold Spring Harbor Laboratory Press. pp. 51–169. ISBN 0-87969-477-7.
17. Watson, J. D.; Crick, FH (1953). "Molecular Structure of Nucleic Acids: A Structure for Deoxyribose Nucleic Acid" (PDF). Nature. 171 *(4356): 737–8. Bibcode*:1953Natur.171..737W. doi:10.1038/171737a0. PMID 13054692.
18. Benzer S (1955). "FINE STRUCTURE OF A GENETIC REGION IN BACTERIOPHAGE". Proc. Natl. Acad. Sci. U.S.A. 41 *(6): 344–54. PMC 528093*. PMID 16589677.
19. Benzer S (1959). "ON THE TOPOLOGY OF THE GENETIC FINE STRUCTURE". Proc. Natl. Acad. Sci. U.S.A. 45 *(11): 1607–20. PMC 22276*. PMID 16590553.
20. Min Jou W, Haegeman G, Ysebaert M, Fiers W (May 1972). "Nucleotide sequence of the gene coding for the bacteriophage MS2 coat protein". Nature. 237 *(5350): 82–8. Bibcode*:1972Natur.237...82J. doi:10.1038/237082a0. PMID 4555447.
21. Sanger, F; Nicklen, S; Coulson, AR (1977). "DNA sequencing with chain-terminating inhibitors". Proceedings of the National Academy of Sciences of the United States of America. 74 *(12): 5463–7. Bibcode*:1977PNAS...74.5463S. doi:10.1073/pnas.74.12.5463. PMC 431766. PMID 271968.
22. Adams, Jill U. (2008). "DNA Sequencing Technologies". Nature Education Knowledge. SciTable. Nature Publishing Group. 1 *(1): 193.*
23. Huxley, Julian (1942). Evolution: the modern synthesis (Definitive ed.). Cambridge, Mass.: MIT Press. ISBN 978-0262513661.
24. Williams, George C. (2001). Adaptation and Natural Selection a Critique of Some Current Evolutionary Thought. ([Online-Ausg.]. ed.). Princeton: Princeton University Press. ISBN 9781400820108.
25. Dawkins, Richard (1977). The selfish gene (Repr. (with corr.) ed.). London: Oxford Univ. Press. ISBN 0-19-857519-X.
26. Dawkins, Richard (1989). The extended phenotype. (Pbk. ed.). Oxford: Oxford University Press. ISBN 0-19-286088-7.
27. Stryer L, Berg JM, Tymoczko JL (2002). Biochemistry (5th ed.). San Francisco: W.H. Freeman. ISBN 0-7167-4955-6.
28. Bolzer, Andreas; Kreth, Gregor; Solovei, Irina; Koehler, Daniela; Saracoglu, Kaan; Fauth, Christine; Müller, Stefan; Eils, Roland; Cremer, Christoph; Speicher, Michael R.; Cremer, Thomas (2005). "Three-Dimensional Maps of All

Chromosomes in Human Male Fibroblast Nuclei and Prometaphase Rosettes". PLoS Biology. 3 *(5): e157. doi:10.1371/journal.pbio.0030157. PMC 1084335.* PMID 15839726.

29. Braig M, Schmitt CA (March 2006). "Oncogene-induced senescence: putting the brakes on tumor development". Cancer Research. 66 *(6): 2881–4. doi:10.1158/0008-5472.CAN-05-4006. PMID 16540631.*
30. Bennett, PM (March 2008). "Plasmid encoded antibiotic resistance: acquisition and transfer of antibiotic resistance genes in bacteria.". *British Journal of Pharmacology. 153* Suppl 1: S347–57. doi:10.1038/sj.bjp.0707607. PMC 2268074. PMID 18193080.
31. International Human Genome Sequencing Consortium (October 2004). "Finishing the euchromatic sequence of the human genome". Nature. 431 *(7011): 931–45. Bibcode*:2004Natur.431..931H. doi:10.1038/nature03001. PMID 15496913.
32. Mortazavi A, Williams BA, McCue K, Schaeffer L, Wold B (July 2008). "Mapping and quantifying mammalian transcriptomes by RNA-Seq". Nature Methods. 5 *(7): 621–8. doi:10.1038/nmeth.1226. PMID 18516045.*
33. Pennacchio, L. A.; Bickmore, W.; Dean, A.; Nobrega, M. A.; Bejerano, G. (2013). "Enhancers: Five essential questions". Nature Reviews Genetics. 14 *(4): 288–95.* doi:10.1038/nrg3458. PMID 23503198.
34. Maston, G. A.; Evans, S. K.; Green, M. R. (2006). "Transcriptional Regulatory Elements in the Human Genome". Annual Review of Genomics and Human Genetics. 7: 29–59. doi:10.1146/annurev.genom.7.080505.115623. PMID 16719718.
35. Mignone, Flavio; Gissi, Carmela; Liuni, Sabino; Pesole, Graziano (2002-02-28). "Untranslated regions of mRNAs". Genome Biology. 3 *(3): reviews0004.* doi:10.1186/gb-2002-3-3-reviews0004. ISSN 1465-6906. PMID 11897027.
36. Bicknell AA, Cenik C, Chua HN, Roth FP, Moore MJ (December 2012). "Introns in UTRs: why we should stop ignoring them.". BioEssays. 34 *(12): 1025–34. doi:10.1002/bies.201200073. PMID 23108796.*
37. Salgado, H.; Moreno-Hagelsieb, G.; Smith, T.; Collado-Vides, J. (2000). "Operons in Escherichia coli: Genomic analyses and predictions". Proceedings of the National Academy of Sciences. 97 *(12): 6652–6657. Bibcode*:2000PNAS...97.6652S. doi:10.1073/pnas.110147297. PMC 18690. PMID 10823905.
38. Blumenthal, Thomas (November 2004). "Operons in eukaryotes". Briefings in Functional Genomics & Proteomics. 3 *(3): 199–211. doi:10.1093/bfgp/3.3.199. ISSN 2041-2649. PMID 15642184.*
39. JACOB F, MONOD J (1961). "Genetic regulatory mechanisms in the synthesis of proteins". J. Mol. Biol. 3: 318–56. doi:10.1016/S0022-2836(61)80072-7. PMID 13718526.
40. Spilianakis CG, Lalioti MD, Town T, Lee GR, Flavell RA (June 2005). "Interchromosomal associations between alternatively expressed loci". Nature. 435 *(7042): 637–45. Bibcode*:2005Natur.435..637S. doi:10.1038/nature03574. PMID 15880101.

41. Williams, A; Spilianakis, CG; Flavell, RA (April 2010). "Interchromosomal association and gene regulation in trans.". *Trends in* genetics : TIG. 26 *(4): 188–97. doi:10.1016/j.tig.2010.01.007. PMID 20236724.*

42. Beadle GW, Tatum EL (1941). "Genetic Control of Biochemical Reactions in Neurospora". Proc. Natl. Acad. Sci. U.S.A. 27 *(11): 499–506. PMC 1078370.* PMID 16588492.

43. Horowitz NH, Berg P, Singer M, Lederberg J, Susman M, Doebley J, Crow JF (2004). "A centennial: George W. Beadle, 1903-1989". Genetics. 166 *(1): 1–10. doi:10.1534/genetics.166.1.1. PMC 1470705.* PMID 15020400.

44. Marande W, Burger G (October 2007). "Mitochondrial DNA as a genomic jigsaw puzzle". Science. AAAS. 318 *(5849): 415. Bibcode*:2007Sci...318..415M. doi:10.1126/science.1148033. PMID 17947575.

45. Parra G, Reymond A, Dabbouseh N, Dermitzakis ET, Castelo R, Thomson TM, Antonarakis SE, Guigó R (January 2006). "Tandem chimerism as a means to increase protein complexity in the human genome". Genome Research. 16 *(1): 37–44. doi:10.1101/gr.4145906. PMC 1356127.* PMID 16344564.

46. Eddy SR (December 2001). "Non-coding RNA genes and the modern RNA world". Nat. Rev. Genet. 2 *(12): 919–29.* doi:10.1038/35103511. PMID 11733745.

47. CRICK FH, BARNETT L, BRENNER S, WATTS-TOBIN RJ (1961). "General nature of the genetic code for proteins". Nature. 192: 1227–32. doi:10.1038/1921227a0. PMID 13882203.

48. Crick, Francis (1962). The genetic code. WH Freeman and Company. PMID 13882204.

49. Woodson SA (May 1998). "Ironing out the kinks: splicing and translation in bacteria". Genes & Development. 12 *(9): 1243–7. doi:10.1101/gad.12.9.1243. PMID 9573040.*

50. Jacob F; Monod J (June 1961). "Genetic regulatory mechanisms in the synthesis of proteins". J Mol Biol. 3 *(3): 318–56.* doi:10.1016/S0022-2836(61)80072-7. PMID 13718526.

51. Koonin, Eugene V.; Dolja, Valerian V.; Morris, T. Jack (January 1993). "Evolution and Taxonomy of Positive-Strand RNA Viruses: Implications of Comparative Analysis of Amino Acid Sequences". Critical Reviews in Biochemistry and Molecular Biology. 28 *(5): 375–430.* doi:10.3109/10409239309078440. PMID 8269709.

52. Domingo, Esteban (2001). "RNA Virus Genomes". ELS. doi:10.1002/9780470015902.a0001488.pub2. ISBN 0470016175.

53. Domingo, E; Escarmís, C; Sevilla, N; Moya, A; Elena, SF; Quer, J; Novella, IS; Holland, JJ (June 1996). "Basic concepts in RNA virus evolution.". *FASEB Journal.* 10 *(8): 859–64. PMID 8666162.*

54. Morris, KV; Mattick, JS (June 2014). "The rise of regulatory RNA.". *Nature Reviews Genetics.* 15 *(6): 423–37.* doi:10.1038/nrg3722. PMID 24776770.

55. Miko, Ilona (2008). "Gregor Mendel and the Principles of Inheritance". Nature

Education Knowledge. SciTable. Nature Publishing Group. 1 *(1): 134.*

56. Chial, Heidi (2008). "Mendelian Genetics: Patterns of Inheritance and Single-Gene Disorders". Nature Education Knowledge. SciTable. Nature Publishing Group. 1 *(1): 63.*

57. McCarthy D, Minner C, Bernstein H, Bernstein C (1976). "DNA elongation rates and growing point distributions of wild-type phage T4 and a DNA-delay amber mutant". J. Mol. Biol. 106 *(4): 963–81.* doi:10.1016/0022-2836(76)90346-6. PMID 789903.

58. Lobo, Ingrid; Shaw, Kelly (2008). "Discovery and Types of Genetic Linkage". Nature Education Knowledge. SciTable. Nature Publishing Group. 1 *(1): 139.*

59. Nachman MW, Crowell SL (September 2000). "Estimate of the mutation rate per nucleotide in humans". Genetics. 156 *(1): 297–304. PMC 1461236.* PMID 10978293.

60. Roach JC, Glusman G, Smit AF, et al. (April 2010). "Analysis of genetic inheritance in a family quartet by whole-genome sequencing". Science. 328 *(5978): 636–9. Bibcode*:2010Sci...328..636R. doi:10.1126/science.1186802. PMC 3037280. PMID 20220176.

61. Drake JW, Charlesworth B, Charlesworth D, Crow JF (April 1998). "Rates of spontaneous mutation". Genetics. 148 *(4): 1667–86. PMC 1460098.* PMID 9560386.

62. «What kinds of gene mutations are possible?". *Genetics Home Reference. United States National Library of Medicine. 11 May 2015. Retrieved 19 May 2015.*

63. Andrews, Christine A. (2010). "Natural Selection, Genetic Drift, and Gene Flow Do Not Act in Isolation in Natural Populations". Nature Education Knowledge. SciTable. Nature Publishing Group. 3 *(10): 5.*

64. Patterson, C (November 1988). "Homology in classical and molecular biology.". *Molecular Biology and Evolution.* 5 *(6): 603–25. PMID 3065587.*

65. Studer, RA; Robinson-Rechavi, M (May 2009). "How confident can we be that orthologs are similar, but paralogs differ?". *Trends in* genetics : TIG. 25 *(5): 210–6. doi:10.1016/j.tig.2009.03.004. PMID 19368988.*

66. Altenhoff, AM; Studer, RA; Robinson-Rechavi, M; Dessimoz, C (2012). "Resolving the ortholog conjecture: orthologs tend to be weakly, but significantly, more similar in function than paralogs.". *PLOS Computational Biology.* 8 *(5): e1002514. doi:10.1371/journal.pcbi.1002514. PMC 3355068.* PMID 22615551.

67. NOSIL, PATRIK; FUNK, DANIEL J.; ORTIZ-BARRIENTOS, DANIEL (February 2009). "Divergent selection and heterogeneous genomic divergence". Molecular Ecology. 18 *(3): 375–402.* doi:10.1111/j.1365-294X.2008.03946.x. PMID 19143936.

68. Emery, Laura. "Introduction to Phylogenetics". EMBL-EBI. Retrieved 19 May 2015.

69. Mitchell, Matthew W.; Gonder, Mary Katherine (2013). "Primate Speciation: A

Case Study of African Apes". Nature Education Knowledge. SciTable. Nature Publishing Group. 4 *(2): 1.*

70. Guerzoni, D; McLysaght, A (November 2011). "De novo origins of human genes.". *PLOS Genetics. 7 (11): e1002381. doi:10.1371/journal.pgen.1002381. PMC 3213182*. PMID 22102832.

71. Reams, AB; Roth, JR (2 February 2015). "Mechanisms of gene duplication and amplification.". *Cold Spring Harbor perspectives in biology. 7 (2): a016592. doi:10.1101/cshperspect.a016592. PMID 25646380.*

72. Demuth, JP; De Bie, T; Stajich, JE; Cristianini, N; Hahn, MW (20 December 2006). "The evolution of mammalian gene families". PLoS ONE. 1: e85. Bibcode:2006PLoSO...1...85D. doi:10.1371/journal.pone.0000085. PMC 1762380. PMID 17183716.

73. Knowles, DG; McLysaght, A (October 2009). "Recent de novo origin of human protein-coding genes.". *Genome Research.* 19 *(10): 1752–9. doi:10.1101/ gr.095026.109. PMC 2765279*. PMID 19726446.

74. Wu, DD; Irwin, DM; Zhang, YP (November 2011). "De novo origin of human protein-coding genes.". *PLOS Genetics.* 7 *(11): e1002379. doi:10.1371/journal. pgen.1002379. PMC 3213175*. PMID 22102831.

75. McLysaght, Aoife; Guerzoni, Daniele (31 August 2015). "New genes from non-coding sequence: the role of de novo protein-coding genes in eukaryotic evolutionary innovation". Philosophical Transactions of the Royal Society B: Biological Sciences. 370 *(1678): 20140332. doi:10.1098/rstb.2014.0332.*

76. Neme, Rafik; Tautz, Diethard (2013). "Phylogenetic patterns of emergence of new genes support a model of frequent de novo evolution". BMC Genomics. 14 *(1): 117.* doi:10.1186/1471-2164-14-117.

77. Treangen, TJ; Rocha, EP (27 January 2011). "Horizontal transfer, not duplication, drives the expansion of protein families in prokaryotes.". *PLOS Genetics.* 7 *(1): e1001284. doi:10.1371/journal.pgen.1001284. PMC 3029252*. PMID 21298028.

78. Ochman, H; Lawrence, JG; Groisman, EA (18 May 2000). "Lateral gene transfer and the nature of bacterial innovation.". *Nature.* 405 *(6784): 299–304. Bibcode*:2000Natur.405..299O. doi:10.1038/35012500. PMID 10830951.

79. Keeling, PJ; Palmer, JD (August 2008). "Horizontal gene transfer in eukaryotic evolution.". *Nature Reviews Genetics.* 9 *(8): 605–18.* doi:10.1038/nrg2386. PMID 18591983.

80. Schönknecht, G; Chen, WH; Ternes, CM; Barbier, GG; Shrestha, RP; Stanke, M; Bräutigam, A; Baker, BJ; Banfield, JF; Garavito, RM; Carr, K; Wilkerson, C; Rensing, SA; Gagneul, D; Dickenson, NE; Oesterhelt, C; Lercher, MJ; Weber, AP (8 March 2013). "Gene transfer from bacteria and archaea facilitated evolution of an extremophilic eukaryote.". *Science.* 339 *(6124): 1207–10. Bibcode*:2013Sci...339.1207S. doi:10.1126/science.1231707. PMID 23471408.

81. Ridley, M. (2006). Genome. New York, NY: Harper Perennial. ISBN 0-06-019497-9

82. Watson, JD, Baker TA, Bell SP, Gann A, Levine M, Losick R. (2004). "Ch9-10", Molecular Biology of the Gene, 5th ed., Peason Benjamin Cummings; CSHL Press.

83. "Integr8 – A.thaliana Genome Statistics:".

84. "Understanding the Basics". The Human Genome Project. Retrieved 26 April 2015.

85. "WS227 Release Letter". WormBase. 10 August 2011. Retrieved 19 November 2013.

86. Yu, J. (5 April 2002). "A Draft Sequence of the Rice Genome (Oryza sativa L. ssp. indica)". Science. 296 *(5565): 79–92. Bibcode:*2002Sci...296...79Y. doi:10.1126/science.1068037. PMID 11935017.

87. Anderson, S.; Bankier, A. T.; Barrell, B. G.; de Bruijn, M. H. L.; Coulson, A. R.; Drouin, J.; Eperon, I. C.; Nierlich, D. P.; Roe, B. A.; Sanger, F.; Schreier, P. H.; Smith, A. J. H.; Staden, R.; Young, I. G. (9 April 1981). "Sequence and organization of the human mitochondrial genome". Nature. 290 *(5806): 457–465. Bibcode:*1981Natur.290..457A. doi:10.1038/290457a0. PMID 7219534.

88. Adams, M. D. (24 March 2000). "The Genome Sequence of Drosophila melanogaster". Science. 287 *(5461): 2185–2195. Bibcode:*2000Sci...287.2185.. doi:10.1126/science.287.5461.2185. PMID 10731132.

89. Pertea, Mihaela; Salzberg, Steven L (2010). "Between a chicken and a grape: estimating the number of human genes". Genome Biology. 11 *(5): 206.* doi:10.1186/gb-2010-11-5-206. PMC 2898077. PMID 20441615.

90. Belyi, V. A.; Levine, A. J.; Skalka, A. M. (22 September 2010). "Sequences from Ancestral Single-Stranded DNA Viruses in Vertebrate Genomes: the Parvoviridae and Circoviridae Are More than 40 to 50 Million Years Old". Journal of Virology. 84 *(23): 12458–12462. doi:10.1128/JVI.01789-10. PMC 2976387.* PMID 20861255.

91. Flores, Ricardo; Di Serio, Francesco; Hernández, Carmen (February 1997). "Viroids: The Noncoding Genomes". Seminars in Virology. 8 *(1): 65–73. doi:10.1006/smvy.1997.0107.*

92. Zonneveld, B. J. M. (2010). "New Record Holders for Maximum Genome Size in Eudicots and Monocots". Journal of Botany. 2010: 1–4. doi:10.1155/2010/527357.

93. Yu J, Hu S, Wang J, Wong GK, Li S, Liu B, Deng Y, Dai L, Zhou Y, Zhang X, Cao M, Liu J, Sun J, Tang J, Chen Y, Huang X, Lin W, Ye C, Tong W, Cong L, Geng J, Han Y, Li L, Li W, Hu G, Huang X, Li W, Li J, Liu Z, Li L, Liu J, Qi Q, Liu J, Li L, Li T, Wang X, Lu H, Wu T, Zhu M, Ni P, Han H, Dong W, Ren X, Feng X, Cui P, Li X, Wang H, Xu X, Zhai W, Xu Z, Zhang J, He S, Zhang J, Xu J, Zhang K, Zheng X, Dong J, Zeng W, Tao L, Ye J, Tan J, Ren X, Chen X, He J, Liu D, Tian W, Tian C, Xia H, Bao Q, Li G, Gao H, Cao T, Wang J, Zhao W, Li P, Chen W, Wang X, Zhang Y, Hu J, Wang J, Liu S, Yang J, Zhang G, Xiong Y, Li Z, Mao L, Zhou C, Zhu Z, Chen R, Hao B, Zheng W, Chen S, Guo W, Li G, Liu S, Tao M, Wang J, Zhu L, Yuan L, Yang H (April 2002). "A draft sequence of the rice genome (Oryza sativa L. ssp.

indica)". Science. 296 *(5565): 79–92. Bibcode*:2002Sci...296...79Y. doi:10.1126/science.1068037. PMID 11935017.

94. Perez-Iratxeta C, Palidwor G, Andrade-Navarro MA (December 2007). "Towards completion of the Earth's proteome". EMBO Reports. 8 *(12): 1135–1141. doi:10.1038/sj.embor.7401117. PMC 2267224.* PMID 18059312.

95. Kauffman SA (1969). "Metabolic stability and epigenesis in randomly constructed genetic nets". Journal of Theoretical Biology. Elsevier. 22 *(3): 437–467.* doi:10.1016/0022-5193(69)90015-0. PMID 5803332.

96. Schuler GD, Boguski MS, Stewart EA, Stein LD, Gyapay G, Rice K, White RE, Rodriguez-Tomé P, Aggarwal A, Bajorek E, Bentolila S, Birren BB, Butler A, Castle AB, Chiannilkulchai N, Chu A, Clee C, Cowles S, Day PJ, Dibling T, Drouot N, Dunham I, Duprat S, East C, Edwards C, Fan JB, Fang N, Fizames C, Garrett C, Green L, Hadley D, Harris M, Harrison P, Brady S, Hicks A, Holloway E, Hui L, Hussain S, Louis-Dit-Sully C, Ma J, MacGilvery A, Mader C, Maratukulam A, Matise TC, McKusick KB, Morissette J, Mungall A, Muselet D, Nusbaum HC, Page DC, Peck A, Perkins S, Piercy M, Qin F, Quackenbush J, Ranby S, Reif T, Rozen S, Sanders C, She X, Silva J, Slonim DK, Soderlund C, Sun WL, Tabar P, Thangarajah T, Vega-Czarny N, Vollrath D, Voyticky S, Wilmer T, Wu X, Adams MD, Auffray C, Walter NA, Brandon R, Dehejia A, Goodfellow PN, Houlgatte R, Hudson JR, Ide SE, Iorio KR, Lee WY, Seki N, Nagase T, Ishikawa K, Nomura N, Phillips C, Polymeropoulos MH, Sandusky M, Schmitt K, Berry R, Swanson K, Torres R, Venter JC, Sikela JM, Beckmann JS, Weissenbach J, Myers RM, Cox DR, James MR, Bentley D, Deloukas P, Lander ES, Hudson TJ (October 1996). "A gene map of the human genome". Science. 274 *(5287): 540–6. Bibcode*:1996Sci...274..540S. doi:10.1126/science.274.5287.540. PMID 8849440.

97. Claverie JM (September 2005). "Fewer genes, more noncoding RNA". Science. 309(5740): 1529–30. Bibcode:2005Sci...309.1529C. doi:10.1126/science.1116800. PMID 16141064.

98. Carninci P, Hayashizaki Y (April 2007). "Noncoding RNA transcription beyond annotated genes". Current Opinion in Genetics & Development. 17 *(2): 139–44. doi:10.1016/j.gde.2007.02.008. PMID 17317145.*

99. to:[a] [b] Hutchison, Clyde A.; Chuang, Ray-Yuan; Noskov, Vladimir N.; Assad-Garcia, Nacyra; Deerinck, Thomas J.; Ellisman, Mark H.; Gill, John; Kannan, Krishna; Karas, Bogumil J. (2016-03-25). "Design and synthesis of a minimal bacterial genome". Science. 351(6280): aad6253. Bibcode:2016Sci...351.....H. doi:10.1126/science.aad6253. ISSN 0036-8075. PMID 27013737.

100. Glass, J. I.; Assad-Garcia, N.; Alperovich, N.; Yooseph, S.; Lewis, M. R.; Maruf, M.; Hutchison, C. A.; Smith, H. O.; Venter, J. C. (3 January 2006). "Essential genes of a minimal bacterium". Proceedings of the National Academy of Sciences. 103 *(2): 425–430. Bibcode*:2006PNAS..103..425G. doi:10.1073/pnas.0510013103. PMC 1324956. PMID 16407165.

101. Gerdes, SY; Scholle, MD; Campbell, JW; Balázsi, G; Ravasz, E; Daugherty, MD; Somera, AL; Kyrpides, NC; Anderson, I; Gelfand, MS; Bhattacharya,

A; Kapatral, V; D'Souza, M; Baev, MV; Grechkin, Y; Mseeh, F; Fonstein, MY; Overbeek, R; Barabási, AL; Oltvai, ZN; Osterman, AL (October 2003). "Experimental determination and system level analysis of essential genes in Escherichia coli MG1655.". *Journal of Bacteriology.* 185(19): 5673–84. doi:10.1128/jb.185.19.5673-5684.2003. PMC 193955. PMID 13129938.

102. Baba, T; Ara, T; Hasegawa, M; Takai, Y; Okumura, Y; Baba, M; Datsenko, KA; Tomita, M; Wanner, BL; Mori, H (2006). "Construction of Escherichia coli K-12 in-frame, single-gene knockout mutants: the Keio collection.". *Molecular systems biology.* 2: 2006.0008. doi:10.1038/msb4100050. PMC 1681482. PMID 16738554.

103. Juhas, M; Reuß, DR; Zhu, B; Commichau, FM (November 2014). "Bacillus subtilis and Escherichia coli essential genes and minimal cell factories after one decade of genome engineering.". *Microbiology (Reading, England).* 160 *(Pt 11): 2341–51. doi:10.1099/mic.0.079376-0. PMID 25092907.*

104. Tu, Z; Wang, L; Xu, M; Zhou, X; Chen, T; Sun, F (21 February 2006). "Further understanding human disease genes by comparing with housekeeping genes and other genes". BMC Genomics. 7: 31. doi:10.1186/1471-2164-7-31. PMC 1397819. PMID 16504025.

105. Georgi, B; Voight, BF; Bućan, M (May 2013). "From mouse to human: evolutionary genomics analysis of human orthologs of essential genes.". *PLOS Genetics.* 9 *(5): e1003484. doi:10.1371/journal.pgen.1003484. PMC 3649967.* PMID 23675308.

106. Eisenberg, E; Levanon, EY (October 2013). "Human housekeeping genes, revisited.". *Trends in* genetics : TIG. 29 *(10): 569–74. doi:10.1016/j.tig.2013.05.010. PMID 23810203.*

107. Amsterdam, A; Hopkins, N (September 2006). "Mutagenesis strategies in zebrafish for identifying genes involved in development and disease.". *Trends in* genetics : TIG. 22 *(9): 473–8. doi:10.1016/j.tig.2006.06.011. PMID 16844256.*

108. «About the HGNC". HGNC Database of Human Gene Names. HUGO Gene Nomenclature Committee. Retrieved 14 May 2015.

109. Stanley N. Cohen; Annie C. Y. Chang (1 May 1973). "Recircularization and Autonomous Replication of a Sheared R-Factor DNA Segment in Escherichia coli Transformants — PNAS". Pnas.org. Retrieved 17 July 2010.

110. Esvelt, KM.; Wang, HH. (2013). "Genome-scale engineering for systems and synthetic biology". Mol Syst Biol. 9 *(1): 641. doi:10.1038/msb.2012.66. PMC 3564264.* PMID 23340847.

111. Tan, WS.; Carlson, DF.; Walton, MW.; Fahrenkrug, SC.; Hackett, PB. (2012). "Precision editing of large animal genomes". Adv Genet. Advances in Genetics. 80: 37–97. doi:10.1016/B978-0-12-404742-6.00002-8. ISBN 9780124047426. PMC 3683964. PMID 23084873.

112. Puchta, H.; Fauser, F. (2013). "Gene targeting in plants: 25 years later". Int. J. Dev. Biol. 57 *(6–7–8): 629–637. doi:10.1387/ijdb.130194hp.*

113. Ran FA, Hsu PD, Wright J, Agarwala V, Scott DA, Zhang F (2013). "Genome engineering using the CRISPR-Cas9 system". Nat Protoc. 8 *(11): 2281–308. doi:10.1038/nprot.2013.143. PMC 3969860*. PMID 24157548.

114. Kittleson, Joshua (2012). "Successes and failures in modular genetic engineering". Current Opinion in Chemical Biology. 16 *(3–4): 329–336. doi:10.1016/j.cbpa.2012.06.009. PMID 22818777.*

115. Berg, P.; Mertz, J. E. (2010). "Personal Reflections on the Origins and Emergence of Recombinant DNA Technology". Genetics. 184 *(1): 9–17. doi:10.1534/genetics.109.112144. PMC 2815933*. PMID 20061565.

116. Austin, Christopher P.; Battey, James F.; Bradley, Allan; Bucan, Maja; Capecchi, Mario; Collins, Francis S.; Dove, William F.; Duyk, Geoffrey; Dymecki, Susan (September 2004). "The Knockout Mouse Project". Nature Genetics. 36 *(9): 921–924*. doi:10.1038/ng0904-921. ISSN 1061-4036. PMC 2716027. PMID 15340423.

117. Guan, Chunmei; Ye, Chao; Yang, Xiaomei; Gao, Jiangang (2010). "A review of current large-scale mouse knockout efforts". Genesis: NA. doi:10.1002/dvg.20594.

118. Deng C (2007). "In celebration of Dr. Mario R. Capecchi's Nobel Prize". International Journal of Biological Sciences. 3 *(7): 417–419. doi:10.7150/ijbs.3.417. PMC 2043165*. PMID 17998949.

CHAPTER 9

Plasmid and Protein

Chapter Contents

- **Introduction**
- **Properties and Characteristics**
- **Classifications and Types**
- **Vectors**
- **Episomes**
- **Plasmid Maintenance**
- **Yeast Plasmids**
- **Plasmid DNA Extraction**
- **Conformations**
- **Software for Bioinformatics and Design**
- **Biochemistry**
- **Synthesis**
- **Structure**
- **Cellular Functions**
- **Methods of Study**
- **Nutrition**
- **History and Etymology**

A **plasmid** is a small DNA molecule within a cell that is physically separated from a chromosomal DNA and can replicate independently. They are most commonly found in bacteria as small circular, double-stranded DNA molecules; however, plasmids are sometimes present in archaea and eukaryotic organisms. In nature, plasmids often carry genes that may benefit the survival of the organism, for example antibiotic resistance. While the chromosomes are big and contain all the essential genetic information for living under normal conditions, plasmids usually are very small and contain only additional genes that may be useful to the organism under certain situations or particular conditions. Artificial plasmids are widely used as vectors in molecular cloning, serving to drive the replication of recombinant DNA sequences within host organisms.

INTRODUCTION

Plasmids are considered replicons, a unit of DNA capable of replicating autonomously within a suitable host. However, plasmids, like viruses, are not generally classified as life. Plasmids can be transmitted from one bacterium to another (even of another species) via three main mechanisms: transformation, transduction, and conjugation. This host-to-host transfer of genetic material is called horizontal gene transfer, and plasmids can be considered part of the mobilome. Unlike viruses (which encase their genetic material in a protective protein coat called a capsid), plasmids are "naked" DNA and do not encode genes necessary to encase the genetic material for transfer to a new host. However, some classes of plasmids encode the conjugative "sex" pilus necessary for their own transfer. The size of the plasmid varies from 1 to over 200 kbp, and the number of identical plasmids in a single cell can range anywhere from one to thousands under some circumstances.

The relationship between microbes and plasmid DNA is neither parasitic nor mutualistic, because each implies the presence of an independent species living in a detrimental or commensal state with the host organism. Rather, plasmids provide a mechanism for horizontal gene transfer within a population of microbes and typically provide a selective advantage under a given environmental state. Plasmids may carry genes that provide resistance to naturally occurring antibiotics in a competitive environmental niche, or the proteins produced may act as toxins under similar circumstances, or allow the organism to utilize particular organic compounds that would be advantageous when nutrients are scarce.

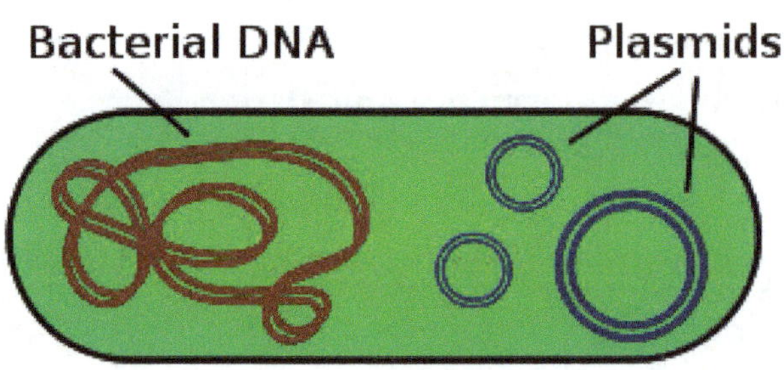

Illustration of a bacterium showing chromosomal DNA and plasmids. Not to scale.

Proteins are large biomolecules, or macromolecules, consisting of one or more long chains of amino acidresidues. Proteins perform a vast array of functions within organisms, including catalysing metabolic reactions, DNA replication, responding to stimuli, and transporting molecules from one location to another. Proteins differ from one another primarily in their sequence of amino acids, which is dictated by the nucleotide sequence of their genes, and which usually results in protein folding into a specific three-dimensional structure that determines its activity.

A linear chain of amino acid residues is called a polypeptide. A protein contains at least one long polypeptide. Short polypeptides, containing less than 20–30 residues, are rarely considered to be proteins and are commonly called peptides, or sometimes oligopeptides. The individual amino acid residues are bonded together by peptide bonds and adjacent amino acid residues. The sequence of amino acid residues in a protein is defined by the sequence of a gene, which is encoded in the genetic code. In general, the genetic code specifies 20 standard amino acids; however, in certain organisms the genetic code can include selenocysteine and—in certain archaea—pyrrolysine. Shortly after or even during synthesis, the residues in a protein are often chemically modified by post-translational modification, which alters the physical and chemical properties, folding, stability, activity, and ultimately, the function of the proteins. Sometimes proteins have non-peptide groups attached, which can be called prosthetic groups or cofactors. Proteins can also work together to achieve a particular function, and they often associate to form stable protein complexes.

Once formed, proteins only exist for a certain period of time and are then degraded and recycled by the cell's machinery through the process of protein turnover. A protein's lifespan is measured in terms of its half-life and covers a wide range. They can exist for minutes or years with an average lifespan of 1–2 days in mammalian cells. Abnormal and or misfolded proteins are degraded more rapidly either due to being targeted for destruction or due to being unstable.

Like other biological macromolecules such as polysaccharides and nucleic acids, proteins are essential parts of organisms and participate in virtually every process within cells. Many proteins are enzymes that catalyse biochemical reactions and are vital to metabolism. Proteins also have structural or mechanical functions, such as actin and myosin in muscle and the proteins in the cytoskeleton, which form a system of scaffolding that maintains cell shape. Other proteins are important in cell signaling, immune responses, cell adhesion, and the cell cycle. In animals, proteins are needed in the diet to provide the essential amino acids that cannot be synthesized. Digestion breaks the proteins down for use in the metabolism.

Proteins may be purified from other cellular components using a variety of techniques such as ultracentrifugation, precipitation, electrophoresis, and chromatography; the advent of genetic engineering has made possible a number of methods to facilitate purification. Methods commonly used to study protein structure and function include immunohistochemistry, site-directed mutagenesis, X-ray crystallography, nuclear magnetic resonance and mass spectrometry.

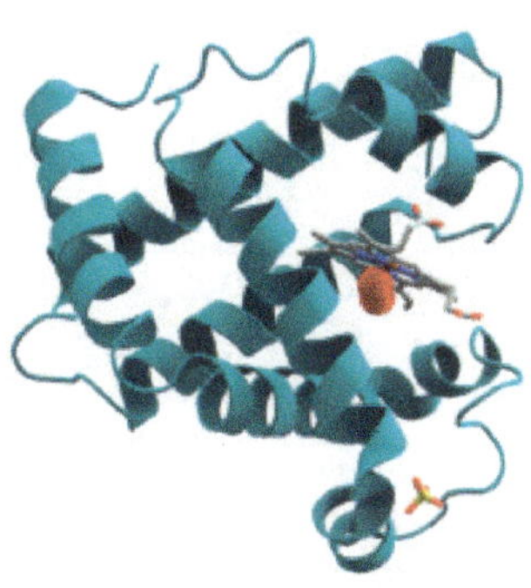

A representation of the 3D structure of the protein myoglobin showing turquoise α-helices. This protein was the first to have its structure solved by X-ray crystallography. Towards the right-center among the coils, a prosthetic group called a heme group(shown in gray) with a bound oxygen molecule (red).

PROPERTIES AND CHARACTERISTICS

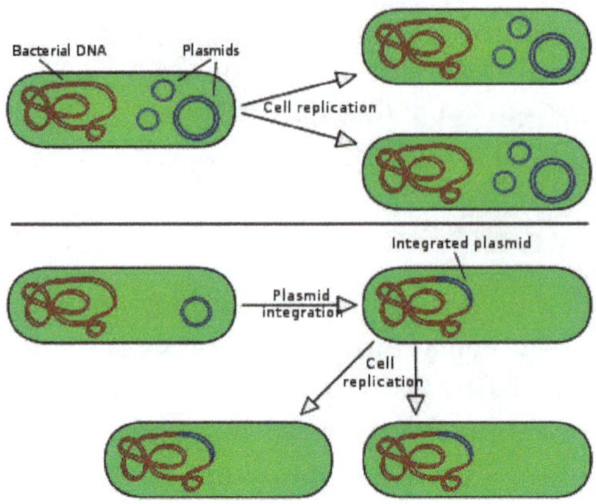

There are two types of plasmid integration into a host bacteria: Non-integrating plasmids replicate as with the top instance, whereas episomes, the lower example, can integrate into the host chromosome.

The American molecular biologist Joshua Lederberg first introduced the term plasmid in 1952 - originally to describe any bacterial genetic material that exists in an extrachromosomal state for at least part of its replication cycle. Later in 1968, it was decided that the term plasmid should be adopted as the term for extrachromosomal genetic element, and to distinguish it from viruses, the definition was narrowed to genetic elements that exist exclusively or predominantly outside of the chromosome and can replicate autonomously.

In order for plasmids to replicate independently within a cell, they must possess a stretch of DNA that can act as an origin of replication. The self-replicating unit, in this case the plasmid, is called a replicon. A typical bacterial replicon may consist of a number of elements, such as the gene for plasmid-specific replication initiation protein (Rep), repeating units called iterons, DnaA boxes, and an adjacent AT-rich region. Smaller plasmids make use of the host replicative enzymes to make copies of themselves, while larger plasmids may carry genes specific for the replication of those plasmids. A few types of plasmids can also insert into the host chromosome, and these integrative plasmids are sometimes referred to as episomes in prokaryotes.

Plasmids almost always carry at least one gene. Many of the genes carried by a plasmid are beneficial for the host cells, for example: enabling the host cell to survive in an environment that would otherwise be lethal or restrictive for growth. Some of these genes encode traits for antibiotic resistance or resistance to heavy metal, while others may produce virulence factors that enable a bacterium to colonize a host and overcome its defences, or have specific metabolic functions that allow the bacterium to utilize a particular nutrient, including the ability to degrade recalcitrant or toxic organic compounds. Plasmids can also provide bacteria with the ability to fix nitrogen. Some plasmids, however, have no observable effect on the phenotype of the host cell or its benefit to the host cells cannot be determined, and these plasmids are called cryptic plasmids.

Naturally occurring plasmids vary greatly in their physical properties. Their size can range from very small mini-plasmids of less than a 1 kilobase pairs (Kbp), to very large megaplasmids of several megabase pairs (Mbp). At the upper end, little can differentiate between a megaplasmid and a minichromosome. Plasmids are generally circular, however examples of linear plasmids are also known. These linear plasmids require specialized mechanisms to replicate their ends.

Plasmids may be present in an individual cell in varying number, ranging from one to several hundreds. The normal number of copies of plasmid that may be found in a single cell is called the copy number, and is determined by how the replication initiation is regulated and the size of the molecule. Larger plasmids tend to have lower copy numbers. Low-copy-number plasmids that exist only as one or a few copies in each bacterium are, upon cell division, in danger of being lost in one of the segregating bacteria. Such single-copy plasmids have systems that attempt to actively distribute a copy to both daughter cells. These systems, which include the parABS system and parMRC system, are often referred to as the partition system or partition function of a plasmid.

CLASSIFICATIONS AND TYPES

Plasmids may be classified in a number of ways. Plasmids can be broadly classified into conjugative plasmids and non-conjugative plasmids. Conjugative plasmids contain a set of transfer or tra genes which promote sexual conjugation between different cells. In the complex process of conjugation, plasmid may be transferred from one bacterium to another via sex pili encoded by some of the tra genes (see figure). Non-conjugative plasmids are incapable of initiating conjugation, hence they can be transferred only with the assistance of conjugative plasmids. An intermediate class of plasmids are mobilizable, and carry only a subset of the genes required for transfer. They can parasitize a conjugative plasmid, transferring at high frequency only in its presence.

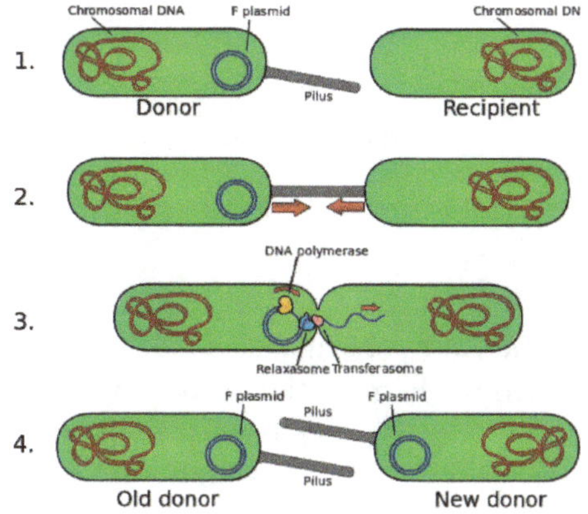

Overview of bacterial conjugation

Plasmids can also be classified into incompatibility group. A microbe can harbour different types of plasmids, however, different plasmids can only exist in a single bacterial cell if they are compatible. If two plasmids are not compatible, one or the other will be rapidly lost from the cell. Different plasmids may therefore be assigned to different incompatibility group depending on whether they can coexist together. Incompatible plasmids normally share the same replication or partition mechanisms.

Another way to classify plasmids is by function. There are five main classes:

- Fertility F-plasmids, which contain tra genes. They are capable of conjugation and result in the expression of sex pili.
- Resistance (R) plasmids, which contain genes that provide resistance against antibiotics or poisons. Historically known as R-factors, before the nature of plasmids was understood.
- Col plasmids, which contain genes that code for bacteriocins, proteins that can kill other bacteria.
- Degradative plasmids, which enable the digestion of unusual substances, e.g. toluene and salicylic acid.
- Virulence plasmids, which turn the bacterium into a pathogen.

Plasmids can belong to more than one of these functional groups.

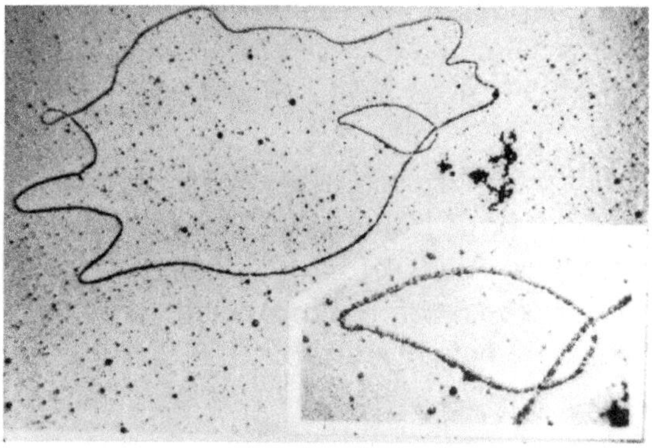

Electron micrograph of a DNA fiber bundle, presumably of a single bacterial chromosome loop.

VECTORS

Artificially constructed plasmids may be used as vectors in genetic engineering. These plasmids serve as important tools in genetics and biotechnology labs, where they are commonly used to clone and amplify (make many copies of) or express particular genes. A wide variety of plasmids are commercially available for such uses. The gene to be replicated is normally inserted into a plasmid that typically contains a number of features for their use. These include a gene that confers resistance to particular antibiotics (ampicillin is most frequently used for bacterial strains), an origin of replication to allow the bacterial cells to replicate the plasmid DNA, and a suitable site for cloning.

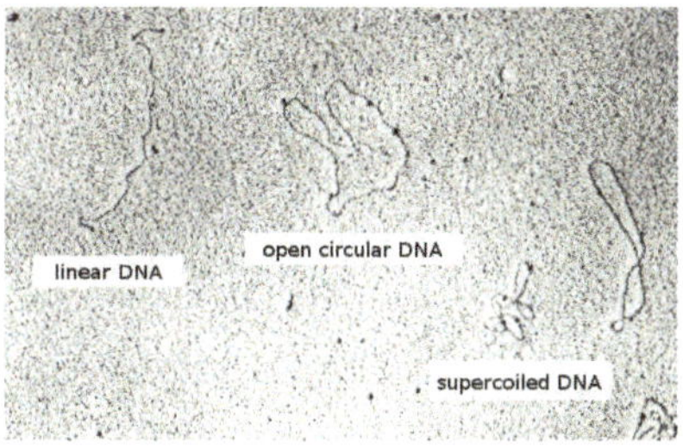

Electron micrograph of a bacterial DNA plasmid (chromosome fragment).

Cloning

Plasmids are the most-commonly used bacterial cloning vectors. These cloning vectors contain a site that allows DNA fragments to be inserted, for example a multiple cloning site or polylinker which has several commonly used restriction sites to which DNA fragments may be ligated. After the gene of interest is inserted, the plasmids are introduced into bacteria by a process called transformation. These plasmids contain a selectable marker, usually an antibiotic resistance gene, which confer on the bacteria an ability to survive and proliferate in a selective growth medium containing the particular antibiotics. The cells after transformation are exposed to the selective media, and only cells containing the plasmid may survive. In this way, the antibiotics act as a filter to select only the bacteria containing the plasmid DNA. The vector may also contain other marker genes or reporter genes to facilitate selection of plasmid with cloned insert. Bacteria containing the plasmid can then be grown in large amounts, harvested, and the plasmid of interest may then be isolated using various methods of plasmid preparation.

A plasmid cloning vector is typically used to clone DNA fragments of up to 15 kbp. To clone longer lengths of DNA, lambda phage with lysogeny genes deleted, cosmids, bacterial artificial chromosomes, or yeast artificial chromosomes are used.

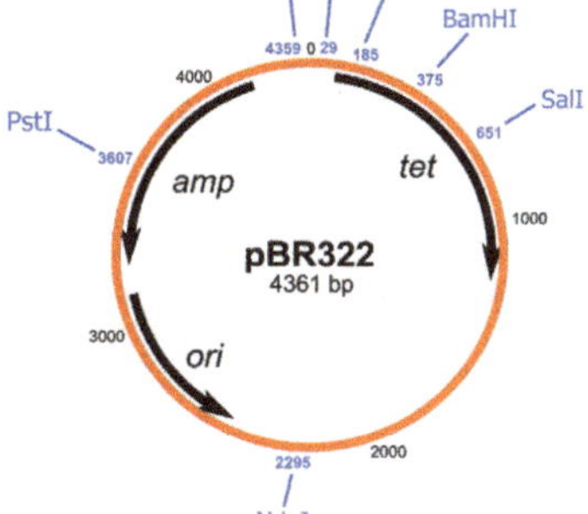

A schematic representation of the pBR322 plasmid, one of the first plasmids to be used widely as a cloning vector. Shown on the plasmid diagram are the genes encoded (amp and tet for ampicillin and tetracycline resistance respectively), its origin of replication (ori), and various restriction sites(indicated in blue).

Protein Production

Another major use of plasmids is to make large amounts of proteins. In this case, researchers grow bacteria containing a plasmid harboring the gene of interest. Just as the bacterium produces proteins to confer its antibiotic resistance, it can also be induced to produce large amounts of proteins from the inserted gene. This is a cheap and easy way of mass-producing the protein the gene codes for, for example, insulin.

Gene Therapy

Plasmid may also be used for gene transfer into human cells as potential treatment in gene therapy so that it may express the protein that is lacking in the cells. Some strategies of gene therapy require the insertion of therapeutic genes at pre-selected chromosomal target sites within the human genome. Plasmid vectors are one of many approaches that could be used for this purpose. Zinc finger nucleases (ZFNs) offer a way to cause a site-specific double-strand break to the DNA genome and cause homologous recombination. Plasmids encoding ZFN could help deliver a therapeutic gene to a specific site so that cell damage, cancer-causing mutations, or an immune response is avoided.

Disease Models

Plasmids were historically used to genetically engineer the embryonic stem cells of rats in order to create rat genetic disease models. The limited efficiency of plasmid-based techniques precluded their use in the creation of more accurate human cell models. However, developments in Adeno-associated virus recombination techniques, and Zinc finger nucleases, have enabled the creation of a new generation of isogenic human disease models.

EPISOMES

The term episome was proposed by François Jacob and Élie Wollman in 1958 to describe extra-chromosomal genetic material that may replicate autonomously or become integrated into the chromosome. The use of the term, however, has diverged since it was first coined as plasmid became the preferred word for autonomously replicating extrachromosomal DNA as proposed in the symposium in 1968 – it was suggested by some that the use of the term episome be abandoned, although others continued to use the term with a shift in meaning. In prokaryotes, episome is now used by some to refer to plasmid that is capable of integrating into the chromosome. The integrative plasmids may be replicated and stably maintained in a cell through multiple generations, but always at some stage they exist as an independent plasmid molecule.

In eukaryotes, episomes are used to mean non-integrated extrachromosomal closed circular DNA molecule that may be replicated in the nucleus. Viruses are the most common examples of this, such as herpesviruses, adenoviruses, and polyomaviruses, but some are plasmids. Other examples include aberrant chromosomal fragments, such as double minute chromosomes, that can arise during artificial gene amplifications or in pathologic processes (e.g., cancer cell transformation). Episomes in eukaryotes behave similarly to plasmids in prokaryotes in that the DNA is stably maintained and replicated with the host cell. Cytoplasmic viral episomes (as in poxvirus infections) can also occur. Some episomes, such as herpesviruses, replicate in a rolling circle mechanism, similar to bacterial phage viruses. Others replicate through a bidirectional replication mechanism (Theta type plasmids). In either case, episomes remain physically separate from host cell chromosomes. Several cancer viruses, including Epstein-Barr virus and Kaposi's sarcoma-associated herpesvirus, are maintained as latent, chromosomally distinct episomes in cancer cells, where the viruses express oncogenes that promote cancer cell proliferation. In cancers, these episomes passively replicate together with host chromosomes when the cell divides. When these viral episomes initiate lytic replication to generate multiple virus particles, they in general activate cellular innate immunity defense mechanisms that kill the host cell.

PLASMID MAINTENANCE

Some plasmids or microbial hosts include an addiction system or postsegregational killing system (PSK), such as the hok/sok (host killing/suppressor of killing) system of plasmid R1 in Escherichia coli. This variant produces both a long-lived poison and a short-lived antidote. Several types of plasmid addiction systems (toxin/ antitoxin, metabolism-based, ORT systems) were described in the literature and used in biotechnical (fermentation) or biomedical (vaccine therapy) applications. Daughter cells that retain a copy of the plasmid survive, while a daughter cell that fails to inherit the plasmid dies or suffers a reduced growth-rate because of the lingering poison from the parent cell. Finally, the overall productivity could be enhanced.

In contrast, virtually all biotechnologically used plasmids (such as pUC18, pBR322 and derived vectors) do not contain toxin-antitoxin addiction systems and thus need to be kept under antibiotic pressure to avoid plasmid loss.

YEAST PLASMIDS

Yeast are organisms that naturally harbour plasmids. Notable plasmids are 2 μm plasmids - small circular plasmids often used for genetic engineering of yeast, and linear pGKL plasmids from Kluyveromyces lactis, that are responsible for killer phenotypes.

Other types of plasmids are often related to yeast cloning vectors that include:

- Yeast integrative plasmid (YIp), yeast vectors that rely on integration into the host chromosome for survival and replication, and are usually used when studying the functionality of a solo gene or when the gene is toxic. Also connected with the gene URA3, that codes an enzyme related to the biosynthesis of pyrimidine nucleotides (T, C);
- Yeast Replicative Plasmid (YRp), which transport a sequence of chromosomal DNA that includes an origin of replication. These plasmids are less stable, as they can get lostduring the budding.

PLASMID DNA EXTRACTION

As alluded to above, plasmids are often used to purify a specific sequence, since they can easily be purified away from the rest of the genome. For their use as vectors, and for molecular cloning, plasmids often need to be isolated.

There are several methods to isolate plasmid DNA from bacteria, the archetypes of which are the **miniprep** and the maxiprep/bulkprep. The former can be used to quickly find out whether the plasmid is correct in any of several bacterial clones. The yield is a small amount of impure plasmid DNA, which is sufficient for analysis by restriction digestand for some cloning techniques.

In the latter, much larger volumes of bacterial suspension are grown from which a maxi-prep can be performed. In essence, this is a scaled-up miniprep followed by additional purification. This results in relatively large amounts (several hundreds micrograms) of very pure plasmid DNA.

In recent times, many commercial kits have been created to perform plasmid extraction at various scales, purity, and levels of automation. Commercial services can prepare plasmid DNA at quoted prices below \$300/mg in milligram quantities and \$15/mg in gram quantities (early 2007).

CONFORMATIONS

Plasmid DNA may appear in one of five conformations, which (for a given size) run at different speeds in a gel during electrophoresis. The conformations are listed below in order of electrophoretic mobility (speed for a given applied voltage) from slowest to fastest:

- Nicked open-circular DNA has one strand cut.
- Relaxed circular DNA is fully intact with both strands uncut, but has been enzymatically relaxed (supercoils removed). This can be modeled by letting a twisted extension cord unwind and relax and then plugging it into itself.
- Linear DNA has free ends, either because both strands have been cut or because the DNA was linear in vivo. This can be modeled with an electrical extension cord that is not plugged into itself.
- Supercoiled (or covalently closed-circular) DNA is fully intact with both strands uncut, and with an integral twist, resulting in a compact form. This can be modeled by twisting an extension cord and then plugging it into itself.
- Supercoiled denatured DNA is like supercoiled DNA, but has unpaired regions that make it slightly less compact; this can result from excessive alkalinity during plasmid preparation.

The rate of migration for small linear fragments is directly proportional to the voltage applied at low voltages. At higher voltages, larger fragments migrate at continuously increasing yet different rates. Thus, the resolution of a gel decreases with increased voltage.

At a specified, low voltage, the migration rate of small linear DNA fragments is a function of their length. Large linear fragments (over 20 kb or so) migrate at a certain fixed rate regardless of length. This is because the molecules 'resperate', with the bulk of the molecule following the leading end through the gel matrix. Restriction digests are frequently used to analyse purified plasmids. These enzymes specifically break the DNA at certain short sequences. The resulting linear fragments form ‹bands' after gel electrophoresis. It is possible to purify certain fragments by cutting the bands out of the gel and dissolving the gel to release the DNA fragments.

Because of its tight conformation, supercoiled DNA migrates faster through a gel than linear or open-circular DNA.

SOFTWARE FOR BIOINFORMATICS AND DESIGN

The use of plasmids as a technique in molecular biology is supported by bioinformatics software. These programs record the DNA sequence of plasmid vectors, help to predict cut sites of restriction enzymes, and to plan manipulations. Examples of software packages that handle plasmid maps are ApE, Clone Manager, GeneConstructionKit, Geneious, Genome Compiler, LabGenius, Lasergene, MacVector, pDraw32, Serial Cloner, VectorFriends, Vector NTI, and WebDSV. These software help conduct entire experiments in silico before doing wet experiments.

BIOCHEMISTRY

Most proteins consist of linear polymers built from series of up to 20 different L-α-amino acids. All proteinogenic amino acidspossess common structural features, including an α-carbon to which an amino group, a carboxyl group, and a variable side chain are bonded. Only proline differs from this basic structure as it contains an unusual ring to the N-end amine group, which forces the CO–NH amide moiety into a fixed conformation. The side chains of the standard amino acids, detailed in the list of standard amino acids, have a great variety of chemical structures and properties; it is the combined effect of all of the amino acid side chains in a protein that ultimately determines its three-dimensional structure and its chemical reactivity.The amino acids in a polypeptide chain are linked by peptide bonds. Once linked in the protein chain, an individual amino acid is called a residue, and the linked series of carbon, nitrogen, and oxygen atoms are known as the main chain or protein backbone.

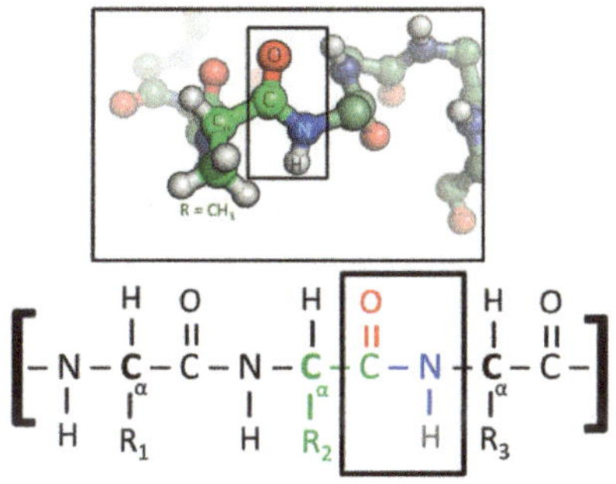

Chemical structure of the peptide bond (bottom) and the three-dimensional structure of a peptide bond between an alanine and an adjacent amino acid (top/inset)

The peptide bond has two resonance forms that contribute some double-bond character and inhibit rotation around its axis, so that the alpha carbons are roughly coplanar. The other two dihedral angles in the peptide bond determine the local shape assumed by the protein backbone. The end of the protein with a free carboxyl group is known as the C-terminus or carboxy terminus, whereas the end with a free amino group is known as the N-terminus or amino terminus. The words protein, polypeptide, and peptide are a little ambiguous and can overlap in meaning. Protein is generally used to refer to the complete biological molecule in a stable conformation, whereas peptide is generally reserved for a short amino acid oligomers often lacking a stable three-dimensional structure. However, the boundary between the two is not well defined and usually lies near 20–30 residues. Polypeptide can refer to any single linear chain of amino acids, usually regardless of length, but often implies an absence of a defined conformation.

Resonance structures of the peptide bond that links individual amino acids to form a protein polymer

Abundance in Cells

It has been estimated that average-sized bacteria contain about 2 million proteins per cell (e.g. E. coli and Staphylococcus aureus). Smaller bacteria, such as Mycoplasma or spirochetes contain fewer molecules, namely on the order of 50,000 to 1 million. By contrast, eukaryotic cells are larger and thus contain much more protein. For instance, yeastcells were estimated to contain about 50 million proteins and human cells on the order of 1 to 3 billion. Bacterial genomes encode about 10 times fewer proteins than humans (e.g. small bacteria ~1,000, E. coli: ~4,000, yeast: ~6,000, human: ~20,000).

The concentration of individual proteins ranges from a few molecules per cell to hundreds of thousands, and about a third of all proteins is not produced in most cells or only induced under certain circumstances. For instance, of the 20,000 or so proteins encoded by the human genome only 6,000 are detected in lymphoblastoid cells.

SYNTHESIS

Biosynthesis

Proteins are assembled from amino acids using information encoded in genes. Each protein has its own unique amino acid sequence that is specified by the nucleotide sequence of the gene encoding this protein. The genetic code is a set of three-nucleotide sets called codons and each three-nucleotide combination designates an amino acid, for example AUG (adenine-uracil-guanine) is the code for methionine. Because DNA contains four nucleotides, the total number of possible codons is 64; hence, there is some redundancy in the genetic code, with some amino acids specified by more than one codon. Genes encoded in DNA are first transcribed into pre-messenger RNA (mRNA) by proteins such as RNA polymerase. Most organisms then process the pre-mRNA (also known as a primary transcript) using various forms of Post-transcriptional modification to form the mature mRNA, which is then used as a template for protein synthesis by the ribosome. In prokaryotes the mRNA may either be used as soon as it is produced, or be bound by a ribosome after having moved away from the nucleoid. In contrast, eukaryotes make mRNA in the cell nucleus and then translocate it across the nuclear membrane into the cytoplasm, where protein synthesis then takes place. The rate of protein synthesis is higher in prokaryotes than eukaryotes and can reach up to 20 amino acids per second.

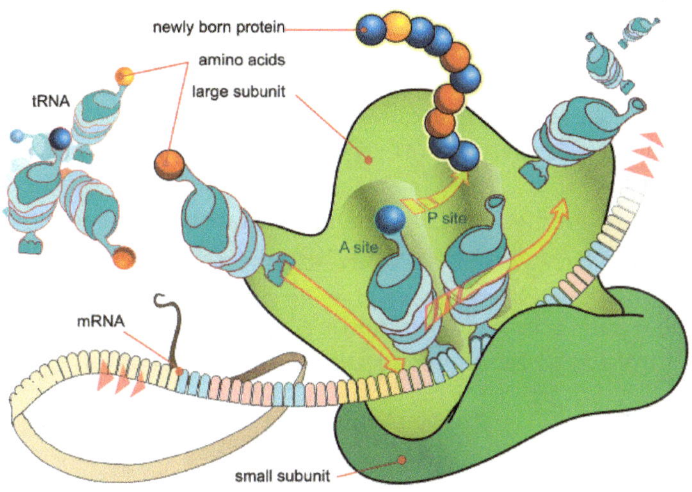

A ribosome produces a protein using mRNA as template

The process of synthesizing a protein from an mRNA template is known as translation. The mRNA is loaded onto the ribosome and is read three nucleotides at a time by matching each codon to its base pairing anticodon located on a transfer RNA molecule, which carries the amino acid corresponding to the codon it recognizes. The enzyme aminoacyl tRNA synthetase "charges" the tRNA molecules with the correct amino acids. The growing polypeptide is often termed the nascent chain. Proteins are always biosynthesized from N-terminus to C-terminus.

The size of a synthesized protein can be measured by the number of amino acids it contains and by its total molecular mass, which is normally reported in units of daltons (synonymous with atomic mass units), or the derivative unit kilodalton (kDa). Yeast proteins are on average 466 amino acids long and 53 kDa in mass. The largest known proteins are the titins, a component of the muscle sarcomere, with a molecular mass of almost 3,000 kDa and a total length of almost 27,000 amino acids.

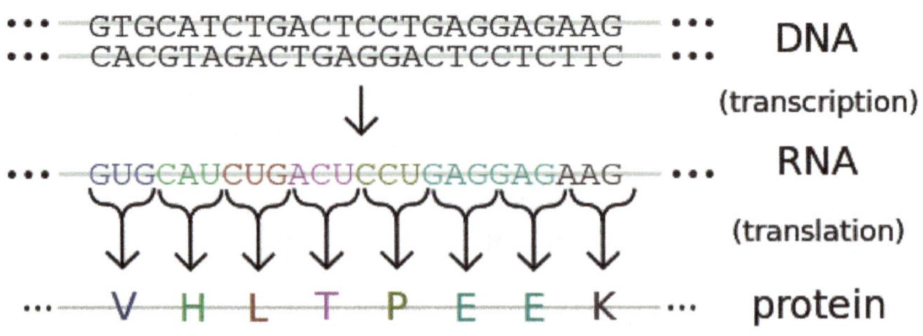

The DNA sequence of a gene encodes the amino acid sequence of a protein

Chemical Synthesis

Short proteins can also be synthesized chemically by a family of methods known as peptide synthesis, which rely on organic synthesis techniques such as chemical ligation to produce peptides in high yield. Chemical synthesis allows for the introduction of non-natural amino acids into polypeptide chains, such as attachment of fluorescent probes to amino acid side chains. These methods are useful in laboratory biochemistry and cell biology, though generally not for commercial applications. Chemical synthesis is inefficient for polypeptides longer than about 300 amino acids, and the synthesized proteins may not readily assume their native tertiary structure. Most chemical synthesis methods proceed from C-terminus to N-terminus, opposite the biological reaction.

STRUCTURE

Most proteins fold into unique 3-dimensional structures. The shape into which a protein naturally folds is known as its native conformation. Although many proteins can fold unassisted, simply through the chemical properties of their amino acids, others require the aid of molecular chaperones to fold into their native states. Biochemists often refer to four distinct aspects of a protein's structure:

- Primary structure: the amino acid sequence. A protein is a polyamide.
- Secondary structure: regularly repeating local structures stabilized by hydrogen bonds. The most common examples are the α-helix, β-sheet and turns. Because secondary structures are local, many regions of different secondary structure can be present in the same protein molecule.
- Tertiary structure: the overall shape of a single protein molecule; the spatial relationship of the secondary structures to one another. Tertiary structure is generally stabilized by nonlocal interactions, most commonly the formation of a hydrophobic core, but also through salt bridges, hydrogen bonds, disulfide bonds, and even posttranslational modifications. The term "tertiary structure" is often used as synonymous with the term fold. The tertiary structure is what controls the basic function of the protein.

- Quaternary structure: the structure formed by several protein molecules (polypeptide chains), usually called protein subunits in this context, which function as a single protein complex.

Proteins are not entirely rigid molecules. In addition to these levels of structure, proteins may shift between several related structures while they perform their functions. In the context of these functional rearrangements, these tertiary or quaternary structures are usually referred to as "conformations", and transitions between them are called conformational changes. Such changes are often induced by the binding of a substrate molecule to an enzyme's active site, or the physical region of the protein that participates in chemical catalysis. In solution proteins also undergo variation in structure through thermal vibration and the collision with other molecules.

The crystal structure of the chaperonin, a huge protein complex. A single protein subunit is highlighted. Chaperonins assist protein folding.

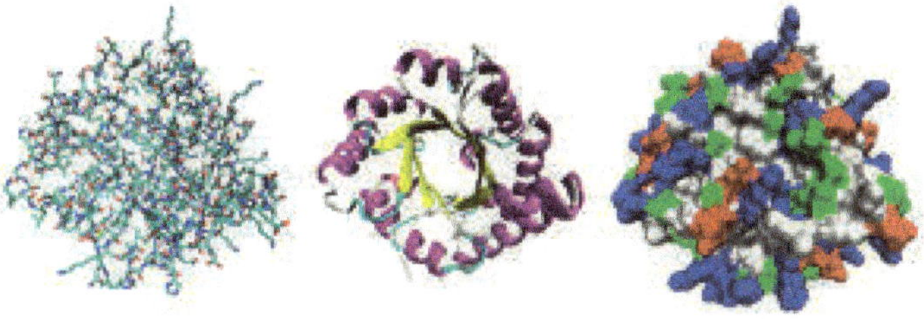

Three possible representations of the three-dimensional structure of the protein triose phosphate isomerase. **Left**: All-atom representation colored by atom type. **Middle:** Simplified representation illustrating the backbone conformation, colored by secondary structure. **Right**: Solvent-accessible surface representation colored by residue type (acidic residues red, basic residues blue, polar residues green, nonpolar residues white).

Proteins can be informally divided into three main classes, which correlate with typical tertiary structures: globular proteins, fibrous proteins, and membrane proteins. Almost all globular proteins are soluble and many are enzymes. Fibrous proteins are often structural, such as collagen, the major component of connective tissue, or keratin, the protein component of hair and nails. Membrane proteins often serve as receptors or provide channels for polar or charged molecules to pass through the cell membrane.

A special case of intramolecular hydrogen bonds within proteins, poorly shielded from water attack and hence promoting their own dehydration, are called dehydrons.

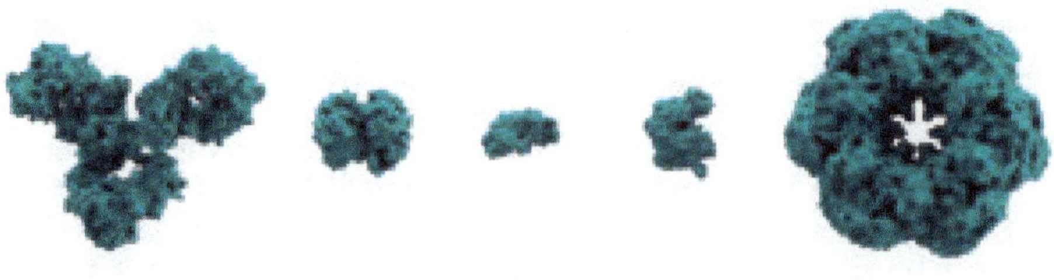

Molecular surface of several proteins showing their comparative sizes. From left to right are: immunoglobulin G (IgG, an antibody), hemoglobin, insulin (a hormone), adenylate kinase (an enzyme), and glutamine synthetase (an enzyme).

Structure Determination

Discovering the tertiary structure of a protein, or the quaternary structure of its complexes, can provide important clues about how the protein performs its function. Common experimental methods of structure determination include X-ray crystallography and NMR spectroscopy, both of which can produce information at atomic resolution. However, NMR experiments are able to provide information from which a subset of distances between pairs of atoms can be estimated, and the final possible conformations for a protein are determined by solving a distance geometry problem. Dual polarisation interferometry is a quantitative analytical method for measuring the overall protein conformation and conformational changes due to interactions or other stimulus. Circular dichroism is another laboratory technique for determining internal β-sheet / α-helical composition of proteins. Cryoelectron microscopy is used to produce lower-resolution structural information about very large protein complexes, including assembled viruses; a variant known as electron crystallography can also produce high-resolution information in some cases, especially for two-dimensional crystals of membrane proteins. Solved structures are usually deposited in the Protein Data Bank (PDB), a freely available resource from which structural data about thousands of proteins can be obtained in the form of Cartesian coordinates for each atom in the protein.

Many more gene sequences are known than protein structures. Further, the set of solved structures is biased toward proteins that can be easily subjected to the conditions required in X-ray crystallography, one of the major structure determination methods. In particular, globular proteins are comparatively easy to crystallize in preparation for X-ray crystallography. Membrane proteins, by contrast, are difficult to crystallize and are underrepresented in the PDB. Structural genomics initiatives have attempted to remedy these deficiencies by systematically solving representative structures of major fold classes. Protein structure prediction methods attempt to provide a means of generating a plausible structure for proteins whose structures have not been experimentally determined.

CELLULAR FUNCTIONS

Proteins are the chief actors within the cell, said to be carrying out the duties specified by the information encoded in genes. With the exception of certain types of RNA, most other biological molecules are relatively inert elements upon which proteins act. Proteins make up half the dry weight of an Escherichia coli cell, whereas other macromolecules such as DNA and RNA make up only 3% and 20%, respectively. The set of proteins expressed in a particular cell or cell type is known as its proteome.

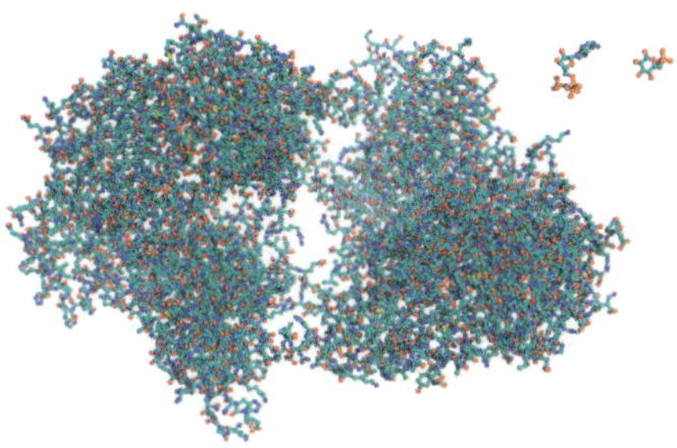

The enzyme hexokinase is shown as a conventional ball-and-stick molecular model. To scale in the top right-hand corner are two of its substrates, ATP and glucose.

The chief characteristic of proteins that also allows their diverse set of functions is their ability to bind other molecules specifically and tightly. The region of the protein responsible for binding another molecule is known as the binding site and is often a depression or "pocket" on the molecular surface. This binding ability is mediated by the tertiary structure of the protein, which defines the binding site pocket, and by the chemical properties of the surrounding amino acids' side chains. Protein binding can be extraordinarily tight and specific; for example, the ribonuclease inhibitor protein binds to human angiogenin with a sub-femtomolar dissociation constant ($<10^{-15}$M) but does not bind at all to its amphibian homolog onconase (>1 M). Extremely minor chemical changes such as the addition of a single methyl group to a binding partner can sometimes suffice to nearly eliminate binding; for example, the aminoacyl tRNA synthetasespecific to the amino acid valine discriminates against the very similar side chain of the amino acid isoleucine.

Proteins can bind to other proteins as well as to small-molecule substrates. When proteins bind specifically to other copies of the same molecule, they can oligomerize to form fibrils; this process occurs often in structural proteins that consist of globular monomers that self-associate to form rigid fibers. Protein–protein interactions also regulate enzymatic activity, control progression through the cell cycle, and allow the assembly of large protein complexes that carry out many closely related reactions with a common biological function. Proteins can also bind to, or even be integrated into, cell membranes. The ability of binding partners to induce conformational changes in proteins allows the construction of enormously complex signaling networks. As interactions between proteins are reversible, and depend heavily on the availability of different groups of partner proteins to form aggregates that are capable to carry out discrete sets of function, study of the interactions between specific proteins is a key to understand important aspects of cellular function, and ultimately the properties that distinguish particular cell types.

Enzymes

The best-known role of proteins in the cell is as enzymes, which catalyse chemical reactions. Enzymes are usually highly specific and accelerate only one or a few chemical reactions. Enzymes carry out most of the reactions involved in metabolism, as well as manipulating DNA in processes such as DNA replication, DNA repair, and transcription. Some enzymes act on other proteins to add or remove chemical groups in a process known as posttranslational modification. About 4,000 reactions are known to be catalysed by enzymes. The rate acceleration conferred by enzymatic catalysis is often enormous—as much as 10^{17}-fold increase in rate over the uncatalysed reaction in the case of orotate decarboxylase (78 million years without the enzyme, 18 milliseconds with the enzyme).

The molecules bound and acted upon by enzymes are called substrates. Although enzymes can consist of hundreds of amino acids, it is usually only a small fraction of the residues that come in contact with the substrate, and an even smaller fraction—three to four residues on average—that are directly involved in catalysis. The region of the enzyme that binds the substrate and contains the catalytic residues is known as the active site.

Dirigent proteins are members of a class of proteins that dictate the stereochemistry of a compound synthesized by other enzymes.

Cell Signaling and Ligand Binding

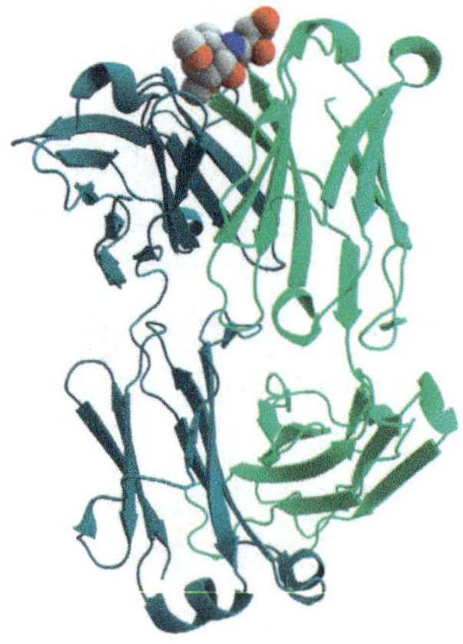

Ribbon diagram of a mouse antibody against cholera that binds a carbohydrate antigen

Many proteins are involved in the process of cell signaling and signal transduction. Some proteins, such as insulin, are extracellular proteins that transmit a signal from the cell in which they were synthesized to other cells in distant tissues. Others are membrane proteins that act as receptorswhose main function is to bind a signaling molecule and induce a biochemical response in the cell. Many receptors have a binding site exposed on the cell surface and an effector domain within the cell, which may have enzymatic activity or may undergo a conformational change detected by other proteins within the cell.

Antibodies are protein components of an adaptive immune system whose main function is to bind antigens, or foreign substances in the body, and target them for destruction. Antibodies can be secreted into the extracellular environment or anchored in the membranes of specialized B cells known as plasma cells. Whereas enzymes are limited in their binding affinity for their substrates by the necessity of conducting their reaction, antibodies have no such constraints. An antibody's binding affinity to its target is extraordinarily high. Many ligand transport proteins bind particular small biomolecules and transport them to other locations in the body of a multicellular organism. These proteins must have a high binding affinity when their ligand is present in high concentrations, but must also release the ligand when it is present at low concentrations in the target tissues. The canonical example of a ligand-binding protein is haemoglobin, which transports oxygenfrom the lungs to other organs and tissues in all vertebrates and has close homologs in every biological kingdom. Lectins are sugar-binding proteins which are highly specific for their sugar moieties. Lectins typically play a role in biological recognition phenomena involving cells and proteins. Receptors and hormones are highly specific binding proteins.

Transmembrane proteins can also serve as ligand transport proteins that alter the permeability of the cell membrane to small molecules and ions. The membrane alone has a hydrophobic core through which polar or charged molecules cannot diffuse. Membrane proteins contain internal channels that allow such molecules to enter and exit the cell. Many ion channel proteins are specialized to select for only a particular ion; for example, potassium and sodium channels often discriminate for only one of the two ions.

Structural Proteins

Structural proteins confer stiffness and rigidity to otherwise-fluid biological components. Most structural proteins are fibrous proteins; for example, collagen and elastin are critical components of connective tissue such as cartilage, and keratin is found in hard or filamentous structures such as hair, nails, feathers, hooves, and some animal shells. Some globular proteins can also play structural functions, for example, actin and tubulin are globular and soluble as monomers, but polymerize to form long, stiff fibers that make up the cytoskeleton, which allows the cell to maintain its shape and size. Other proteins that serve structural functions are motor proteins such as myosin, kinesin, and dynein, which are capable of generating mechanical forces. These proteins are crucial for cellular motility of single celled organisms and the sperm of many multicellular organisms which reproduce sexually. They also generate the forces exerted by contracting muscles and play essential roles in intracellular transport.

METHODS OF STUDY

The activities and structures of proteins may be examined in vitro, in vivo, and in silico. ***In vitro*** studies of purified proteins in controlled environments are useful for learning how a protein carries out its function: for example, enzyme kinetics studies explore the chemical mechanism of an enzyme's catalytic activity and its relative affinity for various possible substrate molecules. By contrast, ***in vivo*** experiments can provide information about the physiological role of a protein in the context of a cell or even a whole organism. ***In silico*** studies use computational methods to study proteins.

Protein Purification

To perform in vitro analysis, a protein must be purified away from other cellular components. This process usually begins with cell lysis, in which a cell's membrane is disrupted and its internal contents released into a solution known as a crude lysate. The resulting mixture can be purified using ultracentrifugation, which fractionates the various cellular components into fractions containing soluble proteins; membrane lipids and proteins; cellular organelles, and nucleic acids. Precipitation by a method known as salting out can concentrate the proteins from this lysate. Various types of chromatography are then used to isolate the protein or proteins of interest based on properties such as molecular weight, net charge and binding affinity. The level of purification can be monitored using various types of gel electrophoresis if the desired protein's molecular weight and isoelectric point are known, by spectroscopy if the protein has distinguishable spectroscopic features, or by enzyme assays if the protein has enzymatic activity. Additionally, proteins can be isolated according their charge using electrofocusing.

For natural proteins, a series of purification steps may be necessary to obtain protein sufficiently pure for laboratory applications. To simplify this process, genetic engineering is often used to add chemical features to proteins that make them easier to purify without affecting their structure or activity. Here, a "tag" consisting of a specific amino acid sequence, often a series of histidine residues (a "His-tag"), is attached to one terminus of the protein. As a result, when the lysate is passed over a chromatography column containing nickel, the histidine residues ligate the nickel and attach to the column while the untagged components of the lysate pass unimpeded. A number of different tags have been developed to help researchers purify specific proteins from complex mixtures.

Cellular Localization

The study of proteins in vivo is often concerned with the synthesis and localization of the protein within the cell. Although many intracellular proteins are synthesized in the cytoplasm and membrane-bound or secreted proteins in the endoplasmic reticulum, the specifics of how proteins are targeted to specific organelles or cellular structures is often unclear. A useful technique for assessing cellular localization uses genetic engineering to express in a cell a fusion protein or chimeraconsisting of the natural protein of interest linked to a "reporter" such as green fluorescent protein (GFP). The fused protein's position within the cell can be cleanly and efficiently visualized using microscopy, as shown in the figure opposite.

Other methods for elucidating the cellular location of proteins requires the use of known compartmental markers for regions such as the ER, the Golgi, lysosomes or vacuoles, mitochondria, chloroplasts, plasma membrane, etc. With the use of fluorescently tagged versions of these markers or of antibodies to known markers, it becomes much simpler to identify the localization of a protein of interest. For example, indirect immunofluorescence will allow for fluorescence colocalization and demonstration of location. Fluorescent dyes are used to label cellular compartments for a similar purpose.

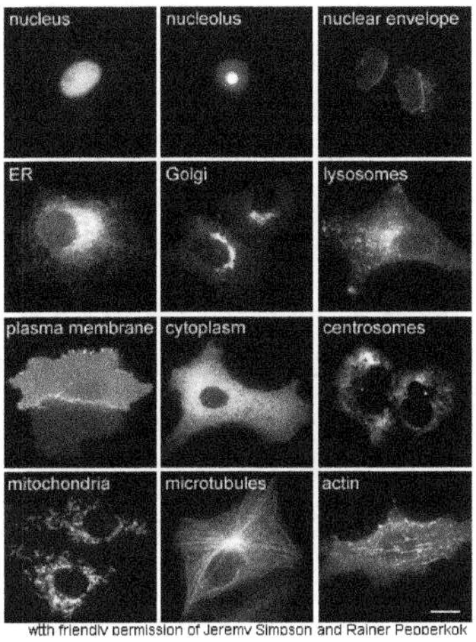

Proteins in different cellular compartments and structures tagged with green fluorescent protein (here, white)

Other possibilities exist, as well. For example, immunohistochemistry usually utilizes an antibody to one or more proteins of interest that are conjugated to enzymes yielding either luminescent or chromogenic signals that can be compared between samples, allowing for localization information. Another applicable technique is cofractionation in sucrose (or other material) gradients using isopycnic centrifugation. While this technique does not prove colocalization of a compartment of known density and the protein of interest, it does increase the likelihood, and is more amenable to large-scale studies.

Finally, the gold-standard method of cellular localization is immunoelectron microscopy. This technique also uses an antibody to the protein of interest, along with classical electron microscopy techniques. The sample is prepared for normal electron microscopic examination, and then treated with an antibody to the protein of interest that is conjugated to an extremely electro-dense material, usually gold. This allows for the localization of both ultrastructural details as well as the protein of interest.

Through another genetic engineering application known as site-directed mutagenesis, researchers can alter the protein sequence and hence its structure, cellular localization, and susceptibility to regulation. This technique even allows the incorporation of unnatural amino acids into proteins, using modified tRNAs, and may allow the rational designof new proteins with novel properties.

Proteomics

The total complement of proteins present at a time in a cell or cell type is known as its proteome, and the study of such large-scale data sets defines the field of proteomics, named by analogy to the related field of genomics. Key experimental techniques in proteomics include 2D electrophoresis, which allows the separation of a large number of proteins, mass spectrometry, which allows rapid high-throughput identification of proteins and sequencing of peptides (most often after in-gel digestion), protein microarrays, which allow the detection of the relative levels of a large number of proteins present in a cell, and two-hybrid screening, which allows the systematic exploration of protein–protein interactions. The total complement of biologically possible such interactions is known as the interactome. A systematic attempt to determine the structures of proteins representing every possible fold is known as structural genomics.

Bioinformatics

A vast array of computational methods have been developed to analyze the structure, function, and evolution of proteins.

The development of such tools has been driven by the large amount of genomic and proteomic data available for a variety of organisms, including the human genome. It is simply impossible to study all proteins experimentally, hence only a few are subjected to laboratory experiments while computational tools are used to extrapolate to similar proteins. Such homologous proteins can be efficiently identified in distantly related organisms by sequence alignment. Genome and gene sequences can be searched by a variety of tools for certain properties. Sequence profiling tools can find restriction enzyme sites, open reading frames in nucleotide sequences, and predict secondary structures. Phylogenetic trees can be constructed and evolutionary hypotheses developed using special software like ClustalW regarding the ancestry of modern organisms and the genes they express. The field of bioinformatics is now indispensable for the analysis of genes and proteins.

Structure Prediction and Simulation

Complementary to the field of structural genomics, protein structure prediction seeks to develop efficient ways to provide plausible models for proteins whose structures have not yet been determined experimentally. The most successful type of structure prediction, known as homology modeling, relies on the existence of a "template" structure with sequence similarity to the protein being modeled; structural genomics' goal is to provide sufficient representation in solved structures to model most of those that remain. Although producing accurate models remains a challenge when only distantly related template structures are available, it has been suggested that sequence alignment is the bottleneck in this process, as quite accurate models can be produced if a "perfect" sequence alignment is known. Many structure prediction methods have served to inform the emerging field of protein engineering, in which novel protein folds have already been designed. A more complex computational problem is the prediction of intermolecular interactions, such as in molecular docking and protein–protein interaction prediction.

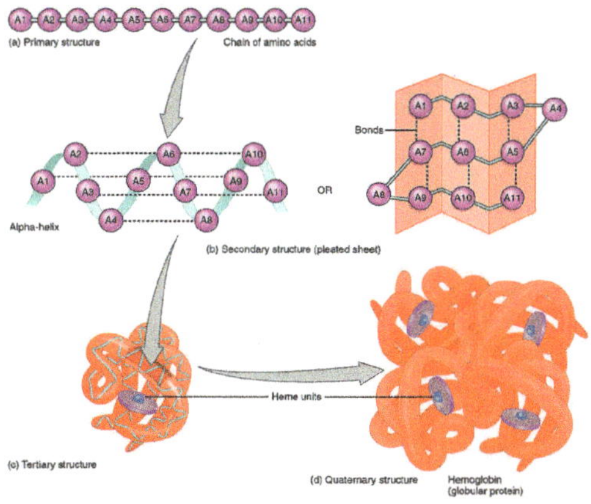

Constituent amino-acids can be analyzed to predict secondary, tertiary and quaternary protein structure, in this case hemoglobin containing heme units

The processes of protein folding and binding can be simulated using such technique as molecular mechanics, in particular, molecular dynamics and Monte Carlo, which increasingly take advantage of parallel and distributed computing (Folding@home project; molecular modeling on GPU). The folding of small α-helical protein domainssuch as the villin headpiece and the HIV accessory protein have been successfully simulated in silico, and hybrid methods that combine standard molecular dynamics with quantum mechanics calculations have allowed exploration of the electronic states of rhodopsins.

Protein Disorder and Unstructure Prediction

Many proteins (in Eucaryota ~33%) contain large unstructured but biologically functional segments and can be classified as intrinsically disordered proteins. Predicting and analysing protein disorder is, therefore, an important part of protein structure characterisation.

NUTRITION

Most microorganisms and plants can biosynthesize all 20 standard amino acids, while animals (including humans) must obtain some of the amino acids from the diet. The amino acids that an organism cannot synthesize on its own are referred to as essential amino acids. Key enzymes that synthesize certain amino acids are not present in animals — such as aspartokinase, which catalyses the first step in the synthesis of lysine, methionine, and threonine from aspartate. If amino acids are present in the environment, microorganisms can conserve energy by taking up the amino acids from their surroundings and downregulating their biosynthetic pathways.

In animals, amino acids are obtained through the consumption of foods containing protein. Ingested proteins are then broken down into amino acids through digestion, which typically involves denaturation of the protein through exposure to acid and hydrolysis by enzymes called proteases. Some ingested amino acids are used for protein biosynthesis, while others are converted to glucose through gluconeogenesis, or fed into the citric acid cycle. This use of protein as a fuel is particularly important under starvation conditions as it allows the body's own proteins to be used to support life, particularly those found in muscle. Amino acids are also an important dietary source of nitrogen.

HISTORY AND ETYMOLOGY

Proteins were recognized as a distinct class of biological molecules in the eighteenth century by Antoine Fourcroy and others, distinguished by the molecules' ability to coagulateor flocculate under treatments with heat or acid. Noted examples at the time included albumin from egg whites, blood serum albumin, fibrin, and wheat gluten.

Proteins were first described by the Dutch chemist Gerardus Johannes Mulder and named by the Swedish chemist Jöns Jacob Berzelius in 1838. Mulder carried out elemental analysis of common proteins and found that nearly all proteins had the same empirical formula, $C_{400}H_{620}N_{100}O_{120}P_1S_1$. He came to the erroneous conclusion that they might be composed of a single type of (very large) molecule. The term "protein" to describe these molecules was proposed by Mulder's associate Berzelius; protein is derived from the Greek word πρώτειος (proteios), meaning "primary", "in the lead", or "standing in front", + -in. Mulder went on to identify the products of protein degradation such as the amino acid leucine for which he found a (nearly correct) molecular weight of 131 Da.

Early nutritional scientists such as the German Carl von Voit believed that protein was the most important nutrient for maintaining the structure of the body, because it was generally believed that "flesh makes flesh." Karl Heinrich Ritthausen extended known protein forms with the identification of glutamic acid. At the Connecticut Agricultural Experiment Station a detailed review of the vegetable proteins was compiled by Thomas Burr Osborne. Working with Lafayette Mendel and applying Liebig's law of the minimumin feeding laboratory rats, the nutritionally essential amino acids were established. The work was continued and communicated by William Cumming Rose. The understanding of proteins as polypeptides came[when?] through the work of Franz Hofmeister and Hermann Emil Fischer. The central role of proteins as enzymes in living organisms was not fully appreciated until 1926, when James B. Sumner showed that the enzyme urease was in fact a protein.

The difficulty in purifying proteins in large quantities made them very difficult for early protein biochemists to study. Hence, early studies focused on proteins that could be purified in large quantities, e.g., those of blood, egg white, various toxins, and digestive/metabolic enzymes obtained from slaughterhouses. In the 1950s, the Armour Hot Dog Co.purified 1 kg of pure bovine pancreatic ribonuclease A and made it freely available to scientists; this gesture helped ribonuclease A become a major target for biochemical study for the following decades.

John Kendrew with model of myoglobin in progress

Linus Pauling is credited with the successful prediction of regular protein secondary structures based on hydrogen bonding, an idea first put forth by William Astbury in 1933. Later work by Walter Kauzmann on denaturation, based partly on previous studies by Kaj Linderstrøm-Lang, contributed an understanding of protein folding and structure mediated by hydrophobic interactions.

The first protein to be sequenced was insulin, by Frederick Sanger, in 1949. Sanger correctly determined the amino acid sequence of insulin, thus conclusively demonstrating that proteins consisted of linear polymers of amino acids rather than branched chains, colloids, or cyclols. He won the Nobel Prize for this achievement in 1958.

The first protein structures to be solved were hemoglobin and myoglobin, by Max Perutz and Sir John Cowdery Kendrew, respectively, in 1958. As of 2016, the Protein Data Bank has over 115,000 atomic-resolution structures of proteins. In more recent times, cryo-electron microscopy of large macromolecular assemblies and computational protein structure prediction of small protein domains are two methods approaching atomic resolution.

REFERENCES

1. Sinkovics, J; Harvath J; Horak A. (1998). "The Origin and evolution of viruses (a review)". Acta Microbiologica et Immunologica Hungarica. ***45*** (3–4): 349–90. PMID 9873943.
2. Thomas, Christopher M; Summers, David (2008). "Bacterial Plasmids". Encyclopedia of Life Sciences. doi:10.1002/9780470015902.a0000468.pub2. ISBN 0470016175.
3. Wolfgang Schumann (2008). "Chapter 1 - Escherichia coli Cloning and Expression Vectors". In Georg Lipps. Plasmids: Current Research and Future Trends. Caister Academic Press. pp. 1–2. ISBN 978-1-904455-35-6.
4. Lederberg J (1952). "Cell genetics and hereditary symbiosis". Physiol. Rev. ***32*** (4): 403–430. PMID 13003535.
5. Stanley Falkow. "Microbial Genomics: Standing on the Shoulders of Giants". Microbiology Society.
6. Finbarr Hayes (2003). "Chapter 1 - The Function and Organization of Plasmids". In Nicola Casali, Andrew Presto. E. Coli Plasmid Vectors: Methods and Applications. Methods in Molecular Biology. ***235***. Humana Press. pp. 1–5.

ISBN 978-1-58829-151-6.

7. Finbarr Hayes (2003). "Chapter 1 - The Function and Organization of Plasmids". In Nicola Casali, Andrew Preston. E. Coli Plasmid Vectors: Methods and Applications. Methods in Molecular Biology, Vol. 235. Humana Press. pp. 5–6. ISBN 978-1-58829-151-6.
8. T. A. Brown (2010). "Chapter 2 - Vectors for Gene Cloning: Plasmids and Bacteriophages". Gene Cloning and DNA Analysis: An Introduction (6th ed.). Wiley-Blackwell. ISBN 978-1405181730.
9. David Summers (1996). "Chapter 1 - The Function and Organization of Plasmids". The Biology of Plasmids. Wiley-Blackwell; First Edition. pp. 21–22. ISBN 978-0632034369.
10. David P. Clark; Nanette Jean Pazdernik (2012). Molecular Biology (2nd ed.). Academic Cell. p. 795. ISBN 978-0123785947.
11. Margaret C. M. Smith and R. Elizabeth Sockett, eds. (1999). Genetic Methods for Diverse Prokaryotes. Methods in Microbiology, vol. 29. Academic Press. pp. 75–77. ISBN 0-12-652340-1.
12. Russell, David W.; Sambrook, Joseph (2001). Molecular cloning: a laboratory manual. Cold Spring Harbor, N.Y: Cold Spring Harbor Laboratory.
13. Uldis N. Streips, Ronald E. Yasbin, eds. (2002). Modern Microbial Genetics (2nd ed.). Wiley-Blackwell. p. 248. ISBN 978-0471386650.
14. Andrew Preston (2003). "Chapter 2 - Choosing a Cloning Vector". In Nicola Casali, Andrew Preston. E. Coli Plasmid Vectors: Methods and Applications. Methods in Molecular Biology, Vol. 235. Humana Press. pp. 19–26. ISBN 978-1-58829-151-6.
15. Kandavelou K, Chandrasegaran S (2008). "Plasmids for Gene Therapy". Plasmids: Current Research and Future Trends. Caister Academic Press. ISBN 978-1-904455-35-6.
16. Morange M (2009). "What history tells us XIX. The notion of the episome" (PDF). Journal of Biosciences. ***34*** (6): 845–8. doi:10.1007/s12038-009-0098-z. PMID 20093737. (subscription required (help)).
17. Jacob F & Wollman EL (1958), "Les épisomes, elements génétiques ajoutés", Comptes Rendus de l'Académie des Sciences de Paris, ***247*** (1): 154–156, PMID 13561654
18. Hayes, W (1969). "What are episomes and plasmids?". In Gordon E. W. Wolstenholme; Maeve O'Connor. Bacterial Episomes and Plasmids (CIBA Foundation Symposium). pp. 4–8. ISBN 978-0700014057.
19. Gordon E. W. Wolstenholme; Maeve O'Connor, eds. (1969). Bacterial Episomes and Plasmids (CIBA Foundation Symposium). pp. 244–245. ISBN 978-0700014057.
20. T. A. Brown (2011). Introduction to Genetics: A Molecular Approach. Garland Science. p. 238. ISBN 978-0815365099.

21. Kathleen Van Craenenbroeck, Peter Vanhoenacker and Guy Haegeman (2000). "Episomal vectors for gene expression in mammalian cells". Eur. J. Biochem. ***267*** (18): 5665–5678. doi:10.1046/j.1432-1327.2000.01645.x. PMID 10971576.
22. Colosimo A1, Goncz KK, Holmes AR, Kunzelmann K, Novelli G, Malone RW, Bennett MJ, Gruenert DC. (2000). "Transfer and expression of foreign genes in mammalian cells" (PDF). Biotechniques. ***29*** (2): 314–8, 320–2, 324 passim. PMID 10948433. Archived from the original (PDF) on 24 July 2011.
23. Gerdes K, Rasmussen PB, Molin S (1986). "Unique type of plasmid maintenance function: postsegregational killing of plasmid-free cells". Proc. Natl. Acad. Sci. U.S.A. ***83***(10): 3116–20. Bibcode:1986PNAS...83.3116G. doi:10.1073/pnas.83.10.3116. PMC 323463. PMID 3517851.
24. Kroll J, Klinter S, Schneider C, Voß I, Steinbüchel A (2010). "Plasmid addiction systems: perspectives and applications in biotechnology". Microb. Biotechnol. ***3*** (6): 634–657. doi:10.1111/j.1751-7915.2010.00170.x. PMC 3815339. PMID 21255361.
25. Gunge, N; Murata, K; Sakaguchi, K (July 1982). "Transformation of Saccharomyces cerevisiae with linear DNA killer plasmids from Kluyveromyces lactis". Journal of Bacteriology. ***151*** (1): 462–4. PMC 220260. PMID 7045080.
26. "Vector NTI feedback video". The DNA Lab.
27. Nelson DL, Cox MM (2005). Lehninger's Principles of Biochemistry (4th ed.). New York, New York: W. H. Freeman and Company.
28. Gutteridge A, Thornton JM (2005). "Understanding nature's catalytic toolkit". Trends in Biochemical Sciences. ***30*** (11): 622–29. doi:10.1016/j.tibs.2005.09.006. PMID 16214343.
29. Murray et al., p. 19.
30. Murray et al., p. 31.
31. Lodish H, Berk A, Matsudaira P, Kaiser CA, Krieger M, Scott MP, Zipurksy SL, Darnell J (2004). Molecular Cell Biology (5th ed.). New York, New York: WH Freeman and Company.
32. Milo, Ron (2013-12-01). "What is the total number of protein molecules per cell volume? A call to rethink some published values". BioEssays. ***35*** (12): 1050–1055. doi:10.1002/bies.201300066. ISSN 1521-1878. PMC 3910158. PMID 24114984.
33. Wu, Linfeng; Candille, Sophie I.; Choi, Yoonha; Xie, Dan; Jiang, Lihua; Li-Pook-Than, Jennifer; Tang, Hua; Snyder, Michael. "Variation and genetic control of protein abundance in humans". Nature. ***499*** (7456): 79–82. doi:10.1038/nature12223. PMC 3789121. PMID 23676674.
34. van Holde and Mathews, pp. 1002–42.
35. Dobson CM (2000). "The nature and significance of protein folding". In Pain RH. Mechanisms of Protein Folding. Oxford, Oxfordshire: Oxford University Press. pp. 1–28. ISBN 0-19-963789-X.

36. Fulton A, Isaacs W (1991). "Titin, a huge, elastic sarcomeric protein with a probable role in morphogenesis". BioEssays. ***13*** (4): 157–61. doi:10.1002/bies.950130403. PMID 1859393.
37. Bruckdorfer T, Marder O, Albericio F (2004). "From production of peptides in milligram amounts for research to multi-tons quantities for drugs of the future". Current Pharmaceutical Biotechnology. ***5*** (1): 29–43. doi:10.2174/1389201043489620. PMID 14965208.
38. Schwarzer D, Cole P (2005). "Protein semisynthesis and expressed protein ligation: chasing a protein's tail". Current Opinion in Chemical Biology. ***9*** (6): 561–69. doi:10.1016/j.cbpa.2005.09.018. PMID 16226484.
39. Kent SB (2009). "Total chemical synthesis of proteins". Chemical Society Reviews. ***38***(2): 338–51. doi:10.1039/b700141j. PMID 19169452.
40. Murray et al., p. 36.
41. Murray et al., p. 37.
42. Murray et al., pp. 30–34.
43. van Holde and Mathews, pp. 368–75.
44. van Holde and Mathews, pp. 165–85.
45. Fernández A, Scott R (2003). "Dehydron: a structurally encoded signal for protein interaction". Biophysical Journal. ***85*** (3): 1914–28. Bibcode:2003BpJ....85.1914F. doi:10.1016/S0006-3495(03)74619-0. PMC 1303363. PMID 12944304.
46. Branden and Tooze, pp. 340–41.
47. Gonen T, Cheng Y, Sliz P, Hiroaki Y, Fujiyoshi Y, Harrison SC, Walz T (2005). "Lipid-protein interactions in double-layered two-dimensional AQP0 crystals". Nature. ***438***(7068): 633–38. Bibcode:2005Natur.438..633G. doi:10.1038/nature04321. PMC 1350984. PMID 16319884.
48. Standley DM, Kinjo AR, Kinoshita K, Nakamura H (2008). "Protein structure databases with new web services for structural biology and biomedical research". Briefings in Bioinformatics. ***9*** (4): 276–85. doi:10.1093/bib/bbn015. PMID 18430752.
49. Walian P, Cross TA, Jap BK (2004). "Structural genomics of membrane proteins". Genome Biology. ***5*** (4): 215. doi:10.1186/gb-2004-5-4-215. PMC 395774. PMID 15059248.
50. Sleator RD. (2012). "Prediction of protein functions". Methods in Molecular Biology. Methods in Molecular Biology. ***815***: 15–24. doi:10.1007/978-1-61779-424-7_2. ISBN 978-1-61779-423-0. PMID 22130980.
51. Voet D, Voet JG. (2004). Biochemistry Vol 1 3rd ed. Wiley: Hoboken, NJ.
52. Sankaranarayanan R, Moras D (2001). "The fidelity of the translation of the genetic code". Acta Biochimica Polonica. ***48*** (2): 323–35. PMID 11732604.
53. van Holde and Mathews, pp. 830–49.
54. Copland JA, Sheffield-Moore M, Koldzic-Zivanovic N, Gentry S, Lamprou G, Tzortzatou-Stathopoulou F, Zoumpourlis V, Urban RJ, Vlahopoulos SA (2009).

"Sex steroid receptors in skeletal differentiation and epithelial neoplasia: is tissue-specific intervention possible?". BioEssays. ***31*** (6): 629–41. doi:10.1002/bies.200800138. PMID 19382224.

55. Samarin S, Nusrat A (2009). "Regulation of epithelial apical junctional complex by Rho family GTPases". Frontiers in Bioscience. ***14*** (14): 1129–42. doi:10.2741/3298. PMID 19273120.
56. Bairoch A (2000). "The ENZYME database in 2000" (PDF). Nucleic Acids Research. ***28*** (1): 304–305. doi:10.1093/nar/28.1.304. PMC 102465. PMID 10592255. Archived from the original (PDF) on June 1, 2011.
57. Radzicka A, Wolfenden R (1995). "A proficient enzyme". Science. ***267*** (5194): 90–93. Bibcode:1995Sci...267...90R. doi:10.1126/science.7809611. PMID 7809611.
58. EBI External Services (2010-01-20). "The Catalytic Site Atlas at The European Bioinformatics Institute". Ebi.ac.uk. Retrieved 2011-01-16.
59. Pickel B, Schaller A (2013). "Dirigent proteins: molecular characteristics and potential biotechnological applications". Applied Microbiology and Biotechnology. ***97*** (19): 8427–8438. doi:10.1007/s00253-013-5167-4. PMID 23989917.
60. Branden and Tooze, pp. 251–81.
61. van Holde and Mathews, pp. 247–50.
62. van Holde and Mathews, pp. 220–29.
63. Rüdiger H, Siebert HC, Solís D, Jiménez-Barbero J, Romero A, von der Lieth CW, Diaz-Mariño T, Gabius HJ (2000). "Medicinal chemistry based on the sugar code: fundamentals of lectinology and experimental strategies with lectins as targets". Current Medicinal Chemistry. ***7*** (4): 389–416. doi:10.2174/0929867003375164. PMID 10702616.
64. Branden and Tooze, pp. 232–34.
65. van Holde and Mathews, pp. 178–81.
66. van Holde and Mathews, pp. 258–64; 272.
67. Murray et al., pp. 21–24.
68. Hey J, Posch A, Cohen A, Liu N, Harbers A (2008). "Fractionation of complex protein mixtures by liquid-phase isoelectric focusing". Methods in Molecular Biology. Methods in Molecular Biology™. ***424***: 225–39. doi:10.1007/978-1-60327-064-9_19. ISBN 978-1-58829-722-8. PMID 18369866.
69. Terpe K (2003). "Overview of tag protein fusions: from molecular and biochemical fundamentals to commercial systems". Applied Microbiology and Biotechnology. ***60*** (5): 523–33. doi:10.1007/s00253-002-1158-6. PMID 12536251.
70. Stepanenko OV, Verkhusha VV, Kuznetsova IM, Uversky VN, Turoverov KK (2008). "Fluorescent proteins as biomarkers and biosensors: throwing color lights on molecular and cellular processes". Current Protein & Peptide Science. ***9*** (4): 338–69. doi:10.2174/138920308785132668. PMC 2904242. PMID 18691124.
71. Yuste R (2005). "Fluorescence microscopy today". Nature Methods. ***2*** (12):

902–904. doi:10.1038/nmeth1205-902. PMID 16299474.

72. Margolin W (2000). "Green fluorescent protein as a reporter for macromolecular localization in bacterial cells". Methods (San Diego, Calif.). ***20*** (1): 62–72. doi:10.1006/meth.1999.0906. PMID 10610805.

73. Walker JH, Wilson K (2000). Principles and Techniques of Practical Biochemistry. Cambridge, UK: Cambridge University Press. pp. 287–89. ISBN 0-521-65873-X.

74. Mayhew TM, Lucocq JM (2008). "Developments in cell biology for quantitative immunoelectron microscopy based on thin sections: a review". Histochemistry and Cell Biology. ***130*** (2): 299–313. doi:10.1007/s00418-008-0451-6. PMC 2491712. PMID 18553098.

75. Hohsaka T, Sisido M (2002). "Incorporation of non-natural amino acids into proteins". Current Opinion in Chemical Biology. ***6*** (6): 809–15. doi:10.1016/S1367-5931(02)00376-9. PMID 12470735.

76. Cedrone F, Ménez A, Quéméneur E (2000). "Tailoring new enzyme functions by rational redesign". Current Opinion in Structural Biology. ***10*** (4): 405–10. doi:10.1016/S0959-440X(00)00106-8. PMID 10981626.

77. Görg A, Weiss W, Dunn MJ (2004). "Current two-dimensional electrophoresis technology for proteomics". Proteomics. ***4*** (12): 3665–85. doi:10.1002/pmic.200401031. PMID 15543535.

78. Conrotto P, Souchelnytskyi S (2008). "Proteomic approaches in biological and medical sciences: principles and applications". Experimental Oncology. ***30*** (3): 171–80. PMID 18806738.

79. Joos T, Bachmann J (2009). "Protein microarrays: potentials and limitations". Frontiers in Bioscience. ***14*** (14): 4376–85. doi:10.2741/3534. PMID 19273356.

80. Koegl M, Uetz P (2007). "Improving yeast two-hybrid screening systems". Briefings in Functional Genomics & Proteomics. ***6*** (4): 302–12. doi:10.1093/bfgp/elm035. PMID 18218650.

81. Plewczyński D, Ginalski K (2009). "The interactome: predicting the protein–protein interactions in cells". Cellular & Molecular Biology Letters. ***14*** (1): 1–22. doi:10.2478/s11658-008-0024-7. PMID 18839074.

82. Zhang C, Kim SH (2003). "Overview of structural genomics: from structure to function". Current Opinion in Chemical Biology. ***7*** (1): 28–32. doi:10.1016/S1367-5931(02)00015-7. PMID 12547423.

83. Zhang Y (2008). "Progress and challenges in protein structure prediction". Current Opinion in Structural Biology. ***18*** (3): 342–48. doi:10.1016/j.sbi.2008.02.004. PMC 2680823. PMID 18436442.

84. Xiang Z (2006). "Advances in homology protein structure modeling". Current Protein and Peptide Science. ***7*** (3): 217–27. doi:10.2174/138920306777452312. PMC 1839925. PMID 16787261.

85. Zhang Y, Skolnick J (2005). "The protein structure prediction problem could be solved using the current PDB library". Proceedings of the National Academy of Sciences of the United States of America. ***102*** (4): 1029–34. Bibcode:2005PNAS..102.1029Z. doi:10.1073/pnas.0407152101. PMC 545829.

PMID 15653774.

86. Kuhlman B, Dantas G, Ireton GC, Varani G, Stoddard BL, Baker D (2003). "Design of a novel globular protein fold with atomic-level accuracy". Science. ***302*** (5649): 1364–68. Bibcode:2003Sci...302.1364K. doi:10.1126/science.1089427. PMID 14631033.

87. Ritchie DW (2008). "Recent progress and future directions in protein–protein docking". Current Protein and Peptide Science. ***9*** (1): 1–15. doi:10.2174/138920308783565741. PMID 18336319.

88. Scheraga HA, Khalili M, Liwo A (2007). "Protein-folding dynamics: overview of molecular simulation techniques". Annual Review of Physical Chemistry. ***58***: 57–83. Bibcode:2007ARPC...58...57S. doi:10.1146/annurev.physchem.58.032806.104614. PMID 17034338.

89. Zagrovic B, Snow CD, Shirts MR, Pande VS (2002). "Simulation of folding of a small alpha-helical protein in atomistic detail using worldwide-distributed computing". Journal of Molecular Biology. ***323*** (5): 927–37. doi:10.1016/S0022-2836(02)00997-X. PMID 12417204.

90. Herges T, Wenzel W (2005). "In silico folding of a three helix protein and characterization of its free-energy landscape in an all-atom force field". Physical Review Letters. ***94*** (1): 018101. Bibcode:2005PhRvL..94a8101H. doi:10.1103/PhysRevLett.94.018101. PMID 15698135.

91. Hoffmann M, Wanko M, Strodel P, König PH, Frauenheim T, Schulten K, Thiel W, Tajkhorshid E, Elstner M (2006). "Color tuning in rhodopsins: the mechanism for the spectral shift between bacteriorhodopsin and sensory rhodopsin II". Journal of the American Chemical Society. ***128*** (33): 10808–18. doi:10.1021/ja062082i. PMID 16910676.

92. Ward, J. J.; Sodhi, J. S.; McGuffin, L. J.; Buxton, B. F.; Jones, D. T. (2004). "Prediction and functional analysis of native disorder in proteins from the three kingdoms of life". Journal of Molecular Biology. ***337*** (3): 635–45. doi:10.1016/j.jmb.2004.02.002. PMID 15019783.

93. Peter Tompa; Alan Fersht (18 November 2009). Structure and Function of Intrinsically Disordered Proteins. CRC Press. ISBN 978-1-4200-7893-0.

94. Brosnan J (June 2003). "Interorgan amino acid transport and its regulation". Journal of Nutrition. ***133*** (6 Suppl 1): 2068S–72S. PMID 12771367.

95. Thomas Burr Osborne (1909): The Vegetable Proteins, History pp 1 to 6, from archive.org

96. Bulletin des Sciences Physiques et Naturelles en Néerlande (1838). pg 104. SUR LA COMPOSITION DE QUELQUES SUBSTANCES ANIMALES

97. Hartley Harold (1951). "Origin of the Word ‹Protein.'". Nature. ***168*** (4267): 244–244. doi:10.1038/168244a0.

98. [c] Perrett D (2007). "From ‹protein' to the beginnings of clinical proteomics". Proteomics: Clinical Applications. ***1*** (8): 720–38. doi:10.1002/prca.200700525. PMID 21136729.

99. New Oxford Dictionary of English

100. Reynolds JA, Tanford C (2003). Nature's Robots: A History of Proteins (Oxford Paperbacks). New York, New York: Oxford University Press. p. 15. ISBN 0-19-860694-X.

101. Bischoff TLW, Voit, C (1860). Die Gesetze der Ernaehrung des Pflanzenfressers durch neue Untersuchungen festgestellt (in German). Leipzig, Heidelberg.

102. Sumner JB (1926). "The isolation and crystallization of the enzyme urease. Preliminary paper" (PDF). Journal of Biological Chemistry. ***69*** (2): 435–41.

103. Pauling L, Corey RB, Branson HR (1951). "The structure of proteins: two hydrogen-bonded helical configurations of the polypeptide chain" (PDF). Proceedings of the National Academy of Sciences of the United States of America. ***37*** (5): 235–40. Bibcode:1951PNAS...37..235P. doi:10.1073/pnas.37.5.235. PMC 1063348. PMID 14834145.

104. Kauzmann W (1956). "Structural factors in protein denaturation". Journal of Cellular Physiology. Supplement. ***47*** (Suppl 1): 113–31. doi:10.1002/jcp.1030470410. PMID 13332017.

105. Kauzmann W (1959). "Some factors in the interpretation of protein denaturation". Advances in Protein Chemistry. Advances in Protein Chemistry. ***14***: 1–63. doi:10.1016/S0065-3233(08)60608-7. ISBN 978-0-12-034214-3. PMID 14404936.

106. Kalman SM, Linderstrom-Lang K, Ottesen M, Richards FM (1955). "Degradation of ribonuclease by subtilisin". Biochimica et Biophysica Acta. ***16*** (2): 297–99. doi:10.1016/0006-3002(55)90224-9. PMID 14363272.

107. Sanger F (1949). "The terminal peptides of insulin". Biochemical Journal. ***45*** (5): 563–74. PMC 1275055. PMID 15396627.

108. Sanger F. (1958), Nobel lecture: The chemistry of insulin (PDF), Nobelprize.org

109. Muirhead H, Perutz M (1963). "Structure of hemoglobin. A three-dimensional fourier synthesis of reduced human hemoglobin at 5.5 Å resolution". Nature. ***199*** (4894): 633–38. Bibcode:1963Natur.199..633M. doi:10.1038/199633a0. PMID 14074546.

110. Kendrew J, Bodo G, Dintzis H, Parrish R, Wyckoff H, Phillips D (1958). "A three-dimensional model of the myoglobin molecule obtained by X-ray analysis". Nature. ***181***(4610): 662–66. Bibcode:1958Natur.181..662K. doi:10.1038/181662a0. PMID 13517261.

111. «RCSB Protein Data Bank". Retrieved 2014-04-05.

112. Zhou ZH (2008). "Towards atomic resolution structural determination by single-particle cryo-electron microscopy". Current Opinion in Structural Biology. ***18*** (2): 218–28. doi:10.1016/j.sbi.2008.03.004. PMC 2714865. PMID 18403197.

113. Keskin O, Tuncbag N, Gursoy A (2008). "Characterization and prediction of protein interfaces to infer protein-protein interaction networks". Current Pharmaceutical Biotechnology. ***9*** (2): 67–76. doi:10.2174/138920108783955191. PMID 18393863.

CHAPTER 10

Basic Techniquses and Tools in Biology Field

Chapter Contents

- **Introduction**
- **Bioreactor Design**
- **Types Of Bioreactors**
- **Cell Culture**
- **Flow Cytometry**
- **Spectroscopy**

The three important techniques of biotechnology are: (1) Recombinant DNA Technology (Genetic Engineering) (2) Plant Tissue Culture and (3) Transgenic (Genetically Modified Organisms).

INTRODUCTION

A bioreactor may refer to any manufactured or engineered device or system that supports a biologically active environment. In one case, a bioreactor is a vessel in which a chemical process is carried out which involves organisms or biochemically active substances derived from such organisms. This process can either be aerobic or anaerobic. These bioreactors are commonly cylindrical, ranging in size from litres to cubic metres, and are often made of stainless steel.

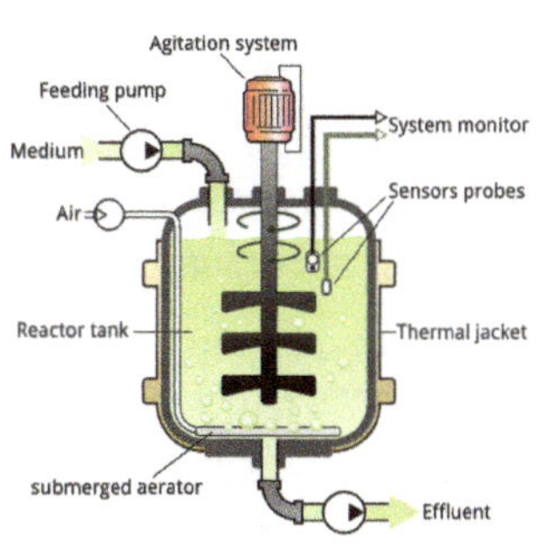

General structure of a continuous stirred-tank type bioreactor

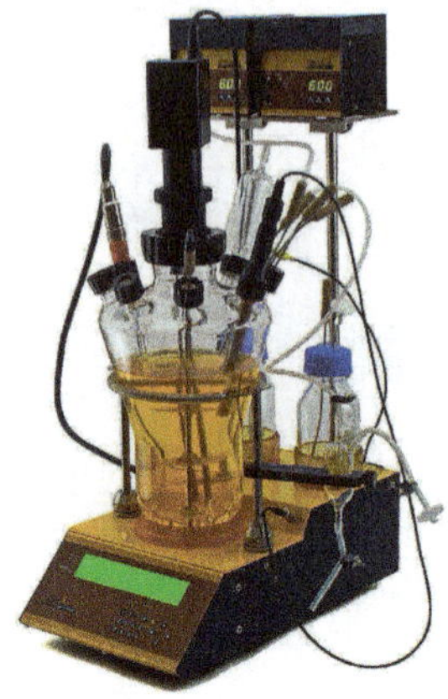

Autoclavable bench-top laboratory bioreactor used for fermentation and cell cultures

A bioreactor may also refer to a device or system meant to grow cells or tissues in the context of cell culture. These devices are being developed for use in tissue engineering or biochemical engineering.

On the basis of mode of operation, a bioreactor may be classified as batch, fed batch or continuous (e.g. a continuous stirred-tank reactor model). An example of a continuous bioreactor is the chemostat.

Organisms growing in bioreactors may be submerged in liquid medium or may be attached to the surface of a solid medium. Submerged cultures may be suspended or immobilized. Suspension bioreactors can use a wider variety of organisms, since special attachment surfaces are not needed, and can operate at much larger scale than immobilized cultures. However, in a continuously operated process the organisms will be removed from the reactor with the effluent. Immobilization is a general term describing a wide variety of cell or particle attachment or entrapment. It can be applied to basically all types of biocatalysis including enzymes, cellular organelles, animal and plant cells.Immobilization is useful for continuously operated processes, since the organisms will not be removed with the reactor effluent, but is limited in scale because the microbes are only present on the surfaces of the vessel.

Large scale immobilized cell bioreactors are:

- moving media, also known as moving bed biofilm reactor (MBBR)
- packed bed
- fibrous bed
- membrane

BIOREACTOR DESIGN

A closed bioreactor used in cellulosic ethanol research

Bioreactor design is a relatively complex engineering task, which is studied in the discipline of biochemical engineering. Under optimum conditions, the microorganisms or cells are able to perform their desired function with limited production of impurities. The environmental conditions inside the bioreactor, such as temperature, nutrient concentrations, pH, and dissolved gases (especially oxygen for aerobic fermentations) affect the growth and productivity of the organisms. The temperature of the fermentation medium is maintained by a cooling jacket, coils, or both. Particularly exothermic fermentations may require the use of external heat exchangers. Nutrients may be continuously added to the fermenter, as in a fed-batch system, or may be charged into the reactor at the beginning of fermentation. The pH of the medium is measured and adjusted with small amounts of acid or base, depending upon the fermentation. For aerobic (and some anaerobic) fermentations, reactant gases (especially oxygen) must be added to the fermentation. Since oxygen is relatively insoluble in water (the basis of nearly all fermentation media), air (or purified oxygen) must be added continuously. The action of the rising bubbles helps mix the fermentation medium and also "strips" out waste gases, such as carbon dioxide. In practice, bioreactors are often pressurized; this increases the solubility of oxygen in water. In an aerobic process, optimal oxygen transfer is sometimes the rate limiting step. Oxygen is poorly soluble in water—even less in warm fermentation broths—and is relatively scarce in air (20.95%). Oxygen transfer is usually helped by agitation, which is also needed to mix nutrients and to keep the fermentation homogeneous. Gas dispersing agitators are used to break up air bubbles and circulate them throughout the vessel.

Fouling can harm the overall efficiency of the bioreactor, especially the heat exchangers. To avoid it, the bioreactor must be easily cleaned. Interior surfaces are typically made of stainless steel for easy cleaning and sanitation. Typically bioreactors are cleaned between batches, or are designed to reduce fouling as much as possible when operated continuously. Heat transfer is an important part of bioreactor design; small vessels can be cooled with a cooling jacket, but larger vessels may require coils or an external heat exchanger.

TYPES OF BIOREACTORS

Photobioreactor

Moss photobioreactor with Physcomitrella patens

A photobioreactor (PBR) is a bioreactor which incorporates some type of light source (that may be natural sunlight or artificial illumination). Virtually any translucent container could be called a PBR, however the term is more commonly used to define a closed system, as opposed to an open tank or pond. Photobioreactors are used to grow small phototrophic organisms such as cyanobacteria, algae, or moss plants.These organisms use light through photosynthesis as their energy source and do not require sugars or lipids as energy source. Consequently, risk of contamination with other organisms like bacteria or fungi is lower in photobioreactors when compared to bioreactors for heterotroph organisms.

Sewage Treatment

Bioreactors are also designed to treat sewage and wastewater. In the most efficient of these systems, there is a supply of a free-flowing, chemically inert medium which acts as a receptacle for the bacteria that break down the raw sewage. Examples of these bioreactors often have separate, sequential tanks and a mechanical separator or cyclone to speed the separation of water and biosolids. Aerators supply oxygen to the sewage and medium, further accelerating breakdown. Submersible mixers provide agitation in anoxic bioreactors to keep the solids in suspension and thereby ensure that the bacteria and the organic materials “meet”. In the process, the liquid’s Biochemical Oxygen Demand (BOD) is reduced sufficiently to render the contaminated water fit for reuse. The biosolids can be collected for further processing, or dried and used as fertilizer. An extremely simple version of a sewage bioreactor is a septic tank whereby the sewage is left in situ, with or without additional media to house bacteria. In this instance, the biosludge itself is the primary host (activated sludge) for the bacteria. Septic systems are best suited where there is sufficient landmass, and the system is not subject to flooding or overly saturated ground, and where time and efficiency are not prioritized.

Because they are the engine that drives biological wastewater treatment, it is critical to closely monitor the quantity and quality of microorganisms in bioreactors. One method for this is via 2nd Generation ATP tests.

Up and Down Agitation Bioreactor

Unique up and down agitation in the bioreactor.

Up and down agitators are useful to avoid shear stress to the cells. These are done by instead of a traditional propeller agitator, which requires an expensive motor and magnetic coupling. Vertical up and down motion is achieved by a motor together with an inexpensive membrane perfectly assure sterility and produce an efficient mixing without formation of a vortex (no baffles needed). At the same time this type of mixing is gentler on cells and produces less foam. Novel biomimicking "fish-tail" stirring discs offer maximum mixing efficiency without cutting edges.

NASA tissue Cloning Bioreactor

A bioreactor used to ferment ethanol from corncob waste being loaded with yeast.

In bioreactors in which the goal is to grow cells or tissues for experimental or therapeutic purposes, the design is significantly different from industrial bioreactors. Many cells and tissues, especially mammalian ones, must have a surface or other structural support in order to grow, and agitated environments are often destructive to these cell types and tissues. Higher organisms, being auxotrophic, also require highly specialized growth media. NASA has developed a new type of bioreactor that artificially grows tissue in cell cultures. NASA's tissue bioreactor can grow heart tissue, skeletal tissue, ligaments, cancer tissue for study, and other types of tissue.

For more information on artificial tissue culture, see tissue engineering.

Modelling of Bioreactor

Mathematical models act as an important tool in various bio-reactor applications including wastewater treatment. These models are useful for planning efficient process control strategies and predicting the future plant performance. Moreover, these models are beneficial in education and research areas.

Bioreactors are generally used in those industries which are concerned with food, beverages and pharmaceuticals. The emergence of Biochemical engineering is of recent origin. Processing of biological materials using biological agents such as cells, enzymes or antibodies are the major pillars of biochemical engineering. Applications of biochemical engineering cover major fields of civilization such as agriculture, food and healthcare, resource recovery and fine chemicals.

Till now, the industries associated with biotechnology have been lagged behind other industries in implementing control over the process and optimization strategies. A main drawback in biotechnological process control is the problem to measure key physical and biochemical parameters.

Operational Stages in a Bio-Process

A bioprocess is composed mainly of three stages — upstream processing, bioreaction, and downstream processing — to convert raw material to finished product.

The raw material can be of biological or non-biological origin. It is first converted to more suitable form for processing. This is done in upstream processing step which involves chemical hydrolysis, preparation of liquid medium, separation of particulate, air purification and many other preparatory operations.

After upstream processing step, the resulting feed is transferred to one or more Bioreaction stages. The Biochemical reactors or bioreactors form the base of the Bioreaction step. This step is mainly consists of three operations namely, production of biomass, metabolize biosynthesis and biotransformation.

Finally, the material produced in the bioreactor must be further processed in the downstream section to convert it into more useful form. The downstream process is mainly consists of physical separation operations which includes, solid liquid separation, adsorption, liquid-liquid extraction, distillation, drying etc.

Specifications of a Bioreactor

A typical bioreactor consists of following parts:

Agitator – used for the mixing of the contents of the reactor which keeps the "cells" in the perfect homogenous condition for better transport of nutrients and oxygen to the desired product(s).

Baffle – used to break the vortex formation in the vessel, which is usually highly undesirable as it changes the center of gravity of the system and consumes additional power.

Sparger – In aerobic cultivation process, the purpose of the sparger is to supply adequate oxygen to the growing cells.

Jacket – The jacket provides the annular area for circulation of constant temperature of water which keeps the temperature of the bioreactor at a constant value.

Development of Modelling Equations for Bioreactors

Assumptions –

- The reactor contents are perfectly mixed together.
- The reactor is operating at a constant temperature (i.e., it is isothermal).
- The feed should be clean and pure (i.e., no biomass in the feed stream).
- The feed stream and reactor contents have equal and constant density (ρ).
- The feed and product streams have the same flow rate (F).

Total Continuity Equation

Making overall mass balance, we get the following equation:

$d(\rho V)/dt = F\rho - F\rho = 0$ (1)

Equation(1) states that the reactor volume (V) is constant since $dV/dt = 0$.

Biomass Continuity Equation

We know,

Flow rate of biomass into the reactor = Fx_i

Flow rate of biomass out of the reactor = Fx

Rate of generation of biomass by reaction = Vr_1

Rate of accumulation of biomass within the reactor = $d(Vx)/dt$

Now, apply general mass balance equation i.e,

Rate of Mass In – Rate of Mass Out + Rate of Generation = Accumulation

$d(Vx)/dt = Fx_i - Fx + Vr_1$ (2)

Where $\mathbf{r_1}$ is the rate of cell generation. Dividing both sides of the above equation by V, we obtain

$dx/dt = (F/V)x_i - (F/V)x + r_1$ (3)

In the chemical reaction engineering, **F/V** is called **space velocity**(s^{-1}) and **V/F** is called the

residence time (s). But in biochemical engineering, **F/V** is known as **Dilution rate (D_r)**. Accordingly, equation(3) yields:

$dx/dt = D_r x_i - D_r x + r_1$ (4)

$dx/dt = D_r(x_i - x) + r_1$ (5)

Substrate Continuity Equation

For substrate balance,

Flow rate of substrate into the bioreactor = FS_i

Flow of the substrate out of the bioreactor = FS

Rate of generation of substrate by reaction = $-Vr_2$

Rate of accumulation of substrate within the reactor = d(VS)/dt

Now, apply general mass balance equation i.e,

Rate of Mass In - Rate of Mass Out + Rate of Generation = Accumulation

$d(VS)/dt = FS_i - FS - Vr_2$ (6)

rearranging above equation, we get

$dS/dt = D_r(S_i - S) - r_2$ (7)

where $\mathbf{r_2}$ is the rate of substrate consumption.

Biochemical Reaction Kinetics

For the chemical reaction,

A ----> P

We can write

$(-r_A) = k (C_A)^n$ (8)

$(r_A) = - k (C_A)^n$ (9)

Where,

$(-r_A)$ = rate of disappearance of A

(r_A) = rate of formation of A

k = reaction rate constant

C_A = Concentration of reactant A

n = order of reaction with respect to component A

For first order reaction, n = 1 and accordingly,

$-r_A = k C_A$

The reaction kinetics involved in biochemical operations is comparatively difficult to obtain than the chemical reaction kinetics. In biochemical operations, the cell kinetics is used for the unstructured models where balanced growth condition is assumed.

The following equation is used to represent the net rate of cell mass growth:

$r_1 = \mu x$ (10)

where μ is the specific growth rate or specific growth rate coefficient(s^{-1}). Here, μ is analogous to first order rate constant k but however, μ is not a constant.

In biochemical engineering, **yield** is defined as the ratio of mass or moles of product formed to the mass or moles of the reactants consumed. The yield (Y) of product (P) with respect to reactant A is defined as:

Y = (mass of P formed)/(mass of A consumed) (11)

In case of bioreactor,

Y = (mass of cells formed)/(mass of substrate consumed) (12)

Thus,

$Y = r_1 / r_2$

Or,

$r_2 = r_1/Y$

Or,

$r_2 = \mu x/Y$ (from 10) (13)

Final form of Equation Of Modelling

By substituting equations (10) & (13) in equations (5) & (7) respectively, we get,

$\mathbf{dx/dt = D_r(x_i - x) + \mu x}$ (14)

$\mathbf{dS/dt = S(S_i - S) - (\mu x/Y)}$ (15)

Since we have assumed that the feed stream does not contain any biomass i.e., $\mathbf{x_i = 0}$, then, bioreactor modelling equation finally get the following form:

$\mathbf{dx/dt = (\mu - D_r)x}$ (16)

$\mathbf{dS/dt = S(S_i - S) - (\mu x/Y)}$ (from 15)

Thus,Equations (15) and (16) are the basic equations which are used for the modelling of any bioreactor.

CELL CULTURE

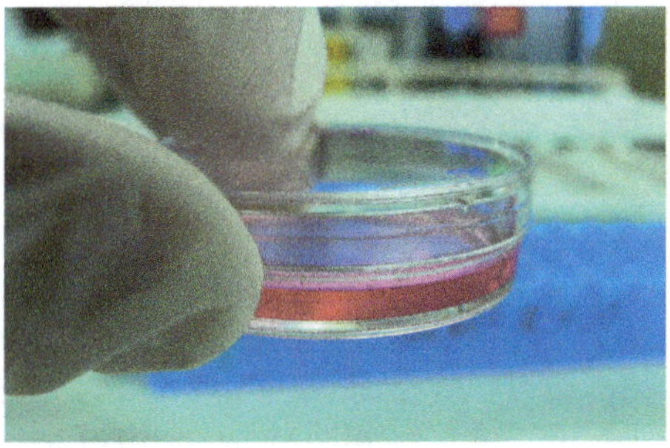

Cell culture in a special tissue culture dish

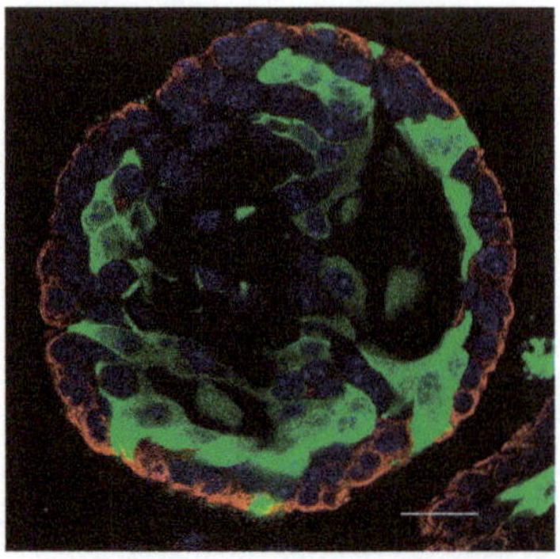

Epithelial cells in culture, stained for keratin (red) and DNA (green)

Cell culture is the process by which cells are grown under controlled conditions, generally outside of their natural environment. Cell culture conditions can vary for each cell type, but artificial environments consist of a suitable vessel with substrate or medium that supplies the essential nutrients (amino acids, carbohydrates, vitamins, minerals), growth factors, hormones, and gases (CO_2, O_2), and regulates the physio-chemical environment (pH buffer, osmotic pressure, temperature). Most cells require a surface or an artificial substrate (adherent or monolayer culture) whereas others can be grown free floating in culture medium (suspension culture).

In practice, the term "cell culture" now refers to the culturing of cells derived from multicellular eukaryotes, especially animal cells, in contrast with other types of culture that also grow cells, such as plant tissue culture, fungal culture, and microbiological culture (of microbes). The historical development and methods of cell culture are closely interrelated to those of tissue culture and organ culture. Viral culture is also related, with cells as hosts for the viruses.

The laboratory technique of maintaining live **cell lines** (a population of cells descended from a single cell and containing the same genetic makeup) separated from their original tissue source became more robust in the middle 20th century.

The 19th-century English physiologist Sydney Ringer developed salt solutions containing the chlorides of sodium, potassium, calcium and magnesium suitable for maintaining the beating of an isolated animal heart outside of the body. In 1885, Wilhelm Roux removed a portion of the medullary plate of an embryonic chicken and maintained it in a warm saline solution for several days, establishing the principle of tissue culture. Ross Granville Harrison, working at Johns Hopkins Medical School and then at Yale University, published results of his experiments from 1907 to 1910, establishing the methodology of tissue culture.

Cell culture techniques were advanced significantly in the 1940s and 1950s to support research in virology. Growing viruses in cell cultures allowed preparation of purified viruses for the manufacture of vaccines. The injectable polio vaccine developed by Jonas Salk was one of the first products mass-produced using cell culture techniques. This vaccine was made possible by the cell culture research of John Franklin Enders, Thomas Huckle Weller, and Frederick Chapman Robbins, who were awarded a Nobel Prize for their discovery of a method of growing the virus in monkey kidney cell cultures.

Concepts in Mammalian Cell Culture

Isolation of Cells

Cells can be isolated from tissues for ex vivo culture in several ways. Cells can be easily purified from blood; however, only the white cells are capable of growth in culture. Mononuclear cells can be released from soft tissues by enzymatic digestion with enzymes such as collagenase, trypsin, or pronase, which break down the extracellular matrix. Alternatively, pieces of tissue can be placed in growth media, and the cells that grow out are available for culture. This method is known as explant culture.

Cells that are cultured directly from a subject are known as primary cells. With the exception of some derived from tumors, most primary cell cultures have limited lifespan.

An established or immortalized cell line has acquired the ability to proliferate indefinitely either through random mutation or deliberate modification, such as artificial expression of the telomerase gene. Numerous cell lines are well established as representative of particular cell types.

Maintaining Cells in Culture

For the majority of isolated primary cells, they undergo the process of senescence and stop dividing after a certain number of population doublings while generally retaining their viability (described as the Hayflick limit).

Cells are grown and maintained at an appropriate temperature and gas mixture (typically, 37 °C, 5% $CO_2$for mammalian cells) in a cell incubator. Culture conditions vary widely for each cell type, and variation of conditions for a particular cell type can result in different phenotypes.

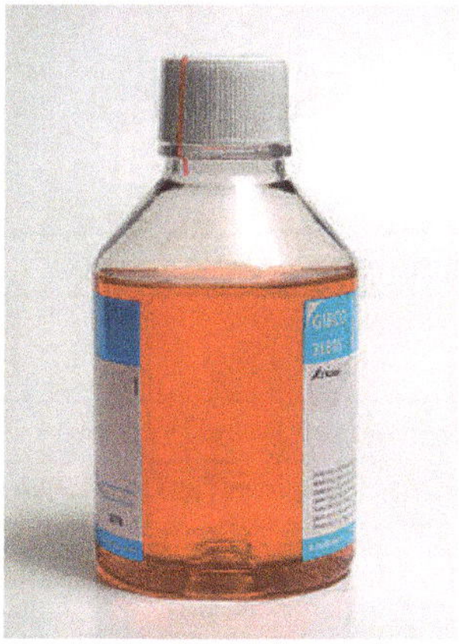

A bottle of DMEM cell culture medium

Aside from temperature and gas mixture, the most commonly varied factor in culture systems is the cell growth medium. Recipes for growth media can vary in pH, glucose concentration, growth factors, and the presence of other nutrients. The growth factors used to supplement media are often derived from the serum of animal blood, such as fetal bovine serum (FBS), bovine calf serum, equine serum, and porcine serum. One complication of these blood-derived ingredients is the potential for contamination of the culture with viruses or prions, particularly in medical biotechnology applications. Current practice is to minimize or eliminate the use of these ingredients wherever possible and use human platelet lysate (hPL). This eliminates the worry of cross-species contamination when using FBS with human cells. hPL has emerged as a safe and reliable alternative as a direct replacement for FBS or other animal serum. In addition, chemically defined media can be used to eliminate any serum trace (human or animal), but this cannot always be accomplished with different cell types. Alternative strategies involve sourcing the animal blood from countries with minimum BSE/TSE risk, such as The United States, Australia and New Zealand, and using purified nutrient concentrates derived from serum in place of whole animal serum for cell culture.

Plating density (number of cells per volume of culture medium) plays a critical role for some cell types. For example, a lower plating density makes granulosa cells exhibit estrogen production, while a higher plating density makes them appear as progesterone-producing theca lutein cells. Cells can be grown either in suspension or adherent cultures. Some cells naturally live in suspension, without being attached to a surface, such as cells that exist in the bloodstream. There are also cell lines that have been modified to be able to survive in suspension cultures so they can be grown to a higher density than adherent conditions would allow. Adherent cells require a surface, such as tissue culture plastic or microcarrier, which may be coated with extracellular matrix (such as collagen and laminin) components to increase adhesion properties and provide other signals needed for growth and differentiation. Most cells derived from solid tissues are adherent. Another type of adherent culture is organotypic culture, which involves growing cells in a three-dimensional (3-D) environment as opposed to two-dimensional culture dishes. This 3D culture system is biochemically and physiologically more similar to in vivo tissue, but is technically challenging to maintain because of many factors (e.g. diffusion).

Components of Cell Culture Media

Component	Function
Carbon source (glucose/ glutamine)	Source of energy
Amino acid	Building blocks of protein
Vitamins	Promote cell survival and growth
Balanced salt solution	It is an isotonic mixture of ions to act as cofactors for enzymatic reactions, cell adhesion etc.
Phenol red dye	pH indicator. The color of phenol red changes from orange/red at pH 7-7.4 to yellow at acidic(lower) pH and purple at basic (higher) pH.
Bicarbonate /HEPES buffer	It is used to maintain a balanced pH in the media

Growth conditions

Parameter	
Temperature	37°C
CO2	5%
Humidity	95%

Cell line cross-Contamination

Cell line cross-contamination can be a problem for scientists working with cultured cells. Studies suggest anywhere from 15–20% of the time, cells used in experiments have been misidentified or contaminated with another cell line. Problems with cell line cross-contamination have even been detected in lines from the NCI-60 panel, which are used routinely for drug-screening studies. Major cell line repositories, including the American Type Culture Collection (ATCC), the European Collection of Cell Cultures (ECACC) and the German Collection of Microorganisms and Cell Cultures (DSMZ), have received cell line submissions from researchers that were misidentified by them. Such contamination poses a problem for the quality of research produced using cell culture lines, and the major repositories are now authenticating all cell line submissions. ATCC uses short tandem repeat (STR) DNA fingerprinting to authenticate its cell lines.

To address this problem of cell line cross-contamination, researchers are encouraged to authenticate their cell lines at an early passage to establish the identity of the cell line. Authentication should be repeated before freezing cell line stocks, every two months during active culturing and before any publication of research data generated using the cell lines. Many methods are used to identify cell lines, including isoenzyme analysis, human lymphocyte antigen (HLA) typing, chromosomal analysis, karyotyping, morphology and STR analysis.

One significant cell-line cross contaminant is the immortal HeLa cell line.

Other Technical Issues

As cells generally continue to divide in culture, they generally grow to fill the available area or volume. This can generate several issues:

- Nutrient depletion in the growth media
- Changes in pH of the growth media
- Accumulation of apoptotic/necrotic (dead) cells
- Cell-to-cell contact can stimulate cell cycle arrest, causing cells to stop dividing, known as contact inhibition.
- Cell-to-cell contact can stimulate cellular differentiation.
- Genetic and epigenetic alterations, with a natural selection of the altered cells potentially leading to overgrowth of abnormal, culture-adapted cells with decreased differentiation and increased proliferative capacity.

Manipulation of Cultured Cells

Among the common manipulations carried out on culture cells are media changes, passaging cells, and transfecting cells. These are generally performed using tissue culture methods that rely on aseptic technique. Aseptic technique aims to avoid contamination with bacteria, yeast, or other cell lines. Manipulations are typically carried out in a biosafety hood or laminar flow cabinet to exclude contaminating micro-organisms. Antibiotics (e.g. penicillin and streptomycin) and antifungals (e.g.amphotericin B) can also be added to the growth media.

As cells undergo metabolic processes, acid is produced and the pH decreases. Often, a pH indicator is added to the medium to measure nutrient depletion.

Media Changes

In the case of adherent cultures, the media can be removed directly by aspiration, and then is replaced. Media changes in non-adherent cultures involve centrifuging the culture and resuspending the cells in fresh media.

Passaging Cells

Passaging (also known as subculture or splitting cells) involves transferring a small number of cells into a new vessel. Cells can be cultured for a longer time if they are split regularly, as it avoids the senescence associated with prolonged high cell density. Suspension cultures are easily passaged with a small amount of culture containing a few cells diluted in a larger volume of fresh media. For adherent cultures, cells first need to be detached; this is commonly done with a mixture of trypsin-EDTA; however, other enzyme mixes are now available for this purpose. A small number of detached cells can then be used to seed a new culture. Some cell cultures, such as RAW cells are mechanically scraped from the surface of their vessel with rubber scrapers.

Transfection and Transduction

Another common method for manipulating cells involves the introduction of foreign DNA by transfection. This is often performed to cause cells to express a gene of interest. More recently, the transfection of RNAi constructs have been realized as a convenient mechanism for suppressing the expression of a particular gene/protein. DNA can also be inserted into cells using viruses, in methods referred to as transduction, infection or transformation. Viruses, as parasitic agents, are well suited to introducing DNA into cells, as this is a part of their normal course of reproduction.

Established Human Cell Lines

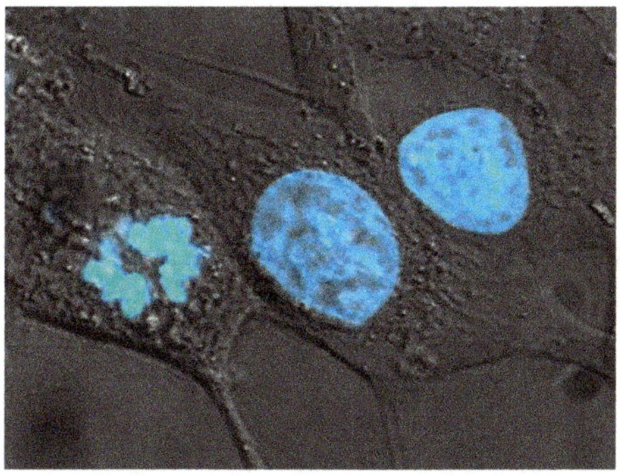

Cultured HeLa cells have been stained with Hoechst turning their nuclei blue, and are one of the earliest human cell lines descended from Henrietta Lacks, who died of cervical cancer from which these cells originated. Cell lines that originate with humans have been somewhat controversial in bioethics, as they may outlive their parent organism and later be used in the discovery of lucrative medical treatments. In the pioneering decision in this area, the Supreme Court of California held in Moore v. Regents of the University of California that human patients have no property rights in cell lines derived from organs removed with their consent.

It is possible to fuse normal cells with an immortalised cell line. This method is used to produce monoclonal antibodies. In brief, lymphocytes isolated from the spleen (or possibly blood) of an immunised animal are combined with an immortal myeloma cell line (B cell lineage) to produce a hybridoma which has the antibody specificity of the primary lymphocyte and the immortality of the myeloma. Selective growth medium (HA or HAT) is used to select against unfused myeloma cells; primary lymphoctyes die quickly in culture and only the fused cells survive. These are screened for production of the required antibody, generally in pools to start with and then after single cloning.

Cell Strains

A cell strain is derived either from a primary culture or a cell line by the selection or cloning of cells having specific properties or characteristics which must be defined. Cell strains are cells that have been adapted to culture but, unlike cell lines, have a finite division potential. Non-immortalized cells stop dividing after 40 to 60 population doublings and, after this, they lose their ability to proliferate (a genetically determined event known as senescence).

Applications of Cell Culture

Mass culture of animal cell lines is fundamental to the manufacture of viral vaccines and other products of biotechnology. Culture of human stem cells is used to expand the number of cells and differentiate the cells into various somatic cell types for transplantation. Stem cell culture is also used to harvest the molecules and exosomes that the stem cells release for the purposes of therapeutic development.

Biological products produced by recombinant DNA (rDNA) technology in animal cell cultures include enzymes, synthetic hormones, immunobiologicals (monoclonal antibodies, interleukins, lymphokines), and anticancer agents. Although many simpler proteins can be produced using rDNA in bacterial cultures, more complex proteins that are glycosylated (carbohydrate-modified) currently must be made in animal cells. An important example of such a complex protein is the hormone erythropoietin. The cost of growing mammalian cell cultures is high, so research is underway to produce such complex proteins in insect cells or in higher plants, use of single embryonic cell and somatic embryos as a source for direct gene transfer via particle bombardment, transit gene expression and confocal microscopy observation is one of its applications. It also offers to confirm single cell origin of somatic embryos and the asymmetry of the first cell division, which starts the process.

Cell culture is also a key technique for cellular agriculture, which aims to provide both new products and new ways of producing existing agricultural products like milk, (cultured) meat, fragrances, and rhino horn from cells and microorganisms. It is therefore considered one means of achieving animal-free agriculture.

Cell culture in two Dimensions

Research in tissue engineering, stem cells and molecular biology primarily involves cultures of cells on flat plastic dishes. This technique is known as two-dimensional (2D) cell culture, and was first developed by Wilhelm Roux who, in 1885, removed a portion of the medullary plate of an embryonic chicken and maintained it in warm saline for several days on a flat glass plate. From the advance of polymer technology arose today's standard plastic dish for 2D cell culture, commonly known as the Petri dish. Julius Richard Petri, a German bacteriologist, is generally credited with this invention while working as an assistant to Robert Koch. Various researchers today also utilize culturing laboratory flasks, conicals, and even disposable bags like those used in single-use bioreactors.

Aside from Petri dishes, scientists have long been growing cells within biologically derived matrices such as collagen or fibrin, and more recently, on synthetic hydrogels such as polyacrylamide or PEG. They do this in order to elicit phenotypes that are not expressed on conventionally rigid substrates. There is growing interest in controlling matrix stiffness, a concept that has led to discoveries in fields such as:

- Stem cell self-renewal
- Lineage specification
- Cancer cell phenotype
- Fibrosis
- Hepatocyte function
- Mechanosensing

Cell culture in three Dimensions

Cell culture in three dimensions has been touted as "Biology's New Dimension". At present, the practice of cell culture remains based on varying combinations of single or multiple cell structures in 2D. Currently, there is an increase in use of 3D cell cultures in research areas including drug discovery, cancer biology, regenerative medicine and basic life science research. There are a variety of platforms used to facilitate the growth of three-dimensional cellular structures including scaffold systems such as hydrogel matrices and solid scaffolds, and scaffold-free systems such as low-adhesion plates, nanoparticle facilitated magnetic levitation, and hanging drop plates.

3D cell culture in Hydrogels

As the natural extracellular matrix (ECM) is important in the survival, proliferation, differentiation and migration of cells, different hydrogel culture matrices mimicking natural ECM structure are seen as potential approaches to in vivo –like cell culturing. Hydrogels are composed of interconnected pores with high water retention, which enables efficient transport of substances such as nutrients and gases. Several different types of hydrogels from natural and synthetic materials are available for 3D cell culture, including animal ECM extract hydrogels, protein hydrogels, peptide hydrogels, polymer hydrogels, and wood-based nanocellulose hydrogel.

3D Cell Culturing by Magnetic Levitation

The 3D Cell Culturing by Magnetic Levitation method (MLM) is the application of growing 3D tissue by inducing cells treated with magnetic nanoparticle assemblies in spatially varying magnetic fields using neodymium magnetic drivers and promoting cell to cell interactions by levitating the cells up to the air/liquid interface of a standard petri dish. The magnetic nanoparticle assemblies consist of magnetic iron oxide nanoparticles, gold nanoparticles, and the polymer polylysine. 3D cell culturing is scalable, with the capability for culturing 500 cells to millions of cells or from single dish to high-throughput low volume systems.

Tissue culture and engineering

Cell culture is a fundamental component of tissue culture and tissue engineering, as it establishes the basics of growing and maintaining cells in vitro. The major application of human cell culture is in stem cell industry, where mesenchymal stem cells can be cultured and cryopreserved for future use. Tissue engineering potentially offers dramatic improvements in low cost medical care for hundreds of thousands of patients annually.

Vaccines

Vaccines for polio, measles, mumps, rubella, and chickenpox are currently made in cell cultures. Due to the H5N1 pandemic threat, research into using cell culture for influenza vaccines is being funded by the United States government. Novel ideas in the field include recombinant DNA-based vaccines, such as one made using human adenovirus (a common cold virus) as a vector, and novel adjuvants.

Culture of Non-Mammalian Cells

Plant cell Culture Methods

Plant cell cultures are typically grown as cell suspension cultures in a liquid medium or as callus cultures on a solid medium. The culturing of undifferentiated plant cells and calli requires the proper balance of the plant growth hormones auxin and cytokinin.

Insect cell Culture

Cells derived from Drosophila melanogaster (most prominently, Schneider 2 cells) can be used for experiments which may be hard to do on live flies or larvae, such as biochemical studies or studies using siRNA. Cell lines derived from the army worm Spodoptera frugiperda, including Sf9 and Sf21, and from the cabbage looper Trichoplusia ni, High Five cells, are commonly used for expression of recombinant proteins using baculovirus.

Bacterial and yeast culture methods

For bacteria and yeasts, small quantities of cells are usually grown on a solid support that contains nutrients embedded in it, usually a gel such as agar, while large-scale cultures are grown with the cells suspended in a nutrient broth.

Viral culture methods

The culture of viruses requires the culture of cells of mammalian, plant, fungal or bacterial origin as hosts for the growth and replication of the virus. Whole wild type viruses, recombinant viruses or viral products may be generated in cell types other than their natural hosts under the right conditions. Depending on the species of the virus, infection and viral replication may result in host cell lysis and formation of a viral plaque.

Common Cell Lines

Human cell lines

- DU145 (prostate cancer)
- H295R (adrenocortical cancer)
- HeLa (cervical cancer)
- KBM-7 (chronic myelogenous leukemia)
- LNCaP (prostate cancer)
- MCF-7 (breast cancer)
- MDA-MB-468 (breast cancer)
- PC3 (prostate cancer)
- SaOS-2 (bone cancer)

- SH-SY5Y (neuroblastoma, cloned from a myeloma)
- T47D (breast cancer)
- THP-1 (acute myeloid leukemia)
- U87 (glioblastoma)
- National Cancer Institute's 60 cancer cell line panel (NCI60)

Primate cell lines

- Vero (African green monkey Chlorocebus kidney epithelial cell line)

Mouse cell lines

- MC3T3 (embryonic calvarium)

Rat tumor cell lines

- GH3 (pituitary tumor)
- PC12 (pheochromocytoma)

Plant cell lines

- Tobacco BY-2 cells (kept as cell suspension culture, they are model system of plant cell)

Other species cell lines

- Dog MDCK kidney epithelial
- Xenopus A6 kidney epithelial
- Zebrafish AB9

List of Cell Lines

Cell line	Meaning	Organism	Origin tissue	Morphology	Links
3T3-L1	"3-day transfer, inoculum 3 x 10^5 cells"	Mouse	Embryo	Fibroblast	ECACCCellosaurus
4T1		Mouse	Mammary gland		ATCCCellosaurus
9L		Rat	Brain	Glioblastoma	ECACCCellosaurus
A172		Human	Brain	Glioblastoma	ECACCCellosaurus
A20		Mouse	B lymphoma	B lymphocyte	Cellosaurus
A253		Human	Submandibular duct	Head and neck carcinoma	ATCCCellosaurus
A2780		Human	Ovary	Ovarian carcinoma	ECACCCellosaurus
A2780ADR		Human	Ovary	Adriamycin-resistant derivative of A2780	ECACCCellosaurus
A2780cis		Human	Ovary	Cisplatin-resistant derivative of A2780	ECACCCellosaurus
A431		Human	Skin epithelium	Squamous cell carcinoma	ECACCCellosaurus
A549		Human	Lung	Lung carcinoma	ECACCCellosaurus

Cell line	Meaning	Organism	Origin tissue	Morphology	Links
AB9		Zebrafish	Fin	Fibroblast	ATCCCellosaurus
AHL-1	Armenian Hamster Lung-1	Hamster	Lung		ECACCCellosaurus
ALC		Mouse	Bone marrow	Stroma	PMID 2435412Cellosaurus
B16		Mouse	Melanoma		ECACCCellosaurus
B35		Rat	Neuroblastoma		ATCCCellosaurus
BCP-1		Human	PBMC	HIV+ primary effusion lymphoma	ATCCCellosaurus
BEAS-2B	Bronchial epithelium + Adenovirus 12-SV40 virus hybrid (Ad12SV40)	Human	Lung	Epithelial	ECACCCellosaurus
bEnd.3	Brain Endothelial 3	Mouse	Brain/cerebral cortex	Endothelium	Cellosaurus
BHK-21	Baby Hamster Kidney-21	Hamster	Kidney	Fibroblast	ECACCCellosaurus
BOSC23	Packaging cell line derived from HEK 293	Human	Kidney (embryonic)	Epithelium	Cellosaurus
BT-20	Breast Tumor-20	Human	Breast epithelium	Breast carcinoma	ATCCCellosaurus
BxPC3	Biopsy xenograft of Pancreatic Carcinoma line 3	Human	Pancreatic adenocarcinoma	Epithelial	ECACCCellosaurus
C2C12		Mouse	Myoblast		ECACCCellosaurus
C3H-10T1/2		Mouse	Embryonic mesenchymal cell line		ECACCCellosaurus
C6		Rat	Brain astrocyte	Glioma	ECACCCellosaurus
C6/36		Insect - Asian tiger mosquito	Larval tissue		ECACCCellosaurus
Caco-2		Human	Colon	Colorectal carcinoma	ECACCCellosaurus
Cal-27		Human	Tongue	Squamous cell carcinoma	ATCCCellosaurus
CGR8		Mouse	Embryonic stem cells		ECACCCellosaurus
CHO	Chinese Hamster Ovary	Hamster	Ovary	Epithelium	ECACCCellosaurus
CML T1	Chronic myeloid leukemia T lymphocyte 1	Human	CML acute phase	T cell leukemia	DSMZCellosaurus
CMT12	Canine Mammary Tumor 12	Dog	Mammary gland	Epithelium	Cellosaurus
COR-L23		Human	Lung	Lung carcinoma	ECACCCellosaurus
COR-L23/5010		Human	Lung	Lung carcinoma	ECACCCellosaurus
COR-L23/CPR		Human	Lung	Lung carcinoma	ECACCCellosaurus
COR-L23/R23-		Human	Lung	Lung carcinoma	ECACCCellosaurus
COS-7	Cercopithecus aethiops, origin-defective SV-40	Old World monkey - Cercopithecus aethiops (Chlorocebus)	Kidney	Fibroblast	ECACCCellosaurus
COV-434		Human	Ovary	Ovarian granulosa cell carcinoma	PMID 8436435ECACC Cellosaurus
CT26		Mouse	Colon	Colorectal carcinoma	Cellosaurus
D17		Dog	Lung metastasis	Osteosarcoma	ATCCCellosaurus
DH82		Dog	Histiocytosis	Monocyte/macrophage	ECACCCellosaurus
DU145		Human	Androgen insensitive prostate carcinoma		ATCCCellosaurus
DuCaP	Dura mater cancer of the Prostate	Human	Metastatic prostate carcinoma	Epithelial	PMID 11317521 Cellosaurus
E14Tg2a		Mouse		Embryonic stem cells	ECACCCellosaurus
EL4		Mouse		T cell leukemia	ECACCCellosaurus
EM-2		Human	CML blast crisis	Ph+ CML line	DSMZCellosaurus
EM-3		Human	CML blast crisis	Ph+ CML line	DSMZCellosaurus

Cell line	Meaning	Organism	Origin tissue	Morphology	Links
EMT6/AR1		Mouse	Mammary gland	Epithelial-like	ECACCCellosaurus
EMT6/AR10.0		Mouse	Mammary gland	Epithelial-like	ECACCCellosaurus
FM3		Human	Lymph node metastasis	Melanoma	ECACCCellosaurus
GL261	Glioma 261	Mouse	Brain	Glioma	Cellosaurus
H1299		Human	Lung	Lung carcinoma	ATCCCellosaurus
HaCaT		Human	Skin	Keratinocyte	CLSCellosaurus
HCA2		Human	Colon	Adenocarcinoma	ECACCCellosaurus
HEK 293	Human Embryonic Kidney 293	Human	Kidney (embryonic)	Epithelium	ECACCCellosaurus
HEK 293T	HEK 293 derivative	Human	Kidney (embryonic)	Epithelium	ECACCCellosaurus
HeLa	"Henrietta Lacks"	Human	Cervix epithelium	Cervical carcinoma	ECACCCellosaurus
Hepa1c1c7	Clone 7 of clone 1 hepatoma line 1	Mouse	Hepatoma	Epithelial	ECACCCellosaurus
Hep G2		Human	Liver	Hepatoblastoma	ECACCCellosaurus
High Five		Insect (moth) - Trichoplusia ni	Ovary		Cellosaurus
HL-60	Human Leukemia-60	Human	Blood	Myeloblast	ECACCCellosaurus
HT-1080		Human		Fibrosarcoma	ECACCCellosaurus
HT-29		Human	Colon epithelium	Adenocarcinoma	ECACCCellosaurus
J558L		Mouse	Myeloma	B lymphocyte cell	ECACCCellosaurus
Jurkat		Human	White blood cells	T cell leukemia	ECACCCellosaurus
JY		Human	Lymphoblastoid	EBV-transformed B cell	ECACCCellosaurus
K562		Human	Lymphoblastoid	CML blast crisis	ECACCCellosaurus
KBM-7		Human	Lymphoblastoid	CML blast crisis	Cellosaurus
KCL-22		Human	Lymphoblastoid	CML	DSMZCellosaurus
KG1		Human	Lymphoblastoid	AML	ECACCCellosaurus
Ku812		Human	Lymphoblastoid	Erythroleukemia	ECACCCellosaurus
KYO-1	Kyoto-1	Human	Lymphoblastoid	CML	DSMZCellosaurus
L1210		Mouse	Lymphocytic leukemia	Ascitic fluid	ECACCCellosaurus
L243		Mouse	Hybridoma	Secretes L243 mAb (against HLA-DR)	ATCCCellosaurus
LNCaP	Lymph Node Cancer of the Prostate	Human	Prostatic adenocarcinoma	Epithelial	ECACCCellosaurus
MA2.1		Mouse	Hybridoma	Secretes MA2.1 mAb (against HLA-A2 and HLA-B17)	ATCCCellosaurus
Ma-Mel 1, 2, 3....48		Human	Skin	A range of melanoma cell lines	ECACCCellosaurus
MC-38	Mouse Colon-38	Mouse	Colon	Adenocarcinoma	Cellosaurus
MCF-7	Michigan Cancer Foundation-7	Human	Breast	Invasive breast ductal carcinoma ER+, PR+	ECACCCellosaurus
MCF-10A	Michigan Cancer Foundation-10A	Human	Breast epithelium		ATCCCellosaurus
MDA-MB-157	M.D. Anderson - Metastatic Breast-157	Human	Pleural effusion metastasis	Breast carcinoma	ECACCCellosaurus
MDA-MB-231	M.D. Anderson - Metastatic Breast-231	Human	Pleural effusion metastasis	Breast carcinoma	ECACCCellosaurus
MDA-MB-361	M.D. Anderson - Metastatic Breast-361	Human		Melanoma (contaminated by M14)	ECACCCellosaurus
MDA-MB-468	M.D. Anderson - Metastatic Breast-468	Human	Pleural effusion metastasis	Breast carcinoma	ATCCCellosaurus
MDCK II	Madin Darby Canine Kidney II	Dog	Kidney	Epithelium	ECACCCellosaurus
MG63		Human	Bone	Osteosarcoma	ECACCCellosaurus
MOR/0.2R		Human	Lung	Lung carcinoma	ECACCCellosaurus
Mono-Mac-6		Human	White blood cells	Myeloid metaplasic AML	DSMZCellosaurus

Cell line	Meaning	Organism	Origin tissue	Morphology	Links
MRC-5	Medical Research Council cell strain 5	Human	Lung (fetal)	Fibroblast	ECACCCellosaurus
MTD-1A		Mouse		Epithelium	Cellosaurus
MyEnd	Myocardial Endothelial	Mouse		Endothelium	Cellosaurus
NCI-H69		Human	Lung	Lung carcinoma	ECACCCellosaurus
NCI-H69/CPR		Human	Lung	Lung carcinoma	ECACCCellosaurus
NCI-H69/ LX10		Human	Lung	Lung carcinoma	ECACCCellosaurus
NCI-H69/ LX20		Human	Lung	Lung carcinoma	ECACCCellosaurus
NCI-H69/LX4		Human	Lung	Lung carcinoma	ECACCCellosaurus
Neuro-2a		Mouse	Nerve/neuroblastoma	Neuronal stem cells	ECACCCellosaurus
NIH-3T3	NIH, 3-day transfer, inoculum 3 x 10^5 cells	Mouse	Embryo	Fibroblast	ECACCCellosaurus
NALM-1		Human	Peripheral blood	Blast-crisis CML	ATCCCellosaurus
Neuro2a		Mouse	Nerve/neuroblastoma	Neuronal stem cells	Cellosaurus
NK-92		Human	Leukemia/lymphoma		ATCCCellosaurus
NTERA-2		Human	Lung metastasis	Embryonal carcinoma	ECACCCellosaurus
NW-145		Human	Skin	Melanoma	ESTDABCellosaurus
OK	Opossum Kidney	Virginia opossum - Didelphis virginiana	Kidney		ECACCCellosaurus
OPCN / OPCT cell lines		Human	Prostate	Range of prostate tumour lines	Cellosaurus
P3X63Ag8		Mouse	Myeloma		ECACCCellosaurus
PC12		Rat	Adrenal medulla	Pheochromocytoma	ECACCCellosaurus
PC-3	Prostate Cancer-3	Human	Bone metastasis	Prostate carcinoma	ECACCCellosaurus
Peer		Human	T cell leukemia		DSMZCellosaurus
PNT1A		Human	Prostate	SV40-transformed tumour line	ECACCCellosaurus
PNT2		Human	Prostate	SV40-transformed tumour line	ECACCCellosaurus
Pt K2	The second cell line derived from Potorous tridactylis	Long-nosed potoroo - Potorous tridactylus	Kidney	Epithelial	ECACCCellosaurus
Raji		Human	B lymphoma	Lymphoblast-like	ECACCCellosaurus
RBL-1	Rat Basophilic Leukemia-1	Rat	Leukemia	Basophil cell	ECACCCellosaurus
RenCa	Renal Carcinoma	Mouse	Kidney	Renal carcinoma	ATCCCellosaurus
RIN-5F		Mouse	Pancreas		ECACCCellosaurus
RMA-S		Mouse		T cell tumour	Cellosaurus
S2	Schneider 2	Insect - Drosophila melanogaster	Late stage (20–24 hours old) embryos		ATCCCellosaurus
SaOS-2	Sarcoma OSteogenic-2	Human	Bone	Osteosarcoma	ECACCCellosaurus
Sf21	Spodoptera frugiperda 21	Insect (moth) - Spodoptera frugiperda	Ovary		ECACCCellosaurus

Cell line	Meaning	Organism	Origin tissue	Morphology	Links
Sf9	Spodoptera frugiperda 9	Insect (moth) - Spodoptera frugiperda	Ovary		ECACCCellosaurus
SH-SY5Y		Human	Bone marrow metastasis	Neuroblastoma	ECACCCellosaurus
SiHa		Human	Cervix epithelium	Cervical carcinoma	ATCCCellosaurus
SK-BR-3	Sloan-Kettering Breast cancer 3	Human	Breast	Breast carcinoma	DSMZCellosaurus
SK-OV-3	Sloan-Kettering Ovarian cancer 3	Human	Ovary	Ovarian carcinoma	ECACCCellosaurus
T2		Human		T cell leukemia/B cell line hybridoma	ATCCCellosaurus
T-47D		Human	Breast	Breast ductal carcinoma	ECACCCellosaurus
T84		Human	Lung metastasis	Colorectal carcinoma	ECACCCellosaurus
T98G		Human	Glioblastoma-astrocytoma	Epithelium	ECACCCellosaurus
THP-1		Human	Monocyte	Acute monocytic leukemia	ECACCCellosaurus
U373		Human	Glioblastoma-astrocytoma	Epithelium	ECACCCellosaurus
U87		Human	Glioblastoma-astrocytoma	Epithelial-like	ECACCCellosaurus
U937		Human	Leukemic monocytic lymphoma		ECACCCellosaurus
VCaP	Vertebral Cancer of the Prostate	Human	Vertebra metastasis	Prostate carcinoma	ECACCCellosaurus
Vero	Vero (truth)	African green monkey - Chlorocebus sabaeus	Kidney epithelium		ECACCCellosaurus
VG-1		Human		Primary effusion lymphoma	Cellosaurus
WM39		Human	Skin	Melanoma	ESTDABCellosaurus
WT-49		Human	Lymphoblastoid		ECACCCellosaurus
YAC-1		Mouse	Lymphoma		ECACCCellosaurus
YAR		Human	Lymphoblastoid	EBV-transformed B cell	Human Immunology ECACC Cellosaurus

FLOW CYTOMETRY

In biotechnology, **flow cytometry** is a laser- or impedance-based, biophysical technology employed in cell counting, cell sorting, biomarker detection and protein engineering, by suspending cells in a stream of fluid and passing them by an electronic detection apparatus. It allows simultaneous multiparametric analysis of the physical and chemical characteristics of up to thousands of particles per second. Flow cytometry is routinely used in the diagnosis of health disorders, especially blood cancers, but has many other applications in basic research, clinical practice and clinical trials. A common variation is to physically sort particles based on their properties, so as to purify populations of interest. The first impedance-based flow cytometry device, using the Coulter principle, was disclosed in U.S. Patent 2,656,508, issued in 1953, to Wallace H. Coulter. Mack Fulwyler was the inventor of the forerunner to today's flow cytometers - particularly the cell sorter. Fulwyler developed this in 1965 with his publication in Science. The first fluorescence-based flow cytometry device (ICP 11) was developed in 1968 by Wolfgang Göhde from the University of Münster, filed for patent on 18 December 1968 and first commercialized in 1968/69 by German developer and manufacturer Partec through Phywe AG in Göttingen. At that time, absorption methods were still widely favored by other scientists over fluorescence methods. Soon after, flow cytometry instruments were developed, including the Cytofluorograph (1971) from Bio/Physics Systems Inc. (later: Ortho Diagnostics), the PAS 8000 (1973) from Partec, the first FACS (Fluorescence-activated cell sorting) instrument from Becton Dickinson (1974), the ICP 22 (1975) from Partec/Phywe and the Epics from Coulter (1977/78). The first label-free high-frequency impedance flow cytometer based on a patented microfluidic "lab-on-chip", Ampha Z30, was introduced by Amphasys (2012).

Name of the Technology

The original name of the fluorescence-based flow cytometry technology was "pulse cytophotometry" (German: Impulszytophotometrie), based on the first patent application on fluorescence-based flow cytometry. At the 5th American Engineering Foundation Conference on Automated Cytology in Pensacola (Florida) in 1976 - eight years after the introduction of the first fluorescence-based flow cytometer (1968) - it was agreed to commonly use the name "flow cytometry", a term that quickly became popular.

Flow Cytometers

Modern flow cytometers are able to analyze several thousand particles every second, in "real time," and can actively separate and isolate particles having specified properties. A flow cytometer is similar to a microscope, except that, instead of producing an image of the cell, flow cytometry offers "high-throughput" (for a large number of cells) automated quantification of set parameters. To analyze solid tissues, a single-cell suspension must first be prepared. A flow cytometer has five main components:

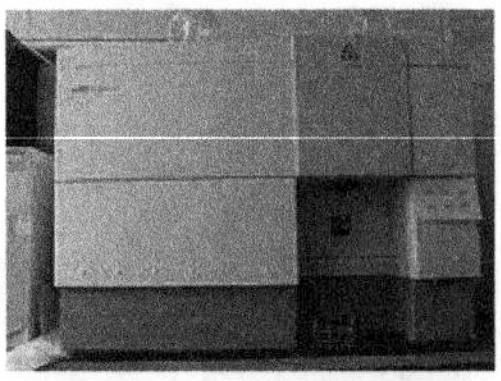

Front view of a desktop flow cytometer - the Becton-Dickinson Fluorescence activated cell sorter (FACSCalibur)

- a flow cell - liquid stream (sheath fluid), which carries and aligns the cells so that they pass single file through the light beam for sensing

- a measuring system - commonly used are measurement of impedance (or conductivity) and optical systems - lamps (mercury, xenon); high-power water-cooled lasers (argon, krypton, dye laser); low-power air-cooled lasers (argon (488 nm), red-HeNe (633 nm), green-HeNe, HeCd (UV)); diode lasers (blue, green, red, violet) resulting in light signals
- a detector and Analog-to-Digital Conversion (ADC) system - which converts analog measurements of forward-scattered light (FSC) and side-scattered light (SSC) as well as dye-specific fluorescence signals into digital signals that can be processed by a binary computer
- an amplification system - linear or logarithmic
- a computer for analysis of the signals.

The process of collecting data from samples using the flow cytometer is termed 'acquisition'. Acquisition is mediated by a computer physically connected to the flow cytometer, and the software which handles the digital interface with the cytometer. The software is capable of adjusting parameters (e.g., voltage, compensation) for the sample being tested, and also assists in displaying initial sample information while acquiring sample data to ensure that parameters are set correctly. Early flow cytometers were, in general, experimental devices, but technological advances have enabled widespread applications for use in a variety of both clinical and research purposes. Due to these developments, a considerable market for instrumentation, analysis software, as well as the reagents used in acquisition such as fluorescently labeled antibodies has developed.

Modern instruments usually have multiple lasers and fluorescence detectors. The current record for a commercial instrument is ten lasers and 18 fluorescence detectors. Increasing the number of lasers and detectors allows for multiple antibody labeling, and can more precisely identify a target population by their phenotypic markers. Certain instruments can even take digital images of individual cells, allowing for the analysis of fluorescent signal location within or on the surface of cells.

Data Analysis

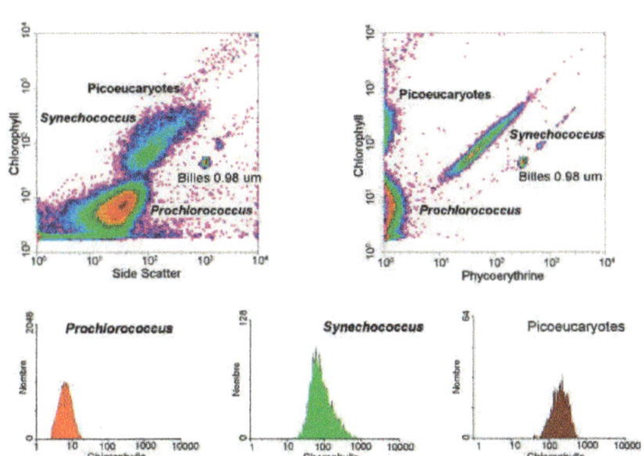

Analysis of a marine sample of photosyntheticpicoplankton by flow cytometry showing three different populations (Prochlorococcus, Synechococcus, and picoeukaryotes)

Gating

The data generated by flow-cytometers can be plotted in a single dimension, to produce a histogram, or in two-dimensional dot plots or even in three dimensions. The regions on these plots can be sequentially separated, based on fluorescence intensity, by creating a series of subset extractions, termed "gates." Specific gating protocols exist for diagnostic and clinical purposes especially in relation to hematology.

The plots are often made on logarithmic scales. Because different fluorescent dyes' emission spectra overlap, signals at the detectors have to be compensated electronically as well as computationally. Data accumulated using the flow cytometer can be analyzed using software, e.g., JMP_(statistical_software), WinMDI, Flowing Software, and web-based Cytobank (all freeware), Cellcion, FCS Express, FlowJo, FACSDiva, CytoPaint (aka Paint-A-Gate), VenturiOne, CellQuest Pro, Infinicyt or Cytospec. Once the data is collected, there is no need to stay connected to the flow cytometer and analysis is most often performed on a separate computer. This is especially necessary in core facilities where usage of these machines is in high demand.

Computational analysis

Recent progress on automated population identification using computational methods has offered an alternative to traditional gating strategies. Automated identification systems could potentially help findings of rare and hidden populations. Representative automated methods include FLOCK in Immunology Database and Analysis Portal (ImmPort), SamSPECTRAL and flowClust in Bioconductor, and FLAME in GenePattern. T-Distributed Stochastic Neighbor Embedding (tSNE) is an algorithm designed to perform dimensionality reduction, to allow visualization of complex multi-dimensional data in a two-dimensional "map". Collaborative efforts have resulted in an open project called FlowCAP (Flow Cytometry: Critical Assessment of Population Identification Methods,) to provide an objective way to compare and evaluate the flow cytometry data clustering methods, and also to establish guidance about appropriate use and application of these methods.

Fluorescence-activated Cell Sorting (FACS)

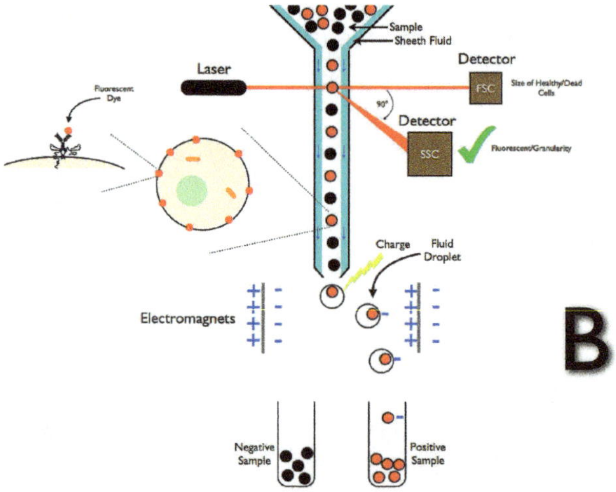

Fluorescence-Activated Cell Sorting (FACS)

Fluorescence-activated cell sorting (FACS) is a specialized type of flow cytometry. It provides a method for sorting a heterogeneous mixture of biological cells into two or more containers, one cell at a time, based upon the specific light scattering and fluorescent characteristics of each cell. It is a useful scientific instrument as it provides fast, objective and quantitative recording of fluorescent signals from individual cells as well as physical separation of cells of particular interest. The technique was expanded by Len Herzenberg, who was responsible for coining the term FACS.Herzenberg won the Kyoto Prize in 2006 for his seminal work in flow cytometry.

The cell suspension is entrained in the center of a narrow, rapidly flowing stream of liquid. The flow is arranged so that there is a large separation between cells relative to their diameter. A vibrating mechanism causes the stream of cells to break into individual droplets. The system is adjusted so that there is a low probability of more than one cell per droplet. Just before the stream breaks into droplets, the flow passes through a fluorescence measuring station where the fluorescent character of each cell of interest is measured. An electrical charging ring is placed just at the point where the stream breaks into droplets. A charge is placed on the ring based immediately prior to fluorescence intensity being measured, and the opposite charge is trapped on the droplet as it breaks from the stream. The charged droplets then fall through an electrostatic deflection system that diverts droplets into containers based upon their charge. In some systems, the charge is applied directly to the stream, and the droplet breaking off retains charge of the same sign as the stream. The stream is then returned to neutral after the droplet breaks off.

The acronym FACS is trademarked and owned by Becton, Dickinson and Company. Among the large majority of researchers who use this technology for sorting or analysis, this term has become genericized, much like Xerox for photocopy or Kleenex for facial tissue.

Labels

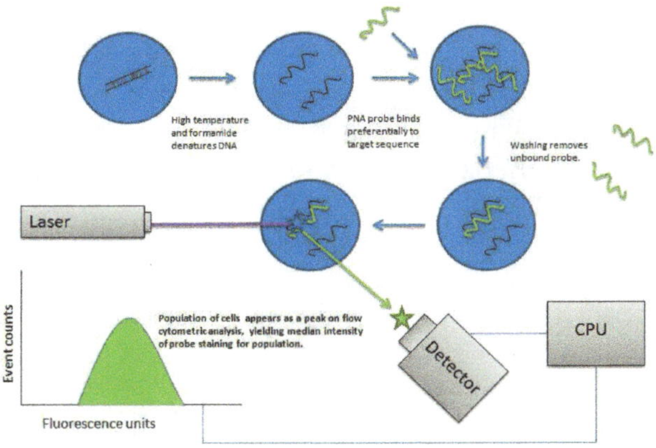

Use of flow cytometry to measure copy number variation of a specific DNA sequence (Flow-FISH)

Fluorescent Labels

A wide range of fluorophores can be used as labels in flow cytometry. Fluorophores, or simply "fluors", are typically attached to an antibody that recognizes a target feature on or in the cell; they may also be attached to a chemical entity with affinity for the cell membrane or another cellular structure. Each fluorophore has a characteristic peak excitation and emission wavelength, and the emission spectra often overlap. Consequently, the combination of labels which can be used depends on the wavelength of the lamp(s) or laser(s) used to excite the fluorochromes and on the detectors available. The maximum number of distinguishable fluorescent labels is thought to be 17 or 18, and this level of complexity necessitates laborious optimization to limit artifacts, as well as complex deconvolution algorithms to separate overlapping spectra. Flow cytometry uses fluorescence as a quantitative tool; the utmost sensitivity of flow cytometry is unmatched by other fluorescent detection platforms such as confocal microscopy. Absolute fluorescence sensitivity is generally lower in confocal microscopy because out-of-focus signals are rejected by the confocal optical system and because the image is built up serially from individual measurements at every location across the cell, reducing the amount of time available to collect signal.

Quantum Dots

Quantum dots are sometimes used in place of traditional fluorophores because of their narrower emission peaks.

Isotope Labeling

Mass cytometry overcomes the fluorescent labeling limit by utilizing lanthanide isotopes attached to antibodies. This method could theoretically allow the use of 40 to 60 distinguishable labels and has been demonstrated for 30 labels. Mass cytometry is fundamentally different from flow cytometry: cells are introduced into a plasma, ionized, and associated isotopes are quantified via time-of-flight mass spectrometry. Although this method permits the use of a large number of labels, it currently has lower throughput capacity than flow cytometry. It also destroys the analysed cells, precluding their recovery by sorting.

Cytometric Bead Array

In addition to the ability to label and identify individual cells via fluorescent antibodies, cellular products such as cytokines, proteins, and other factors may also be measured as well. Similar to ELISA sandwich assays, cytometric bead array (CBA) assays use multiple bead populations typically differentiated by size and different levels of fluorescence intensity to distinguish multiple analytes in a single assay. The amount of the analyte captured is detected via a biotinylated antibody against a secondary epitope of the protein, followed by a streptavidin-R-phycoerythrin treatment. The fluorescent intensity of R-phycoerythrin on the beads is quantified on a flow cytometer equipped with a 488 nm excitation source. Concentrations of a protein of interest in the samples can be obtained by comparing the fluorescent signals to those of a standard curve generated from a serial dilution of a known concentration of the analyte. Commonly also referred to as cytokine bead array (CBA).

Impedance Flow Cytometry

Impedance-based single cell analysis systems are commonly known as Coulter counters. They represent a well-established method for counting and sizing virtually any kind of cells and particles. The label-free technology has recently been enhanced by a "lab-on-a-chip" based approach and by applying high frequency alternating current (AC) in the radio frequency range (from 100 kHz to 30 MHz) instead of a static direct current (DC) or low frequency AC field. This patented technology allows a highly accurate cell analysis and provides additional information like membrane capacitance and viability. The relatively small size and robustness allow battery powered on-site use in the field.

Measurable Parameters

This list is very long and constantly expanding.

- Apoptosis (quantification, measurement of DNA degradation, mitochondrial membrane potential, permeability changes, caspase activity)
- Cell adherence (for instance pathogen-host cell adherence)
- Cell pigments such as chlorophyll or phycoerythrin
- Cell surface antigens (Cluster of differentiation (CD) markers)
- Cell viability
- Characterising multidrug resistance (MDR) in cancer cells
- Chromosome analysis and sorting (library construction, chromosome paint)
- DNA copy number variation (by Flow-FISH or BACs-on-Beads technology)
- Enzymatic activity
- Glutathione
- Intracellular antigens (various cytokines, secondary mediators, etc.)
- Membrane fluidity
- Monitoring electropermeabilization of cells
- Nuclear antigens
- Oxidative burst
- pH, intracellular ionized calcium, magnesium, membrane potential
- Protein expression and localization
- Protein modifications, phospho-proteins
- Scattering of light can be used to measure volume (by forward scatter) and morphological complexity (by side scatter) of cells or other particles, even those that are non-fluorescent. These are conventionally abbreviated as FSC and SSC respectively.

- Total DNA content (cell cycle analysis, cell kinetics, proliferation, ploidy, aneuploidy, endoreduplication, etc.)
- Total RNA content
- Transgenic products in vivo, particularly the Green fluorescent protein or related Fluorescent Proteins
- Various combinations (DNA/surface antigens, etc.)

SPECTROSCOPY

Spectroscopy is the study of the interaction between matter and electromagnetic radiation. Historically, spectroscopy originated through the study of visible light dispersed according to its wavelength, by a prism. Later the concept was expanded greatly to include any interaction with radiative energy as a function of its wavelength or frequency. Spectroscopic data is often represented by an emission spectrum, a plot of the response of interest as a function of wavelength or frequency.

Analysis of white light by dispersing it with a prism is an example of spectroscopy.

Spectroscopy and spectrography are terms used to refer to the measurement of radiation intensity as a function of wavelength and are often used to describe experimental spectroscopic methods. Spectral measurement devices are referred to as spectrometers, spectrophotometers, spectrographs or spectral analyzers.

Daily observations of color can be related to spectroscopy. Neon lighting is a direct application of atomic spectroscopy. Neon and other noble gases have characteristic emission frequencies (colors). Neon lamps use collision of electrons with the gas to excite these emissions. Inks, dyes and paints include chemical compounds selected for their spectral characteristics in order to generate specific colors and hues. A commonly encountered molecular spectrum is that of nitrogen dioxide. Gaseous nitrogen dioxide has a characteristic red absorption feature, and this gives air polluted with nitrogen dioxide a reddish-brown color. Rayleigh scattering is a spectroscopic scattering phenomenon that accounts for the color of the sky. Spectroscopic studies were central to the development of quantum mechanics and included Max Planck's explanation of blackbody radiation, Albert Einstein's explanation of the photoelectric effect and Niels Bohr's explanation of atomic structure and spectra. Spectroscopy is used in physical and analytical chemistry because atoms and molecules have unique spectra. As a result, these spectra can be used to detect, identify and quantify information about the atoms and molecules. Spectroscopy is also used in astronomy and remote sensing on earth. Most research telescopes have spectrographs. The measured spectra are used to determine the chemical composition and physical properties of astronomical objects (such as their temperature and velocity).

Theory

One of the central concepts in spectroscopy is a resonance and its corresponding resonant frequency. Resonances were first characterized in mechanical systems such as pendulums. Mechanical systems that vibrate or oscillate will experience large amplitude oscillations when they are driven at their resonant frequency. A plot of amplitude vs. excitation frequency will have a peak centered at the resonance frequency. This plot is one type of spectrum, with the peak often referred to as a spectral line, and most spectral lines have a similar appearance. In quantum mechanical systems, the analogous resonance is a coupling of two quantum mechanical stationary states of one system, such as an atom, via an oscillatory source of energy such as a photon. The coupling of the two states is strongest when the energy of the source matches the energy difference between the two states. The energy (E) of a photon is related to its frequency (ν) by $E = h\nu$ where h is Planck's constant, and so a spectrum of the system response vs. photon frequency will peak at the resonant frequency or energy. Particles such as electrons and neutrons have a comparable relationship, the de Broglie relations, between their kinetic energy and their wavelength and frequency and therefore can also excite resonant interactions. Spectra of atoms and molecules often consist of a series of spectral lines, each one representing a resonance between two different quantum states. The explanation of these series, and the spectral patterns associated with them, were one of the experimental enigmas that drove the development and acceptance of quantum mechanics. The hydrogen spectral series in particular was first successfully explained by the Rutherford-Bohr quantum model of the hydrogen atom. In some cases spectral lines are well separated and distinguishable, but spectral lines can also overlap and appear to be a single transition if the density of energy states is high enough. Named series of lines include the principal, sharp, diffuse and fundamental series.

Classification of Methods

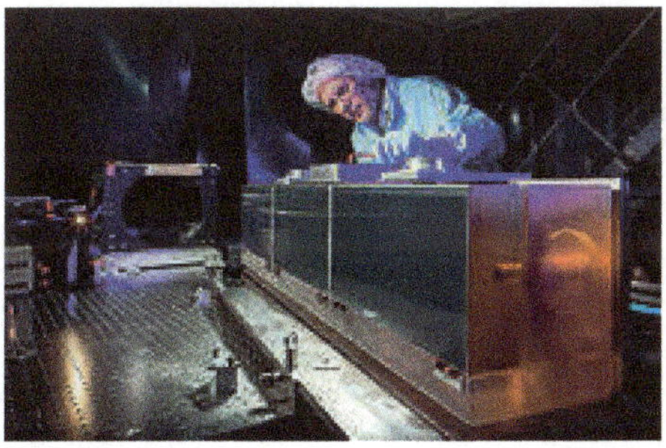

A huge diffraction grating at the heart of the ultra-precise ESPRESSO spectrograph.

Spectroscopy is a sufficiently broad field that many sub-disciplines exist, each with numerous implementations of specific spectroscopic techniques. The various implementations and techniques can be classified in several ways.

Type of Radiative Energy

Types of spectroscopy are distinguished by the type of radiative energy involved in the interaction. In many applications, the spectrum is determined by measuring changes in the intensity or frequency of this energy. The types of radiative energy studied include:

- Electromagnetic radiation was the first source of energy used for spectroscopic studies. Techniques that employ electromagnetic radiation are typically classified by the wavelength region of the spectrum and include microwave, terahertz, infrared, near infrared, visible and ultraviolet, x-ray and gamma spectroscopy.
- Particles, due to their de Broglie wavelength, can also be a source of radiative energy and both electrons and neutrons are commonly used. For a particle, its kinetic energy determines its wavelength.
- Acoustic spectroscopy involves radiated pressure waves.
- Mechanical methods can be employed to impart radiating energy, similar to acoustic waves, to solid materials.

Nature of the Interaction

Types of spectroscopy can also be distinguished by the nature of the interaction between the energy and the material. These interactions include:

- Absorption occurs when energy from the radiative source is absorbed by the material. Absorption is often determined by measuring the fraction of energy transmitted through the material; absorption will decrease the transmitted portion.
- Emission indicates that radiative energy is released by the material. A material's blackbody spectrum is a spontaneous emission spectrum determined by its temperature; this feature can be measured in the infrared by instruments such as the Atmospheric Emitted Radiance Interferometer (AERI).Emission can also be induced by other sources of energy such as flames or sparks or electromagnetic radiation in the case of fluorescence.
- Elastic scattering and reflection spectroscopy determine how incident radiation is reflected or scattered by a material. Crystallography employs the scattering of high energy radiation, such as x-rays and electrons, to examine the arrangement of atoms in proteins and solid crystals.
- Impedance spectroscopy studies the ability of a medium to impede or slow the transmittance of energy. For optical applications, this is characterized by the index of refraction.
- Inelastic scattering phenomena involve an exchange of energy between the radiation and the matter that shifts the wavelength of the scattered radiation. These include Raman and Compton scattering.

- Coherent or resonance spectroscopy are techniques where the radiative energy couples two quantum states of the material in a coherent interaction that is sustained by the radiating field. The coherence can be disrupted by other interactions, such as particle collisions and energy transfer, and so often require high intensity radiation to be sustained. Nuclear magnetic resonance (NMR) spectroscopy is a widely used resonance method and ultrafast laser methods are also now possible in the infrared and visible spectral regions.

Type of Material

Spectroscopic studies are designed so that the radiant energy interacts with specific types of matter.

Atoms

Atomic spectroscopy was the first application of spectroscopy developed. Atomic absorption spectroscopy (AAS) and atomic emission spectroscopy (AES) involve visible and ultraviolet light. These absorptions and emissions, often referred to as atomic spectral lines, are due to electronic transitions of outer shell electrons as they rise and fall from one electron orbit to another. Atoms also have distinct x-ray spectra that are attributable to the excitation of inner shell electrons to excited states. Atoms of different elements have distinct spectra and therefore atomic spectroscopy allows for the identification and quantitation of a sample's elemental composition. Robert Bunsen and Gustav Kirchhoff discovered new elements by observing their emission spectra. Atomic absorption lines are observed in the solar spectrum and referred to as Fraunhofer lines after their discoverer. A comprehensive explanation of the hydrogen spectrum was an early success of quantum mechanics and explained the Lamb shift observed in the hydrogen spectrum, which further led to the development of quantum electrodynamics. Modern implementations of atomic spectroscopy for studying visible and ultraviolet transitions include flame emission spectroscopy, inductively coupled plasma atomic emission spectroscopy, glow discharge spectroscopy, microwave induced plasma spectroscopy, and spark or arc emission spectroscopy. Techniques for studying x-ray spectra include X-ray spectroscopy and X-ray fluorescence (XRF).

Molecules

The combination of atoms into molecules leads to the creation of unique types of energetic states and therefore unique spectra of the transitions between these states. Molecular spectra can be obtained due to electron spin states (electron paramagnetic resonance), molecular rotations, molecular vibration and electronic states. Rotations are collective motions of the atomic nuclei and typically lead to spectra in the microwave and millimeter-wave spectral regions; rotational spectroscopy and microwave spectroscopy are synonymous. Vibrations are relative motions of the atomic nuclei and are studied by both infrared and Raman spectroscopy. Electronic excitations are studied using visible and ultraviolet spectroscopy as well as fluorescence spectroscopy.

Studies in molecular spectroscopy led to the development of the first maser and contributed to the subsequent development of the laser.

Crystals and Extended Materials

The combination of atoms or molecules into crystals or other extended forms leads to the creation of additional energetic states. These states are numerous and therefore have a high density of states. This high density often makes the spectra weaker and less distinct, i.e., broader. For instance, blackbody radiation is due to the thermal motions of atoms and molecules within a material. Acoustic and mechanical responses are due to collective motions as well. Pure crystals, though, can have distinct spectral transitions, and the crystal arrangement also has an effect on the observed molecular spectra. The regular lattice structure of crystals also scatters x-rays, electrons or neutrons allowing for crystallographic studies.

Nuclei

Nuclei also have distinct energy states that are widely separated and lead to gamma ray spectra. Distinct nuclear spin states can have their energy separated by a magnetic field, and this allows for NMR spectroscopy.

Other Types

Other types of spectroscopy are distinguished by specific applications or implementations:

- Acoustic resonance spectroscopy is based on sound waves primarily in the audible and ultrasonic regions
- Auger spectroscopy is a method used to study surfaces of materials on a micro-scale. It is often used in connection with electron microscopy.
- Cavity ring down spectroscopy
- Circular Dichroism spectroscopy
- Coherent anti-Stokes Raman spectroscopy (CARS) is a recent technique that has high sensitivity and powerful applications for in vivo spectroscopy and imaging.
- Cold vapour atomic fluorescence spectroscopy
- Correlation spectroscopy encompasses several types of two-dimensional NMR spectroscopy.
- Deep-level transient spectroscopy measures concentration and analyzes parameters of electrically active defects in semiconducting materials
- Dual polarisation interferometry measures the real and imaginary components of the complex refractive index
- Electron phenomenological spectroscopy measures physicochemical properties and characteristics of electronic structure of multicomponent and complex molecular systems.
- EPR spectroscopy
- Force spectroscopy
- Fourier transform spectroscopy is an efficient method for processing spectra data

obtained using interferometers. Fourier transform infrared spectroscopy (FTIR) is a common implementation of infrared spectroscopy. NMR also employs Fourier transforms.

- Hadron spectroscopy studies the energy/mass spectrum of hadrons according to spin, parity, and other particle properties. Baryon spectroscopy and meson spectroscopy are both types of hadron spectroscopy.
- Hyperspectral imaging is a method to create a complete picture of the environment or various objects, each pixel containing a full visible, VNIR, NIR, or infrared spectrum.
- Inelastic electron tunneling spectroscopy (IETS) uses the changes in current due to inelastic electron-vibration interaction at specific energies that can also measure optically forbidden transitions.
- Inelastic neutron scattering is similar to Raman spectroscopy, but uses neutrons instead of photons.
- Laser-Induced Breakdown Spectroscopy (LIBS), also called Laser-induced plasma spectrometry (LIPS)
- Laser spectroscopy uses tunable lasers and other types of coherent emission sources, such as optical parametric oscillators, for selective excitation of atomic or molecular species.
- Mass spectroscopy is an historical term used to refer to mass spectrometry. Current recommendations are to use the latter term. Use of the term mass spectroscopy originated in the use of phosphor screens to detect ions.
- Mössbauer spectroscopy probes the properties of specific isotopic nuclei in different atomic environments by analyzing the resonant absorption of gamma-rays. See also Mössbauer effect.
- Neutron spin echo spectroscopy measures internal dynamics in proteins and other soft matter systems
- Photoacoustic spectroscopy measures the sound waves produced upon the absorption of radiation.
- Photoemission spectroscopy
- Photothermal spectroscopy measures heat evolved upon absorption of radiation.
- Pump-probe spectroscopy can use ultrafast laser pulses to measure reaction intermediates in the femtosecond timescale.
- Raman optical activity spectroscopy exploits Raman scattering and optical activity effects to reveal detailed information on chiral centers in molecules.
- Raman spectroscopy
- Saturated spectroscopy
- Scanning tunneling spectroscopy
- Spectrophotometry
- Time-resolved spectroscopy measures the decay rate(s) of excited states using various spectroscopic methods.

- Time-Stretch Spectroscopy
- Thermal infrared spectroscopy measures thermal radiation emitted from materials and surfaces and is used to determine the type of bonds present in a sample as well as their lattice environment. The techniques are widely used by organic chemists, mineralogists, and planetary scientists.
- Ultraviolet photoelectron spectroscopy (UPS)
- Ultraviolet–visible spectroscopy
- Vibrational circular dichroism spectroscopy
- Video spectroscopy
- X-ray photoelectron spectroscopy (XPS)

Applications

UVES is a high-resolution spectrograph on the Very Large Telescope.

- Cure monitoring of composites using optical fibers.
- Estimate weathered wood exposure times using near infrared spectroscopy.
- Measurement of different compounds in food samples by absorption spectroscopy both in visible and infrared spectrum.

REFERENCES

1. "Some landmarks in the development of tissue and cell culture". Retrieved 2006-04-19.
2. "Cell Culture". Retrieved 2006-04-19.
3. "Whonamedit - Ringer's solution". whonamedit.com. Retrieved 2014-06-09.
4. "Animals and alternatives in testing". Archived from the original on 2006-02-25. Retrieved 2006-04-19.
5. Schiff, JudithAnn. "An unsung hero of medical research". Retrieved 2006-04-19. Yale Alumni Magazine, February 2002.
6. "Post - Blog | Boval BioSolutions, LLC". bovalco.com. Retrieved 2014-12-02.
7. "LipiMAX purified lipoprotein solution from bovine serum". Selborne Biological

Services. 2006. Retrieved 2010-02-02.

8. Portela VM, Zamberlam G, Price CA (April 2010). "Cell plating density alters the ratio of estrogenic to progestagenic enzyme gene expression in cultured granulosa cells". Fertil. Steril. 93 (6): 2050–5. doi:10.1016/j.fertnstert.2009.01.151. PMID 19324349.

9. Jill Neimark (27 February 2015). "Line of attack". Science. 347 (6225): 938–940. doi:10.1126/science.347.6225.938.

10. Drexler, HG; Dirks, WG; MacLeod, RA (Oct 1999). "False human hematopoietic cell lines: cross-contaminations and misinterpretations". Leukemia. 13 (10): 1601–7. doi:10.1038/sj/leu/2401510. ISSN 0887-6924. PMID 10516762.

11. Drexler, HG; MacLeod, RA; Dirks, WG (Dec 2001). "Cross-contamination: HS-Sultan is not a myeloma but a Burkitt lymphoma cell line" (Free full text). Blood. 98 (12): 3495–6. doi:10.1182/blood.V98.12.3495. ISSN 0006-4971. PMID 11732505.

12. Cabrera, CM; Cobo, F; Nieto, A; Cortés, JL; Montes, RM; Catalina, P; Concha, A (Jun 2006). "Identity tests: determination of cell line cross-contamination". Cytotechnology. 51 (2): 45–50. doi:10.1007/s10616-006-9013-8. ISSN 0920-9069. PMC 3449683. PMID 19002894.

13. Chatterjee, R (Feb 2007). "Cell biology. Cases of mistaken identity". Science. 315 (5814): 928–31. doi:10.1126/science.315.5814.928. ISSN 0036-8075. PMID 17303729.

14. Liscovitch, M; Ravid, D (Jan 2007). "A case study in misidentification of cancer cell lines: MCF-7/AdrR cells (re-designated NCI/ADR-RES) are derived from OVCAR-8 human ovarian carcinoma cells". Cancer Letters. 245 (1–2): 350–2. doi:10.1016/j.canlet.2006.01.013. ISSN 0304-3835. PMID 16504380.

15. Macleod, RA; Dirks, WG; Matsuo, Y; Kaufmann, M; Milch, H; Drexler, HG (Nov 1999). "Widespread intraspecies cross-contamination of human tumor cell lines arising at source". International Journal of Cancer. Journal International Du Cancer. 83 (4): 555–63. doi:10.1002/(SICI)1097-0215(19991112)83:4<555::AID-IJC19>3.0.CO;2-2. ISSN 0020-7136. PMID 10508494.

16. Masters, JR (Apr 2002). "HeLa cells 50 years on: the good, the bad and the ugly". Nature reviews. Cancer. 2 (4): 315–9. doi:10.1038/nrc775. ISSN 1474-175X. PMID 12001993.

17. Dunham, J.H.; Guthmiller, P. (2008). "Doing good science: Authenticating cell line identity" (PDF). Cell Notes. 22: 15–17.

18. Nguyen, H. T.; Geens, M.; Spits, C. (2012). "Genetic and epigenetic instability in human pluripotent stem cells". Human Reproduction Update. 19 (2): 187–205. doi:10.1093/humupd/dms048. PMID 23223511.

19. "Moore v. Regents of University of California (1990) 51 C3d 120". Online.ceb.com. Retrieved 2012-01-27.

20. Hayflick, L. (September 1998). "A brief history of the mortality and immortality of cultured cells". Keio J Med. 3. 47 (3): 174–182. doi:10.2302/kjm.47.174. PMID 9785764.

21. "Worthington tissue guide". Retrieved 2013-04-30.

22. Qiana L, Saltzman WM (2004) "Improving the expansion and neuronal differentiation of mesenchymal stem cells through culture surface modification." Biomaterials 25:

1331-1337.

23. Maguire G (2016) Therapeutics from Adult Stem Cells and the Hype Curve. ACS Med. Chem. Lett. 7 (5): 441–443

24. Discher, D. E.; Janmey, P; Wang, YL (2005). "Tissue Cells Feel and Respond to the Stiffness of Their Substrate". Science. 310 (5751): 1139–43. doi:10.1126/science.1116995. PMID 16293750.

25. Gilbert, PM; Havenstrite, KL; Magnusson, KE; Sacco, A; Leonardi, NA; Kraft, P; Nguyen, NK; Thrun, S; Lutolf, MP; Blau, HM (2010). "Substrate elasticity regulates skeletal muscle stem cell self-renewal in culture". Science. 329 (5995): 1078–81. doi:10.1126/science.1191035. PMC 2929271. PMID 20647425.

26. Chowdhury, Farhan; Li, Yanzhen; Poh, Yeh-Chuin; Yokohama-Tamaki, Tamaki; Wang, Ning; Tanaka, Tetsuya S. (2010). Zhou, Zhongjun, ed. "Soft Substrates Promote Homogeneous Self-Renewal of Embryonic Stem Cells via Downregulating Cell-Matrix Tractions". PLoS ONE. 5 (12): e15655. doi:10.1371/journal.pone.0015655. PMC 3001487 PMID 21179449.

27. Engler, Adam J.; Sen, Shamik; Sweeney, H. Lee; Discher, Dennis E. (2006). "Matrix Elasticity Directs Stem Cell Lineage Specification". Cell. 126 (4): 677–89. doi:10.1016/j.cell.2006.06.044. PMID 16923388.

28. Paszek, Matthew J.; Zahir, Nastaran; Johnson, Kandice R.; Lakins, Johnathon N.; Rozenberg, Gabriela I.; Gefen, Amit; Reinhart-King, Cynthia A.; Margulies, Susan S.; Dembo, Micah; Boettiger, David; Hammer, Daniel A.; Weaver, Valerie M. (2005). "Tensional homeostasis and the malignant phenotype". Cancer Cell. 8 (3): 241–54. doi:10.1016/j.ccr.2005.08.010. PMID 16169468.

29. Levental, Kandice R.; Yu, Hongmei; Kass, Laura; Lakins, Johnathon N.; Egeblad, Mikala; Erler, Janine T.; Fong, Sheri F.T.; Csiszar, Katalin; Giaccia, Amato; Weninger, Wolfgang; Yamauchi, Mitsuo; Gasser, David L.; Weaver, Valerie M. (2009). "Matrix Crosslinking Forces Tumor Progression by Enhancing Integrin Signaling". Cell. 139 (5): 891–906. doi:10.1016/j.cell.2009.10.027. PMC 2788004. PMID 19931152.

30. Tilghman, Robert W.; Cowan, Catharine R.; Mih, Justin D.; Koryakina, Yulia; Gioeli, Daniel; Slack-Davis, Jill K.; Blackman, Brett R.; Tschumperlin, Daniel J.; Parsons, J. Thomas (2010). Hotchin, Neil A, ed. "Matrix Rigidity Regulates Cancer Cell Growth and Cellular Phenotype". PLoS ONE. 5 (9): e12905. doi:10.1371/journal.pone.0012905. PMC 2944843. PMID 20886123.

31. Liu, F; Mih, JD; Shea, BS; Kho, AT; Sharif, AS; Tager, AM; Tschumperlin, DJ (2010). "Feedback amplification of fibrosis through matrix stiffening and COX-2 suppression". The Journal of Cell Biology. 190 (4): 693–706. doi:10.1083/jcb.201004082. PMC 2928007. PMID 20733059.

32. Wipff, PJ; Rifkin, DB; Meister, JJ; Hinz, B (2007). "Myofibroblast contraction activates latent TGF-beta1 from the extracellular matrix". The Journal of Cell Biology. 179 (6): 1311–23. doi:10.1083/jcb.200704042. PMC 2140013. PMID 18086923.

33. Georges, PC; Hui, JJ; Gombos, Z; McCormick, ME; Wang, AY; Uemura, M; Mick, R; Janmey, PA; Furth, EE; Wells, RG (2007). "Increased stiffness of the rat liver precedes matrix deposition: Implications for fibrosis". American Journal of Physiology. Gastrointestinal and Liver Physiology. 293 (6): G1147–54. doi:10.1152/

ajpgi.00032.2007. PMID 17932231.

34. Li, L; Sharma, N; Chippada, U; Jiang, X; Schloss, R; Yarmush, ML; Langrana, NA (2008). "Functional modulation of ES-derived hepatocyte lineage cells via substrate compliance alteration". Annals of Biomedical Engineering. 36 (5): 865–76. doi:10.1007/s10439-008-9458-3. PMID 18266108.

35. Semler, Eric J.; Lancin, Perry A.; Dasgupta, Anouska; Moghe, Prabhas V. (2005). "Engineering hepatocellular morphogenesis and function via ligand-presenting hydrogels with graded mechanical compliance". Biotechnology and Bioengineering. 89 (3): 296–307. doi:10.1002/bit.20328. PMID 15744840.

36. Friedland, JC; Lee, MH; Boettiger, D (2009). "Mechanically activated integrin switch controls alpha5beta1 function". Science. 323 (5914): 642–4. doi:10.1126/science.1168441. PMID 19179533.

37. Chan, CE; Odde, DJ (2008). "Traction dynamics of filopodia on compliant substrates". Science. 322 (5908): 1687–91. doi:10.1126/science.1163595. PMID 19074349.

38. Dupont, Sirio; Morsut, Leonardo; Aragona, Mariaceleste; Enzo, Elena; Giulitti, Stefano; Cordenonsi, Michelangelo; Zanconato, Francesca; Le Digabel, Jimmy; Forcato, Mattia; Bicciato, Silvio; Elvassore, Nicola; Piccolo, Stefano (2011). "Role of YAP/TAZ in mechanotransduction". Nature. 474 (7350): 179–83. doi:10.1038/nature10137. PMID 21654799.

39. «drug discovery@nature.com". Nature.com. Retrieved 2013-03-26.

40. Duell, BL; Cripps, AW; Schembri, MA; Ulett, GC (2011). "Epithelial cell coculture models for studying infectious diseases: Benefits and limitations". Journal of biomedicine & biotechnology. 2011: 852419. doi:10.1155/2011/852419. PMC 3189631. PMID 22007147.

41. Barrila, Jennifer; Radtke, Andrea L.; Crabbé, Aurélie; Sarker, Shameema F.; Herbst-Kralovetz, Melissa M.; Ott, C. Mark; Nickerson, Cheryl A. (2010). "Organotypic 3D cell culture models: Using the rotating wall vessel to study host–pathogen interactions". Nature Reviews Microbiology. 8 (11): 791–801. doi:10.1038/nrmicro2423. PMID 20948552.

42. M. Bhattacharya et al. Nanofibrillar cellulose hydrogel promotes three-dimensional liver cell culture. J Control Release 2012 Dec 28;164(3):291-8

43. Derosa, Maria C.; Monreal, Carlos; Schnitzer, Morris; Walsh, Ryan; Sultan, Yasir (2010). "Nanotechnology in fertilizers". Nature Nanotechnology. 5 (2): 91. doi:10.1038/nnano.2010.2. PMID 20130583.

44. Hsiao, Amy Y.; Tung, Yi-Chung; Qu, Xianggui; Patel, Lalit R.; Pienta, Kenneth J.; Takayama, Shuichi (2012). "384 hanging drop arrays give excellent Z-factors and allow versatile formation of co-culture spheroids". Biotechnology and Bioengineering. 109 (5): 1293–304. doi:10.1002/bit.24399. PMC 3306496. PMID 22161651.

45. M.W. Tibbitt, K.S. Anseth Hydrogels as Extracellular Matrix Mimics for 3D Cell Culture Biotechnol Bioeng. 2009 Jul 1;103(4):655-63

46. Reuters (2006-01-26). "Quickie Bird Flu Vaccine Created". Wired. Wired.com. Retrieved 2010-01-31.

47. Gao W; Soloff AC; Lu X; Montecalvo A; Nguyen DC; Matsuoka Y; Robbins PD; Swayne DE; Donis RO; Katz JM; Barratt-Boyes SM; Gambotto A. (February 2006). "Protection of mice and poultry from lethal H5N1 avian influenza virus through adenovirus-based immunization". Journal of Virology. United States: American Society for Microbiology.

80 (4): 1959–1964. doi:10.1128/JVI.80.4.1959-1964.2006. ISSN 0022-538X. PMC 1367171. PMID 16439551. Retrieved 2010-01-31.

48. «NIAID Taps Chiron to Develop Vaccine Against H9N2 Avian Influenza". National Institute of Allergy and Infectious Diseases (NIAID). 2004-08-17. Retrieved 2010-01-31.

49. Hunt, P; Robertson, D; Weiss, D; Rennick, D; Lee, F; Witte, ON (1987). "A single bone marrow-derived stromal cell type supports the in vitro growth of early lymphoid and myeloid cells". Cell. 48 (6): 997–1007. doi:10.1016/0092-8674(87)90708-2. PMID 2435412.

50. Van Den Berg-Bakker, CA; Hagemeijer, A; Franken-Postma, EM; Smit, VT; Kuppen, PJ; Van Ravenswaay Claasen, HH; Cornelisse, CJ; Schrier, PI (1993). "Establishment and characterization of 7 ovarian carcinoma cell lines and one granulosa tumor cell line: Growth features and cytogenetics". International Journal of Cancer. Journal International Du Cancer. 53 (4): 613–20. doi:10.1002/ijc.2910530415. PMID 8436435.

51. Lee, YG; Korenchuk, S; Lehr, J; Whitney, S; Vessela, R; Pienta, KJ (2001). "Establishment and characterization of a new human prostatic cancer cell line: DuCaP". In vivo (Athens, Greece). 15 (2): 157–62. PMID 11317521.

52. Ou, Dawei; Mitchell, Leslie Ann; Décarie, Diane; Tingle, Aubrey J; Nepom, Gerald T (1998). "Promiscuous T-Cell Recognition of a Rubella Capsid Protein Epitope Restricted by DRB1*0403 and DRB1*0901 Molecules Sharing an HLA DR Supertype". Human Immunology. 59 (3): 149–57. doi:10.1016/S0198-8859(98)00006-8. PMID 9548074.

CHAPTER 11

Basic Techniques and Tools Chemical Field

Chapter Contents

INTRODUCTION

Centrifugation is a process which involves the application of the centripetal force for the sedimentation of heterogeneous mixtures with a centrifuge, and is used in industrial and laboratory settings. This process is used to separate two miscible substances, but also to analyze the hydrodynamic properties of macromolecules. More-dense components of the mixture migrate away from the axis of the centrifuge, while less-dense components of the mixture migrate towards the axis. Chemists and biologists may increase the effective gravitational force on a test tube so as to more rapidly and completely cause the precipitate (pellet) to gather on the bottom of the tube. The remaining solution (supernatant) may be discarded with a pipette.

There is a correlation between the size and density of a particle and the rate that the particle separates from a heterogeneous mixture, when the only force applied is that of gravity. The larger the size and the larger the density of the particles, the faster they separate from the mixture. By applying a larger effective gravitational force to the mixture, like a centrifuge does, the separation of the particles is accelerated. This is ideal in industrial and lab settings because particles that would naturally separate over a long period of time can be separated in much less time.

The rate of centrifugation is specified by the angular velocity usually expressed as revolutions per minute (RPM), or acceleration expressed as *g*. The conversion factor between RPM and *g* depends on the radius of the centrifuge rotor. The particles' settling velocity in centrifugation is a function of their size and shape, centrifugal acceleration, the volume fraction of solids present, the density difference between the particle and the liquid, and the viscosity. The most common application is the separation of solid from highly concentrated suspensions, which is used in the treatment of sewage sludges for dewatering where less consistent sediment is produced.

In the chemical and food industries, special centrifuges can process a continuous stream of particle-laden liquid.

Centrifugation is the most common method used for uranium enrichment, relying on the slight mass difference between atoms of U238 and U235 in uranium hexafluoride gas..

MATHEMATICAL FORMULA

The general formula for calculating the revolutions per minute (RPM) of a centrifuge is

$$RPM = \sqrt{\frac{g}{r}},$$

where g represents the respective force of the centrifuge and r the radius from the center of the rotor to a point in the sample. However, depending on the centrifuge model used, the respective angle of the rotor and the radius may vary, thus the formula gets modified. For example, the Sorvall #SS-34 rotor has a maximum radius of 10.8 cm, so the formula becomes $RPM = 299\sqrt{\frac{g}{r}}$, which can further simplify to $RPM = 91\sqrt{g}$.

CENTRIFUGATION IN BIOLOGICAL RESEARCH

Microcentrifuges

Microcentrifuges are used to process small volumes of biological molecules, cells, or nuclei. Microcentrifuge tubes generally hold 0.5 - 2.0 mL of liquid, and are spun at maximum angular speeds of 12,000–13,000 rpm. Microcentrifuges are small enough to fit on a table-top and have rotors that can quickly change speeds. They may or may not have a refrigeration function.

High-speed Centrifuges

High-speed or superspeed centrifuges can handle larger sample volumes, from a few tens of millilitres to several litres. Additionally, larger centrifuges can also reach higher angular velocities (around 30,000 rpm). The rotors may come with different adapters to hold various sizes of test tubes, bottles, or microtiter plates.

Fractionation Process

General method of fractionation: Cell sample is stored in a suspension which is:

- Buffered - neutral pH, preventing damage to the structure of proteins including enzymes (which could affect ionic bonds)
- Isotonic (of equal water potential) - this prevents water gain or loss by the organelles
- Cool - reducing the overall activity of enzyme released later in the procedure
    - Cells are homogenised in a blender and filtered to remove debris
    - The homogenised sample is placed in an ultracentrifuge and spun in low speed - nuclei settle out, forming a pellet
    - The supernatant (suspension containing remaining organelles) is spun at a higher speed - chloroplasts settle out
    - The supernatant is spun at a higher speed still - mitochondria and lysosomes settle out
    - The supernatant is spun at an even higher speed - ribosomes, membranes settle out

The ribosomes, membranes and Golgi complexes can be separated by another technique called density gradient centrifugation.

Ultracentrifugations

Ultracentrifugation makes use of high centrifugal force for studying properties of biological particles. Compared to microcentrifuges or high-speed centrifuges, ultracentrifuges can isolate much smaller particles, including ribosomes, proteins, and viruses. Ultracentrifuges can also be used in the study of membrane fractionation. This occurs because ultracentrifuges can reach maximum angular velocities in excess of 70,000 rpm. Additionally, while microcentrifuges and supercentrifuges separate particles in batches (limited volumes of samples must be handled manually in test tubes or bottles), ultracentrifuges can separate molecules in batch or continuous flow systems.

In addition to purification, analytical ultracentrifugation (AUC) can be used for determination of the properties of macromolecules such as shape, mass, composition, and conformation. Samples are centrifuged with a high-density solution such as sucrose, caesium chloride, or iodixanol. The high-density solution may be at a uniform concentration throughout the test tube (“cushion”) or a varying concentration (“gradient”). Molecular properties can be modeled through sedimentation velocity analysis or sedimentation equilibrium analysis. During the run, the particle or molecules will migrate through the test tube at different speeds depending on their physical properties and the properties of the solution, and eventually form a pellet at the bottom of the tube, or bands at various heights.

Density Gradient Centrifugation

Density gradient centrifugation Is considered one of the more efficient methods of separating suspended particles. Density gradient centrifugation can be used both as a separation technique and as a method of measuring the densities of particles or molecules in a mixture. A tube, after being centrifuged by this method, has particles in order of density based on height. The object or particle of interest will reside in the position within the tube corresponding to its density.

Linderstorm-Lang, in 1937, discovered that density gradient tubes could be used for density measurements. He discovered this when working with potato yellow-dwarf virus.

This method was also used in Meselson and Stahl's famous experiment in which they proved that DNA replication is semi-conservative by using different isotopes of nitrogen. They used density gradient centrifugation to determine which isotope or isotopes of nitrogen were present in the DNA after cycles of replication.

Nevertheless, some non-ideal sedimentations are still possible when using this method. The first potential issue is the unwanted aggregation of particles, but this can occur in any centrifugation. The second possibility occurs when droplets of solution that contain particles sediment. This is more likely to occur when working with a solution that has a layer of suspension floating on a dense liquid, which in fact have little to no density gradient.

Differential Centrifugation

Differential Centrifugation is a type of centrifugation in which one selectively spins down components of a mixture by a series of increasing centrifugation forces. This method is commonly used to separate organelles and membranes found in cells. Organelles generally differ from each other in density in size, making the use of differential centrifugation, and centrifugation in general, possible. The organelles can then be identified by testing for indicators that are unique to the specific organelles.

Other applications

- Separating chalk powder from water
- Removing fat from milk to produce skimmed milk
- Separating particles from an air-flow using cyclonic separation
- The clarification and stabilization of wine
- Separation of urine components and blood components in forensic and research laboratories
- Aids in separation of proteins using purification techniques such as salting out, e.g. ammonium sulfate precipitation.

HISTORY

By 1923 Theodor Svedberg and his student H. Rinde had successfully analyzed large-grained sols in terms of their gravitational sedimentation. Sols consist of a substance evenly distributed in another substance, also known as a colloid. However, smaller grained sols, such as those containing gold, could not be analyzed. To investigate this problem Svedberg developed an analytical centrifuge, equipped with a photographic absorption system, which would exert a much greater centrifugal effect.In addition, he developed the theory necessary to measure molecular weight. During this time, Svedberg's attention shifted from gold to proteins.

By 1900, it had been generally accepted that proteins were composed of amino acids; however, whether proteins were colloids or macromolecules was still under debate. One protein being investigated at the time was hemoglobin. It was determined to have 712 carbon, 1,130 hydrogen, 243 oxygen, two sulfur atoms, and at least one iron atom. This gave hemoglobin a resulting weight of approximately 16,000 dalton (Da) but it was uncertain whether this value was a multiple of one or four (dependent upon the number of iron atoms present).

Through a series of experiments utilizing the sedimentation equilibrium technique, two important observations were made: hemoglobin has a molecular weight of 68,000 Da, suggesting that there are four iron atoms present rather than one, and that no matter where the hemoglobin was isolated from, it had exactly the same molecular weight. How something of such a large molecular mass could be consistently found, regardless of where it was sampled from in the body, was unprecedented and favored the idea the proteins are macromolecules rather than colloids. In order to investigate this phenomenon, a centrifuge with even higher speeds was needed, and thus the ultracentrifuge was created to apply the theory of sedimentation-diffusion. The same molecular mass was determined, and the presence of a spreading boundary suggested that it was a single compact particle. Further application of centrifugation showed that under different conditions the large homogeneous particles could be broken down into discrete subunits. The development of centrifugation was a great advance in experimental protein science.

CONTINUOUS STIRRED-TANK REACTOR

The continuous flow stirred-tank reactor (CSTR), also known as vat- or backmix reactor, is a common ideal reactor type in chemical engineering. A CSTR often refers to a model used to estimate the key unit operation variables when using a continuous[†] agitated-tank reactor to reach a specified output. The mathematical model works for all fluids: liquids, gases, and slurries.

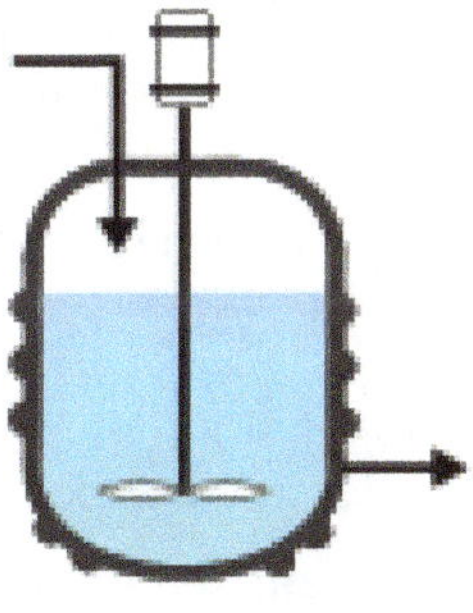

CSTR symbol

The behavior of a CSTR is often approximated or modeled by that of a Continuous Ideally Stirred-Tank Reactor (CISTR). All calculations performed with CISTRs assume perfect mixing. In a perfectly mixed reactor, the output composition is identical to composition of the material inside the reactor, which is a function of residence time and rate of reaction. If the residence time is 5-10 times the mixing time, this approximation is valid for engineering purposes. The CISTR model is often used to simplify engineering calculations and can be used to describe research reactors. In practice it can only be approached, in particular in industrial size reactors.

Assume:

- perfect or ideal mixing, as stated above

Integral mass balance on number of moles N_i of species i in a reactor of volume V. $[accumulation] = [in] - [out] + [generation]$

1. $\frac{dN_i}{dt} = F_{io} - F_i + V\nu_i r_i$

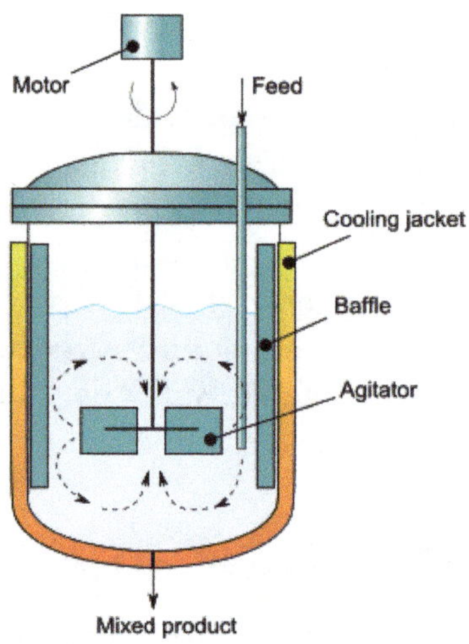

Cross-sectional diagram of Continuous flow stirred-tank reactor

where F_{io} is the molar flow rate inlet of species i, F_i the molar flow rate outlet, and ν_i stoichiometric coefficient. The reaction rate, r, is generally dependent on the reactant concentration and the rate constant (k). The rate constant can be determined by using a known empirical reaction rates that is adjusted for temperature using the Arrhenius temperature dependence. Generally, as the temperature increases so does the rate at which the reaction occurs. Residence time, τ , is the average amount of time a discrete quantity of reagent spends inside the tank.

Assume:

- constant density (valid for most liquids; valid for gases only if there is no net change in the number of moles or drastic temperature change)
- isothermal conditions, or constant temperature (k is constant)
- steady state

- single, irreversible reaction ($v_A = -1$)
- first-order reaction ($r = kC_A$)

A → products

$N_A = C_A V$ (where C_A is the concentration of species A, V is the volume of the reactor, N_A is the number of moles of species A)

2. $$C_A = \frac{C_{Ao}}{1 + k\tau}$$

The values of the variables, outlet concentration and residence time, in Equation 2 are major design criteria.

To model systems that do not obey the assumptions of constant temperature and a single reaction, additional dependent variables must be considered. If the system is considered to be in unsteady-state, a differential equation or a system of coupled differential equations must be solved.

CSTR's are known to be one of the systems which exhibit complex behavior such as steady-state multiplicity, limit cycles and chaos.

CRYSTALLIZATION

Crystallization is the (natural or artificial) process where a solid forms where the atoms or molecules are highly organized in a structure known as a crystal. Some of the ways which crystals form are through precipitating from a solution, melting or more rarely deposition directly from a gas. Crystallization is also a chemical solid–liquid separation technique, in which mass transfer of a solute from the liquid solution to a pure solid crystalline phase occurs. In chemical engineering crystallization occurs in a crystallizer. Crystallization is therefore related to precipitation, although the result is not amorphous or disordered, but a crystal.

Process

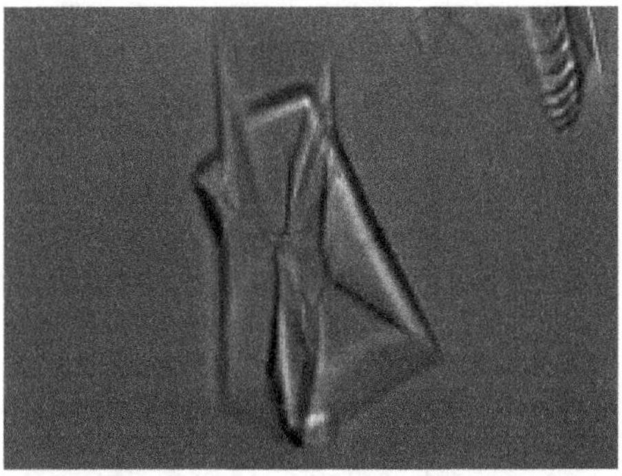

Time-lapse of growth of a citric acid crystal. The video covers an area of 2.0 by 1.5 mm and was captured over 7.2 min.

The crystallization process consists of two major events, *nucleation* and *crystal growth* which are driven by thermodynamic properties as well as chemical properties. In crystallization *Nucleation* is the step where the solute molecules or atoms dispersed in the solvent start to gather into clusters, on the microscopic scale (elevating solute concentration in a small region), that become stable under the current operating conditions. These stable clusters constitute the nuclei. Therefore, the clusters need to reach a critical size in order to become stable nuclei. Such critical size is dictated by many different factors (temperature, supersaturation, etc.). It is at the stage of nucleation that the atoms or molecules arrange in a defined and periodic manner that defines the crystal structure — note that "crystal structure" is a special term that refers to the relative arrangement of the atoms or molecules, not the macroscopic properties of the crystal (size and shape), although those are a result of the internal crystal structure.

The *crystal growth* is the subsequent size increase of the nuclei that succeed in achieving the critical cluster size. Crystal growth is a dynamic process occurring in equilibrium where solute molecules or atoms precipitate out of solution, and dissolve back into solution. Supersaturation is one of the driving forces of crystallization, as the solubility of a species is an equilibrium process quantified by Ksp. Depending upon the conditions, either nucleation or growth may be predominant over the other, dictating crystal size.

Many compounds have the ability to crystallize with some having different crystal structures, a phenomenon called polymorphism. Each polymorph is in fact a different thermodynamic solid state and crystal polymorphs of the same compound exhibit different physical properties, such as dissolution rate, shape (angles between facets and facet growth rates), melting point, etc. For this reason, polymorphism is of major importance in industrial manufacture of crystalline products. Additionally, crystal phases can sometimes be interconverted by varying factors such as temperature.

In Nature

Snowflakes are a very well-known example, where subtle differences in *crystal growth* conditions result in different geometries.

Crystallized honey

There are many examples of natural process that involve crystallization. Geological time scale process examples include:

- Natural (mineral) crystal formation (see also gemstone);
- Stalactite/stalagmite, rings formation.

Usual time scale process examples include:

- Snow flakes formation;
- Honey crystallization (nearly all types of honey crystallize).

Methods

Crystal formation can be divided into two types, where the first type of crystals are composed of a cation and anion, also known as a salt, such as sodium acetate. The second type of crystals are composed of uncharged species, for example menthol.

Crystal formation can be achieved by various methods, such as: cooling, evaporation, addition of a second solvent to reduce the solubility of the solute (technique known as antisolvent or drown-out), solvent layering, sublimation, changing the cation or anion, as well as other methods.

The formation of a supersaturated solution does not guarantee crystal formation, and often a seed crystal or scratching the glass is required to form nucleation sites.

A typical laboratory technique for crystal formation is to dissolve the solid in a solution in which it is partially soluble, usually at high temperatures to obtain supersaturation. The hot mixture is then filtered to remove any insoluble impurities. The filtrate is allowed to slowly cool. Crystals that form are then filtered and washed with a solvent in which they are not soluble, but is miscible with the mother liquor. The process is then repeated to increase the purity in a technique known as recrystallization.

Typical Equipment

Equipment for the main industrial processes for crystallization.

- *Tank crystallizers*. Tank crystallization is an old method still used in some specialized cases. Saturated solutions, in tank crystallization, are allowed to cool in open tanks. After a period of time the mother liquor is drained and the crystals removed. Nucleation and size of crystals are difficult to control. Typically, labor costs are very high.

Thermodynamic View

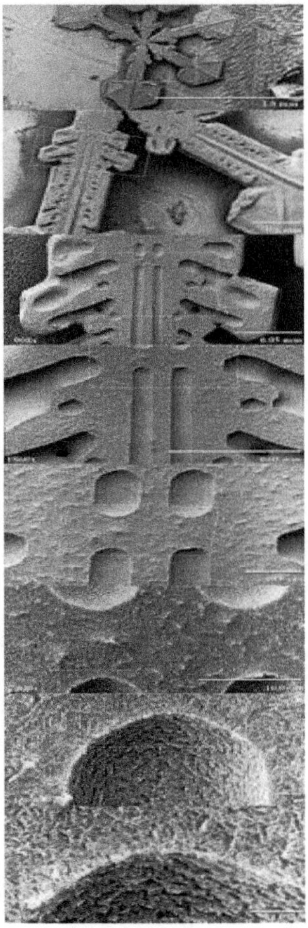

Low-temperature SEM magnification series for a snow crystal. The crystals are captured, stored, and sputter coated with platinum at cryo-temperatures for imaging.

The nature of a crystallization process is governed by both thermodynamic and kinetic factors, which can make it highly variable and difficult to control. Factors such as impurity level, mixing regime, vessel design, and cooling profile can have a major impact on the size, number, and shape of crystals produced.

Now put yourself in the place of a molecule within a pure and *perfect crystal*, being heated by an external source. At some sharply defined temperature, the complicated architecture of the crystal collapses to that of a liquid. Textbook thermodynamics says that melting occurs because the entropy, S, gain in your system by spatial randomization of the molecules has overcome the enthalpy, H, loss due to breaking the crystal packing forces:

$$T(S_{liquid} - S_{solid}) > H_{liquid} - H_{solid}$$
$$G_{liquid} < G_{solid}$$

This rule suffers no exceptions when the temperature is rising. By the same token, on cooling the melt, at the very same temperature the bell should ring again, and molecules should click back into the very same crystalline form. The entropy decrease due to the ordering of molecules within the system is overcompensated by the thermal randomization of the surroundings, due to the release of the heat of fusion; the entropy of the universe increases.

But liquids that behave in this way on cooling are the exception rather than the rule; in spite of the second principle of thermodynamics, crystallization usually occurs at lower temperatures (supercooling). This can only mean that a crystal is more easily destroyed than it is formed. Similarly, it is usually much easier to dissolve a perfect crystal in a solvent than to grow again a good crystal from the resulting solution. The nucleation and growth of a crystal are under kinetic, rather than thermodynamic, control.

Dynamics

As mentioned above, a crystal is formed following a well-defined pattern, or structure, dictated by forces acting at the molecular level. As a consequence, during its formation process the crystal is in an environment where the solute concentration reaches a certain critical value, before changing status. Solid formation, impossible below the solubility threshold at the given temperature and pressure conditions, may then take place at a concentration higher than the theoretical solubility level. The difference between the actual value of the solute concentration at the crystallization limit and the theoretical (static) solubility threshold is called supersaturation and is a fundamental factor in crystallization.

Nucleation

Nucleation is the initiation of a phase change in a small region, such as the formation of a solid crystal from a liquid solution. It is a consequence of rapid local fluctuations on a molecular scale in a homogeneous phase that is in a state of metastable equilibrium. Total nucleation is the sum effect of two categories of nucleation – primary and secondary.

Primary nucleation

Primary nucleation is the initial formation of a crystal where there are no other crystals present or where, if there are crystals present in the system, they do not have any influence on the process. This can occur in two conditions. The first is homogeneous nucleation, which is nucleation that is not influenced in any way by solids. These solids include the walls of the crystallizer vessel and particles of any foreign substance. The second category, then, is heterogeneous nucleation. This occurs when solid particles of foreign substances cause an increase in the rate of nucleation that would otherwise not be seen without the existence of these foreign particles. Homogeneous nucleation rarely occurs in practice due to the high energy necessary to begin nucleation without a solid surface to catalyse the nucleation.

Primary nucleation (both homogeneous and heterogeneous) has been modelled with the following:

$$B = \frac{dN}{dt} = k_n(c - c^*)^n$$

- B is the number of nuclei formed per unit volume per unit time.
- N is the number of nuclei per unit volume.
- k_n is a rate constant.
- c is the instantaneous solute concentration.
- c^* is the solute concentration at saturation.
- $(c–c^*)$ is also known as supersaturation.
- n is an empirical exponent that can be as large as 10, but generally ranges between 3 and 4.

Secondary nucleation

Secondary nucleation is the formation of nuclei attributable to the influence of the existing microscopic crystals in the magma. The first type of known secondary crystallization is attributable to fluid shear, the other due to collisions between already existing crystals with either a solid surface of the crystallizer or with other crystals themselves. Fluid shear nucleation occurs when liquid travels across a Crystal at a high speed, sweeping away nuclei that would otherwise be incorporated into a Crystal, causing the swept-away nuclei to become new crystals. Contact nucleation has been found to be the most effective and common method for nucleation. The benefits include the following

- Low kinetic order and rate-proportional to supersaturation, allowing easy control without unstable operation.
- Occurs at low supersaturation, where growth rate is optimum for good quality.
- Low necessary energy at which crystals strike avoids the breaking of existing crystals into new crystals.
- The quantitative fundamentals have already been isolated and are being incorporated into practice.

The following model, although somewhat simplified, is often used to model secondary nucleation:

$$B = \frac{dN}{dt} = k_1 M_T^j (c - c^*)^b$$

- k_1 is a rate constant.
- M_T is the suspension density.

- j is an empirical exponent that can range up to 1.5, but is generally 1.
- b is an empirical exponent that can range up to 5, but is generally 2.

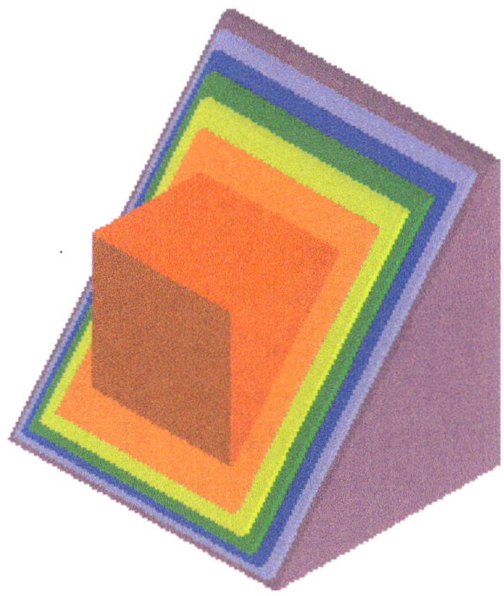

Crystal growth

Growth

Once the first small crystal, the nucleus, forms it acts as a convergence point (if unstable due to supersaturation) for molecules of solute touching – or adjacent to – the crystal so that it increases its own dimension in successive layers. The pattern of growth resembles the rings of an onion, as shown in the picture, where each colour indicates the same mass of solute; this mass creates increasingly thin layers due to the increasing surface area of the growing crystal. The supersaturated solute mass the original nucleus may *capture* in a time unit is called the *growth rate* expressed in kg/(m^2*h), and is a constant specific to the process. Growth rate is influenced by several physical factors, such as surface tension of solution, pressure, temperature, relative crystal velocity in the solution, Reynolds number, and so forth.

The main values to control are therefore:

- Supersaturation value, as an index of the quantity of solute available for the growth of the crystal;
- Total crystal surface in unit fluid mass, as an index of the capability of the solute to fix onto the crystal;
- Retention time, as an index of the probability of a molecule of solute to come into contact with an existing crystal;
- Flow pattern, again as an index of the probability of a molecule of solute to come into contact with an existing crystal (higher in laminar flow, lower in turbulent flow, but the reverse applies to the probability of contact).

The first value is a consequence of the physical characteristics of the solution, while the others define a difference between a well- and poorly designed crystallizer.

Size Distribution

The appearance and size range of a crystalline product is extremely important in crystallization. If further processing of the crystals is desired, large crystals with uniform size are important for washing, filtering, transportation, and storage. The importance lies in the fact that large crystals are easier to filter out of a solution than small crystals. Also, larger crystals have a smaller surface area to volume ratio, leading to a higher purity. This higher purity is due to less retention of mother liquor which contains impurities, and a smaller loss of yield when the crystals are washed to remove the mother liquor. The theoretical crystal size distribution can be estimated as a function of operating conditions with a fairly complicated mathematical process called population balance theory (using population balance equations).

Main Crystallization Processes

Some of the important factors influencing solubility are:

- Concentration
- Temperature
- Polarity
- Ionic Strength

So we may identify two main families of crystallization processes:

- Cooling crystallization
- Evaporative crystallization

This division is not really clear-cut, since hybrid systems exist, where cooling is performed through evaporation, thus obtaining at the same time a concentration of the solution.

A crystallization process often referred to in chemical engineering is the fractional crystallization. This is not a different process, rather a special application of one (or both) of the above.

Cooling crystallization

Application

Most chemical compounds, dissolved in most solvents, show the so-called *direct* solubility that is, the solubility threshold increases with temperature.

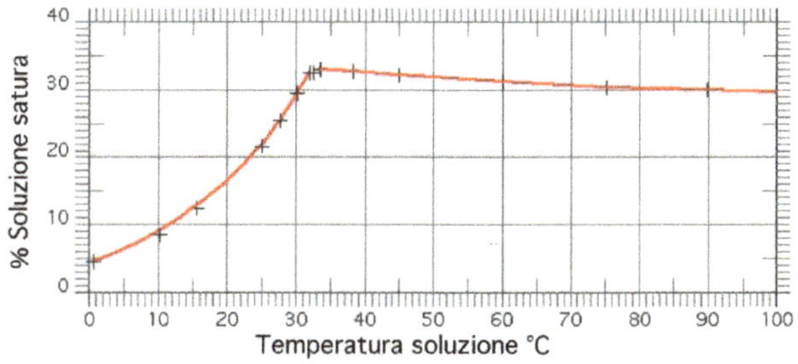

Solubility of the system Na_2SO_4 – H_2O

So, whenever the conditions are favourable, crystal formation results from simply cooling the solution. Here *cooling* is a relative term: austenite crystals in a steel form well above 1000 °C. An example of this crystallization process is the production of Glauber's salt, a crystalline form of sodium sulfate. In the diagram, where equilibrium temperature is on the x-axis and equilibrium concentration (as mass percent of solute in saturated solution) in y-axis, it is clear that sulfate solubility quickly decreases below 32.5 °C. Assuming a saturated solution at 30 °C, by cooling it to 0 °C (note that this is possible thanks to the freezing-point depression), the precipitation of a mass of sulfate occurs corresponding to the change in solubility from 29% (equilibrium value at 30 °C) to approximately 4.5% (at 0 °C) – actually a larger crystal mass is precipitated, since sulfate entrains hydration water, and this has the side effect of increasing the final concentration.

There are of course limitation in the use of cooling crystallization:

- Many solutes precipitate in hydrate form at low temperatures: in the previous example this is acceptable, and even useful, but it may be detrimental when, for example, the mass of water of hydration to reach a stable hydrate crystallization form is more than the available water: a single block of hydrate solute will be formed – this occurs in the case of calcium chloride);
- Maximum supersaturation will take place in the coldest points. These may be the heat exchanger tubes which are sensitive to scaling, and heat exchange may be greatly reduced or discontinued;

- A decrease in temperature usually implies an increase of the viscosity of a solution. Too high a viscosity may give hydraulic problems, and the laminar flow thus created may affect the crystallization dynamics.
- It is of course not applicable to compounds having *reverse* solubility, a term to indicate that solubility increases with temperature decrease (an example occurs with sodium sulfate where solubility is reversed above 32.5 °C).

Cooling Crystallizers

Vertical cooling crystallizer in a beet sugar factory

The simplest cooling crystallizers are tanks provided with a mixer for internal circulation, where temperature decrease is obtained by heat exchange with an intermediate fluid circulating in a jacket. These simple machines are used in batch processes, as in processing of pharmaceuticals and are prone to scaling. Batch processes normally provide a relatively variable quality of product along the batch.

The *Swenson-Walker* crystallizer is a model, specifically conceived by Swenson Co. around 1920, having a semicylindric horizontal hollow trough in which a hollow screw conveyor or some hollow discs, in which a refrigerating fluid is circulated, plunge during rotation on a longitudinal axis. The refrigerating fluid is sometimes also circulated in a jacket around the trough. Crystals precipitate on the cold surfaces of the screw/discs, from which they are removed by scrapers and settle on the bottom of the trough. The screw, if provided, pushes the slurry towards a discharge port.

A common practice is to cool the solutions by flash evaporation: when a liquid at a given T_0 temperature is transferred in a chamber at a pressure P_1 such that the liquid saturation temperature T_1 at P_1 is lower than T_0, the liquid will release heat according to the temperature difference and a quantity of solvent, whose total latent heat of vaporization equals the difference in enthalpy. In simple words, the liquid is cooled by evaporating a part of it.

In the sugar industry vertical cooling crystallizers are used to exhaust the molasses in the last crystallization stage downstream of vacuum pans, prior to centrifugation. The massecuite enters the crystallizers at the top, and cooling water is pumped through pipes in counterflow.

Evaporative Crystallization

Another option is to obtain, at an approximately constant temperature, the precipitation of the crystals by increasing the solute concentration above the solubility threshold. To obtain this, the solute/solvent mass ratio is increased using the technique of evaporation. This process is of course insensitive to change in temperature (as long as hydration state remains unchanged).

All considerations on control of crystallization parameters are the same as for the cooling models.

Evaporative Crystallizers

Most industrial crystallizers are of the evaporative type, such as the very large sodium chloride and sucrose units, whose production accounts for more than 50% of the total world production of crystals. The most common type is the *forced circulation* (FC) model (see evaporator). A pumping device (a pump or an axial flow mixer) keeps the crystal slurry in homogeneous suspension throughout the tank, including the exchange surfaces; by controlling pump flow, control of the contact time of the crystal mass with the supersaturated solution is achieved, together with reasonable velocities at the exchange surfaces. The Oslo, mentioned above, is a refining of the evaporative forced circulation crystallizer, now equipped with a large crystals settling zone to increase the retention time (usually low in the FC) and to roughly separate heavy slurry zones from clear liquid.

DTB Crystallizer

DTB Crystallizer

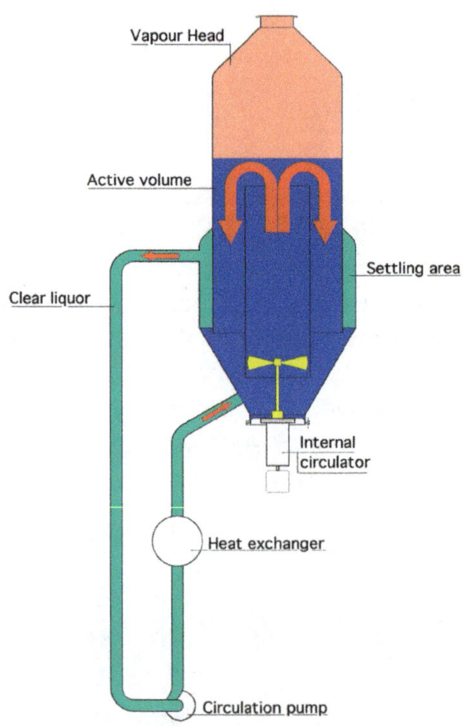

Schematic of DTB

Whichever the form of the crystallizer, to achieve an effective process control it is important to control the retention time and the crystal mass, to obtain the optimum conditions in terms of crystal specific surface and the fastest possible growth. This is achieved by a separation – to put it simply – of the crystals from the liquid mass, in order to manage the two flows in a different way. The practical way is to perform a gravity settling to be able to extract (and possibly recycle separately) the (almost) clear liquid, while managing the mass flow around the crystallizer to obtain a precise slurry density elsewhere. A typical example is the DTB (*Draft Tube and Baffle*) crystallizer, an idea of Richard Chisum Bennett (a Swenson engineer and later President of Swenson) at the end of the 1950s. The DTB crystallizer (see images) has an internal circulator, typically an axial flow mixer – yellow – pushing upwards in a draft tube while outside the crystallizer there is a settling area in an annulus; in it the exhaust solution moves upwards at a very low velocity, so that large crystals settle – and return to the main circulation – while only the fines, below a given grain size are extracted and eventually destroyed by increasing or decreasing temperature, thus creating additional supersaturation. A quasi-perfect control of all parameters is achieved. This crystallizer, and the derivative models (Krystal, CSC, etc.) could be the ultimate solution if not for a major limitation in the evaporative capacity, due to the limited diameter of the vapour head and the relatively low external circulation not allowing large amounts of energy to be supplied to the system.

CHROMATOGRAPHY

Chromatography is the collective term for a set of laboratory techniques for the separation of mixtures. The mixture is dissolved in a fluid called the *mobile phase,* which carries it through a structure holding another material called the *stationary phase.* The various constituents of the mixture travel at different speeds, causing them to separate. The separation is based on differential partitioning between the mobile and stationary phases. Subtle differences in a compound's partition coefficient result in differential retention on the stationary phase and thus changing the separation.

Pictured is a sophisticated gas chromatography system. This instrument records concentrations of acrylonitrile in the air at various points throughout the chemical laboratory.

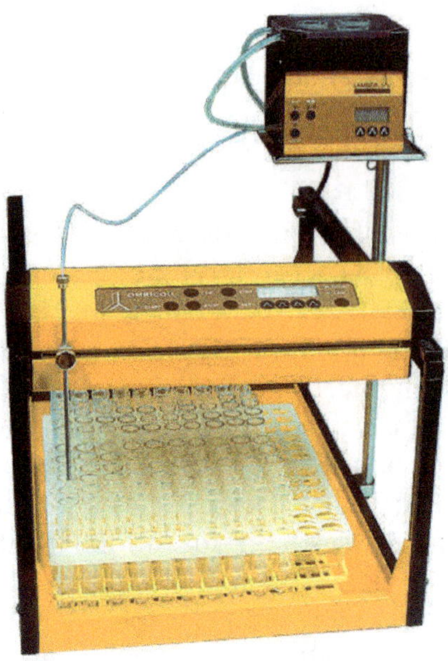

Automated fraction collector and sampler for chromatographic techniques

CHROMATOGRAPHY

Chromatography may be preparative or analytical. The purpose of preparative chromatography is to separate the components of a mixture for more advanced use (and is thus a form of purification). Analytical chromatography is done normally with smaller amounts of material and is for measuring the relative proportions of analytes in a mixture. The two are not mutually exclusive.

Thin layer chromatography is used to separate components of a plant extract, illustrating the experiment with plant pigments that gave chromatography its name

Chromatography was first employed in Russia by the Italian-born scientist Mikhail Tsvet in 1900. He continued to work with chromatography in the first decade of the 20th century, primarily for the separation of plant pigments such as chlorophyll, carotenes, and xanthophylls. Since these components have different colors (green, orange, and yellow, respectively) they gave the technique its name. New types of chromatography developed during the 1930s and 1940s made the technique useful for many separation processes.

Chromatography technique developed substantially as a result of the work of Archer John Porter Martin and Richard Laurence Millington Synge during the 1940s and 1950s, for which they won a Nobel prize. They established the principles and basic techniques of partition chromatography, and their work encouraged the rapid development of several chromatographic methods: paper chromatography, gas chromatography, and what would become known as high performance liquid chromatography. Since then, the technology has advanced rapidly. Researchers found that the main principles of Tsvet's chromatography could be applied in many different ways, resulting in the different varieties of chromatography described below. Advances are continually improving the technical performance of chromatography, allowing the separation of increasingly similar molecules.

Chromatography Terms

- The analyte is the substance to be separated during chromatography. It is also normally what is needed from the mixture.
- Analytical chromatography is used to determine the existence and possibly also the concentration of analyte(s) in a sample.
- A bonded phase is a stationary phase that is covalently bonded to the support particles or to the inside wall of the column tubing.
- A chromatogram is the visual output of the chromatograph. In the case of an optimal separation, different peaks or patterns on the chromatogram correspond to different components of the separated mixture.

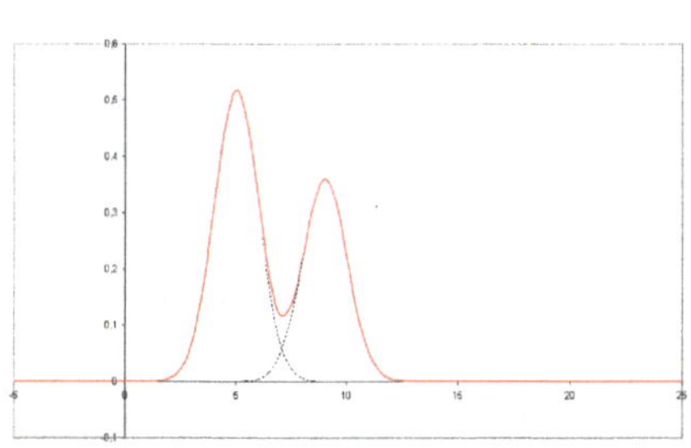

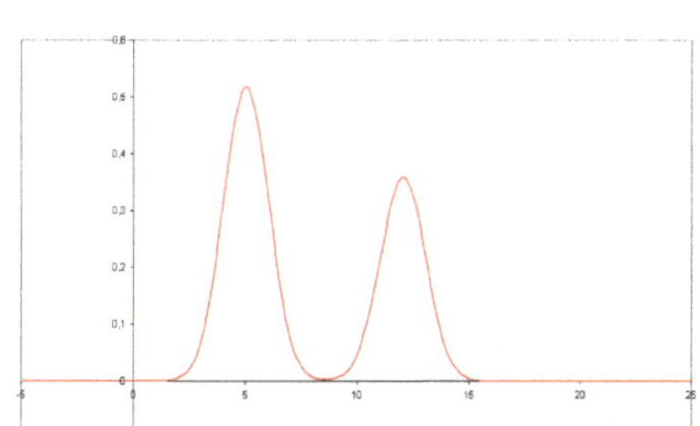

Plotted on the x-axis is the retention time and plotted on the y-axis a signal (for example obtained by a spectrophotometer, mass spectrometer or a variety of other detectors) corresponding to the response created by the analytes exiting the system. In the case of an optimal system the signal is proportional to the concentration of the specific analyte separated.

- A chromatograph is equipment that enables a sophisticated separation, e.g. gas chromatographic or liquid chromatographic separation.
- Chromatography is a physical method of separation that distributes components to separate between two phases, one stationary (stationary phase), the other (the mobile phase) moving in a definite direction.
- The eluate is the mobile phase leaving the column.
- The eluent is the solvent that carries the analyte.
- An eluotropic series is a list of solvents ranked according to their eluting power.
- An immobilized phase is a stationary phase that is immobilized on the support particles, or on the inner wall of the column tubing.
- The mobile phase is the phase that moves in a definite direction. It may be a liquid (LC and Capillary Electrochromatography (CEC)), a gas (GC), or a supercritical fluid (supercritical-fluid chromatography, SFC). The mobile phase consists of the sample being separated/analyzed and the solvent that moves the sample through the column. In the case of HPLC the mobile phase consists of a non-polar solvent(s) such as hexane in normal phase or a polar solvent such as methanol in reverse phase chromatography and the sample being separated. The mobile phase moves through the chromatography column (the stationary phase) where the sample interacts with the stationary phase and is separated.
- Preparative chromatography is used to purify sufficient quantities of a substance for further use, rather than analysis.
- The retention time is the characteristic time it takes for a particular analyte to pass through the system (from the column inlet to the detector) under set conditions. See also: Kovats' retention index
- The sample is the matter analyzed in chromatography. It may consist of a single component or it may be a mixture of components. When the sample is treated in the course of an analysis, the phase or the phases containing the analytes of interest is/are referred to as the sample whereas everything out of interest separated from the sample before or in the course of the analysis is referred to as waste.
- The solute refers to the sample components in partition chromatography.
- The solvent refers to any substance capable of solubilizing another substance, and especially the liquid mobile phase in liquid chromatography.
- The stationary phase is the substance fixed in place for the chromatography procedure. Examples include the silica layer in thin layer chromatography
- The detector refers to the instrument used for qualitative and quantitative detection of analytes after separation.

Chromatography is based on the concept of partition coefficient. Any solute partitions between two immiscible solvents. When we make one solvent immobile (by adsorption on a solid support matrix) and another mobile it results in most common applications of chromatography. If the matrix support, or stationary phase, is polar (e.g. paper, silica etc.) it is forward phase chromatography, and if it is non-polar (C-18) it is reverse phase.

Techniques by Chromatographic Bed Shape

Column Chromatography

Column chromatography is a separation technique in which the stationary bed is within a tube. The particles of the solid stationary phase or the support coated with a liquid stationary phase may fill the whole inside volume of the tube (packed column) or be concentrated on or along the inside tube wall leaving an open, unrestricted path for the mobile phase in the middle part of the tube (open tubular column). Differences in rates of movement through the medium are calculated to different retention times of the sample.

In 1978, W. Clark Still introduced a modified version of column chromatography called flash column chromatography (flash). The technique is very similar to the traditional column chromatography, except for that the solvent is driven through the column by applying positive pressure. This allowed most separations to be performed in less than 20 minutes, with improved separations compared to the old method. Modern flash chromatography systems are sold as pre-packed plastic cartridges, and the solvent is pumped through the cartridge. Systems may also be linked with detectors and fraction collectors providing automation. The introduction of gradient pumps resulted in quicker separations and less solvent usage.

In expanded bed adsorption, a fluidized bed is used, rather than a solid phase made by a packed bed. This allows omission of initial clearing steps such as centrifugation and filtration, for culture broths or slurries of broken cells.

Phosphocellulose chromatography utilizes the binding affinity of many DNA-binding proteins for phosphocellulose. The stronger a protein's interaction with DNA, the higher the salt concentration needed to elute that protein.

Planar Chromatography

Planar chromatography is a separation technique in which the stationary phase is present as or on a plane. The plane can be a paper, serving as such or impregnated by a substance as the stationary bed (paper chromatography) or a layer of solid particles spread on a support such as a glass plate (thin layer chromatography). Different compounds in the sample mixture travel different distances according to how strongly they interact with the stationary phase as compared to the mobile phase. The specific Retention factor (R_f) of each chemical can be used to aid in the identification of an unknown substance.

Paper Chromatography

Paper chromatography is a technique that involves placing a small dot or line of sample solution onto a strip of *chromatography paper*. The paper is placed in a container with a shallow layer of solvent and sealed. As the solvent rises through the paper, it meets the sample mixture, which starts to travel up the paper with the solvent. This paper is made of cellulose, a polar substance, and the compounds within the mixture travel farther if they are non-polar. More polar substances bond with the cellulose paper more quickly, and therefore do not travel as far.

Thin layer Chromatography (TLC)

Thin layer chromatography (TLC) is a widely employed laboratory technique and is similar to paper chromatography. However, instead of using a stationary phase of paper, it involves a stationary phase of a thin layer of adsorbent like silica gel, alumina, or cellulose on a flat, inert substrate. Compared to paper, it has the advantage of faster runs, better separations, and the choice between different adsorbents. For even better resolution and to allow for quantification, high-performance TLC can be used. An older popular use had been to differentiate chromosomes by observing distance in gel (separation of was a separate step).

Displacement chromatography

The basic principle of displacement chromatography is: A molecule with a high affinity for the chromatography matrix (the displacer) competes effectively for binding sites, and thus displace all molecules with lesser affinities. There are distinct differences between displacement and elution chromatography. In elution mode, substances typically emerge from a column in narrow, Gaussian peaks. Wide separation of peaks, preferably to baseline, is desired for maximum purification. The speed at which any component of a mixture travels down the column in elution mode depends on many factors. But for two substances to travel at different speeds, and thereby be resolved, there must be substantial differences in some interaction between the biomolecules and the chromatography matrix. Operating parameters are adjusted to maximize the effect of this difference. In many cases, baseline separation of the peaks can be achieved only with gradient elution and low column loadings. Thus, two drawbacks to elution mode chromatography, especially at the preparative scale, are operational complexity, due to gradient solvent pumping, and low throughput, due to low column loadings. Displacement chromatography has advantages over elution chromatography in that components are resolved into consecutive zones of pure substances rather than "peaks". Because the process takes advantage of the nonlinearity of the isotherms, a larger column feed can be separated on a given column with the purified components recovered at significantly higher concentrations.

Techniques by Physical state of Mobile Phase

Gas Chromatography

Gas chromatography (GC), also sometimes known as gas-liquid chromatography, (GLC), is a separation technique in which the mobile phase is a gas. Gas chromatographic separation is always carried out in a column, which is typically "packed" or "capillary". Packed columns are the routine work horses of gas chromatography, being cheaper and easier to use and often giving adequate performance. Capillary columns generally give far superior resolution and although more expensive are becoming widely used, especially for complex mixtures. Both types of column are made from non-adsorbent and chemically inert materials. Stainless steel and glass are the usual materials for packed columns and quartz or fused silica for capillary columns.

Gas chromatography is based on a partition equilibrium of analyte between a solid or viscous liquid stationary phase (often a liquid silicone-based material) and a mobile gas (most often helium). The stationary phase is adhered to the inside of a small-diameter (commonly 0.53 – 0.18mm inside diameter) glass or fused-silica tube (a capillary column) or a solid matrix inside a larger metal tube (a packed column). It is widely used in analytical chemistry; though the high temperatures used in GC make it unsuitable for high molecular weight biopolymers or proteins (heat denatures them), frequently encountered in biochemistry, it is well suited for use in the petrochemical, environmental monitoring and remediation, and industrial chemical fields. It is also used extensively in chemistry research.

Liquid Chromatography

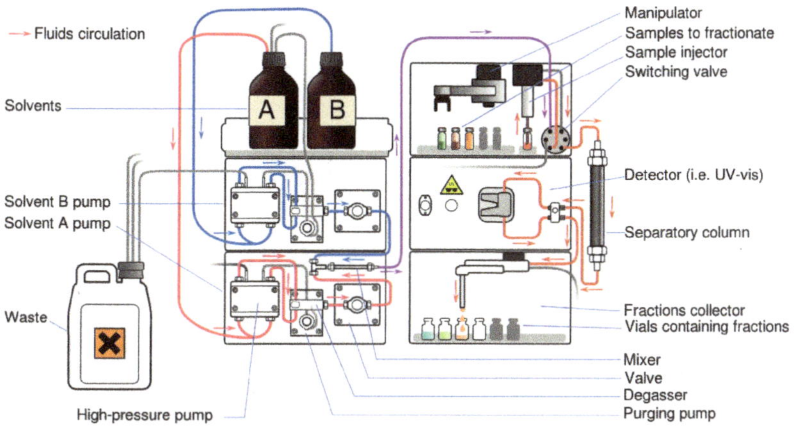

Preparative HPLC apparatus

Liquid chromatography (LC) is a separation technique in which the mobile phase is a liquid. It can be carried out either in a column or a plane. Present day liquid chromatography that generally utilizes very small packing particles and a relatively high pressure is referred to as high performance liquid chromatography (HPLC).

In HPLC the sample is forced by a liquid at high pressure (the mobile phase) through a column that is packed with a stationary phase composed of irregularly or spherically shaped particles, a porous monolithic layer, or a porous membrane. HPLC is historically divided into two different sub-classes based on the polarity of the mobile and stationary phases. Methods in which the stationary phase is more polar than the mobile phase (e.g., toluene as the mobile phase, silica as the stationary phase) are termed normal phase liquid chromatography (NPLC) and the opposite (e.g., water-methanol mixture as the mobile phase and C18 = octadecylsilyl as the stationary phase) is termed reversed phase liquid chromatography (RPLC).

Specific techniques under this broad heading are listed below.

Affinity Chromatography

Affinity chromatography is based on selective non-covalent interaction between an analyte and specific molecules. It is very specific, but not very robust. It is often used in biochemistry in the purification of proteins bound to tags. These fusion proteins are labeled with compounds such as His-tags, biotin or antigens, which bind to the stationary phase specifically. After purification, some of these tags are usually removed and the pure protein is obtained.

Affinity chromatography often utilizes a biomolecule's affinity for a metal (Zn, Cu, Fe, etc.). Columns are often manually prepared. Traditional affinity columns are used as a preparative step to flush out unwanted biomolecules.

However, HPLC techniques exist that do utilize affinity chromatography properties. Immobilized Metal Affinity Chromatography (IMAC) is useful to separate aforementioned molecules based on the relative affinity for the metal (I.e. Dionex IMAC). Often these columns can be loaded with different metals to create a column with a targeted affinity.

Supercritical Fluid Chromatography

Supercritical fluid chromatography is a separation technique in which the mobile phase is a fluid above and relatively close to its critical temperature and pressure.

Techniques by Separation Mechanism

Ion Exchange Chromatography

Ion exchange chromatography (usually referred to as ion chromatography) uses an ion exchange mechanism to separate analytes based on their respective charges. It is usually performed in columns but can also be useful in planar mode. Ion exchange chromatography uses a charged stationary phase to separate charged compounds including anions, cations, amino acids, peptides, and proteins. In conventional methods the stationary phase is an ion exchange resin that carries charged functional groups that interact with oppositely charged groups of the compound to retain. Ion exchange chromatography is commonly used to purify proteins using FPLC.

Size-exclusion chromatography

Size-exclusion chromatography (SEC) is also known as **gel permeation chromatography** (GPC) or **gel filtration chromatography** and separates molecules according to their size (or more accurately according to their hydrodynamic diameter or hydrodynamic volume). Smaller molecules are able to enter the pores of the media and, therefore, molecules are trapped and removed from the flow of the mobile phase. The average residence time in the pores depends upon the effective size of the analyte molecules. However, molecules that are larger than the average pore size of the packing are excluded and thus suffer essentially no retention; such species are the first to be eluted. It is generally a low-resolution chromatography technique and thus it is often reserved for the final, "polishing" step of a purification. It is also useful for determining the tertiary structure and quaternary structure of purified proteins, especially since it can be carried out under native solution conditions.

Expanded bed Adsorption Chromatographic Separation

An expanded bed chromatographic adsorption (EBA) column for a biochemical separation process comprises a pressure equalization liquid distributor having a self-cleaning function below a porous blocking sieve plate at the bottom of the expanded bed, an upper part nozzle assembly having a backflush cleaning function at the top of the expanded bed, a better distribution of the feedstock liquor added into the expanded bed ensuring that the fluid passed through the expanded bed layer displays a state of piston flow. The expanded bed layer displays a state of piston flow. The expanded bed chromatographic separation column has advantages of increasing the separation efficiency of the expanded bed.

Expanded-bed adsorption (EBA) chromatography is a convenient and effective technique for the capture of proteins directly from unclarified crude sample. In EBA chromatography, the settled bed is first expanded by upward flow of equilibration buffer. The crude feed, a mixture of soluble proteins, contaminants, cells, and cell debris, is then passed upward through the expanded bed. Target proteins are captured on the adsorbent, while particulates and contaminants pass through. A change to elution buffer while maintaining upward flow results in desorption of the target protein in expanded-bed mode. Alternatively, if the flow is reversed, the adsorbed particles will quickly settle and the proteins can be desorbed by an elution buffer. The mode used for elution (expanded-bed versus settled-bed) depends on the characteristics of the feed. After elution, the adsorbent is cleaned with a predefined cleaning-in-place (CIP) solution, with cleaning followed by either column regeneration (for further use) or storage.

Special Techniques

Reversed-phase Chromatography

Reversed-phase chromatography (RPC) is any liquid chromatography procedure in which the mobile phase is significantly more polar than the stationary phase. It is so named because in normal-phase liquid chromatography, the mobile phase is significantly less polar than the stationary phase. Hydrophobic molecules in the mobile phase tend to adsorb to the relatively hydrophobic stationary phase. Hydrophilic molecules in the mobile phase will tend to elute first. Separating columns typically comprise a C8 or C18 carbon-chain bonded to a silica particle substrate.

Hydrophobic Interaction Chromatography

Hydrophobic interactions between proteins and the chromatographic matrix can be exploited to purify proteins. In hydrophobic interaction chromatography the matrix material is lightly substituted with hydrophobic groups. These groups can range from methyl, ethyl, propyl, octyl, or phenyl groups.[] At high salt concentrations, non-polar sidechains on the surface on proteins "interact" with the hydrophobic groups; that is, both types of groups are excluded by the polar solvent (hydrophobic effects are augmented by increased ionic strength). Thus, the sample is applied to the column in a buffer which is highly polar. The eluant is typically an aqueous buffer with decreasing salt concentrations, increasing concentrations of detergent (which disrupts hydrophobic interactions), or changes in pH.

In general, Hydrophobic Interaction Chromatography (HIC) is advantageous if the sample is sensitive to pH change or harsh solvents typically used in other types of chromatography but not high salt concentrations. Commonly, it is the amount of salt in the buffer which is varied. In 2012, Müller and Franzreb described the effects of temperature on HIC using Bovine Serum Albumin (BSA) with four different types of hydrophobic resin. The study altered temperature as to effect the binding affinity of BSA onto the matrix. It was concluded that cycling temperature from 50 degrees to 10 degrees would not be adequate to effectively wash all BSA from the matrix but could be very effective if the column would only be used a few times. Using temperature to effect change allows labs to cut costs on buying salt and saves money.

If high salt concentrations along with temperature fluctuations want to be avoided you can use a more hydrophobic to compete with your sample to elute it. [source] This so-called salt independent method of HIC showed a direct isolation of Human Immunoglobulin G (IgG) from serum with satisfactory yield and used Beta-cyclodextrin as a competitor to displace IgG from the matrix. This largely opens up the possibility of using HIC with samples which are salt sensitive as we know high salt concentrations precipitate proteins.

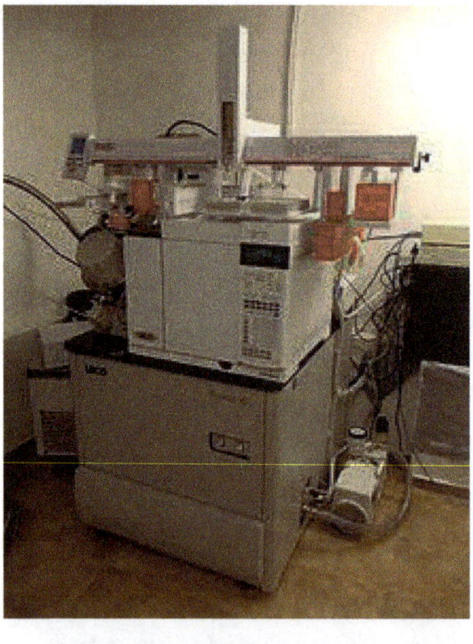

Two-dimensional chromatograph GCxGC-TOFMS at Chemical Faculty of GUT Gdańsk, Poland, 2016

Two-dimensional Chromatography

In some cases, the chemistry within a given column can be insufficient to separate some analytes. It is possible to direct a series of unresolved peaks onto a second column with different physico-chemical (Chemical classification) properties. Since the mechanism of retention on this new solid support is different from the first dimensional separation, it can be possible to separate compounds that are indistinguishable by one-dimensional chromatography. The sample is spotted at one corner of a square plate,developed, air-dried, then rotated by 90° and usually redeveloped in a second solvent system.

Simulated Moving-Bed Chromatography

The simulated moving bed (SMB) technique is a variant of high performance liquid chromatography; it is used to separate particles and/or chemical compounds that would be difficult or impossible to resolve otherwise. This increased separation is brought about by a valve-and-column arrangement that is used to lengthen the stationary phase indefinitely. In the moving bed technique of preparative chromatography the feed entry and the analyte recovery are simultaneous and continuous, but because of practical difficulties with a continuously moving bed, simulated moving bed technique was proposed. In the simulated moving bed technique instead of moving the bed, the sample inlet and the analyte exit positions are moved continuously, giving the impression of a moving bed. True moving bed chromatography (TMBC) is only a theoretical concept. Its simulation, SMBC is achieved by the use of a multiplicity of columns in series and a complex valve arrangement, which provides for sample and solvent feed, and also analyte and waste takeoff at appropriate locations of any column, whereby it allows switching at regular intervals the sample entry in one direction, the solvent entry in the opposite direction, whilst changing the analyte and waste takeoff positions appropriately as well.

Pyrolysis Gas Chromatography

Pyrolysis gas chromatography mass spectrometry is a method of chemical analysis in which the sample is heated to decomposition to produce smaller molecules that are separated by gas chromatography and detected using mass spectrometry.

Pyrolysis is the thermal decomposition of materials in an inert atmosphere or a vacuum. The sample is put into direct contact with a platinum wire, or placed in a quartz sample tube, and rapidly heated to 600–1000 °C. Depending on the application even higher temperatures are used. Three different heating techniques are used in actual pyrolyzers: Isothermal furnace, inductive heating (Curie Point filament), and resistive heating using platinum filaments. Large molecules cleave at their weakest points and produce smaller, more volatile fragments. These fragments can be separated by gas chromatography. Pyrolysis GC chromatograms are typically complex because a wide range of different decomposition products is formed. The data can either be used as fingerprint to prove material identity or the GC/MS data is used to identify individual fragments to obtain structural information. To increase the volatility of polar fragments, various methylating reagents can be added to a sample before pyrolysis.

Besides the usage of dedicated pyrolyzers, pyrolysis GC of solid and liquid samples can be performed directly inside Programmable Temperature Vaporizer (PTV) injectors that provide quick heating (up to 30 °C/s) and high maximum temperatures of 600–650 °C. This is sufficient for some pyrolysis applications. The main advantage is that no dedicated instrument has to be purchased and pyrolysis can be performed as part of routine GC analysis. In this case quartz GC inlet liners have to be used. Quantitative data can be acquired, and good results of derivatization inside the PTV injector are published as well.

Fast protein Liquid Chromatography

Fast protein liquid chromatography (FPLC), is a form of liquid chromatography that is often used to analyze or purify mixtures of proteins. As in other forms of chromatography, separation is possible because the different components of a mixture have different affinities for two materials, a moving fluid (the "mobile phase") and a porous solid (the stationary phase). In FPLC the mobile phase is an aqueous solution, or "buffer". The buffer flow rate is controlled by a positive-displacement pump and is normally kept constant, while the composition of the buffer can be varied by drawing fluids in different proportions from two or more external reservoirs. The stationary phase is a resin composed of beads, usually of cross-linked agarose, packed into a cylindrical glass or plastic column. FPLC resins are available in a wide range of bead sizes and surface ligands depending on the application.

Countercurrent Chromatography

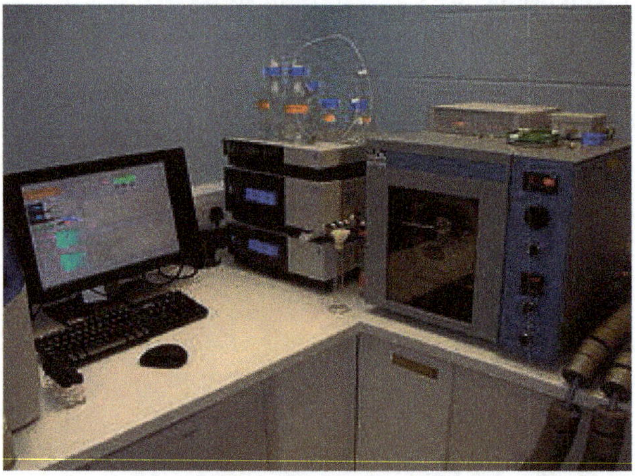

An example of a HPCCC system

Countercurrent chromatography (CCC) is a type of liquid-liquid chromatography, where both the stationary and mobile phases are liquids. The operating principle of CCC equipment requires a column consisting of an open tube coiled around a bobbin. The bobbin is rotated in a double-axis gyratory motion (a cardioid), which causes a variable gravity (G) field to act on the column during each rotation. This motion causes the column to see one partitioning step per revolution and components of the sample separate in the column due to their partitioning coefficient between the two immiscible liquid phases used. There are many types of CCC available today. These include HSCCC (High Speed CCC) and HPCCC (High Performance CCC). HPCCC is the latest and best performing version of the instrumentation available currently.

Chiral Chromatography

Chiral chromatography involves the separation of stereoisomers. In the case of enantiomers, these have no chemical or physical differences apart from being three-dimensional mirror images. Conventional chromatography or other separation processes are incapable of separating them. To enable chiral separations to take place, either the mobile phase or the stationary phase must themselves be made chiral, giving differing affinities between the analytes. Chiral chromatography HPLC columns (with a chiral stationary phase) in both normal and reversed phase are commercially available.

REFERENCES

1. Garrett, Reginald H.; Grisham, Charles M. (2013). Biochemistry (5th ed.). Belmont, CA: Brooks/Cole, Cengage Learning. p. 111. ISBN 9781133106296.
2. Frei, Mark. "Centrifugation Basics". Sigma-Aldrich. Retrieved 10 May 2016.
3. Article on "Centrifugation" retrieved on 15th October 2013 from http://www.lenntech.com/library/clarification/clarification/centrifugation.htm
4. Ballou, David P.; Benore, Marilee; Ninfa, Alexander J. (2008). Fundamental laboratory approaches for biochemistry and biotechnology. (2nd ed.). Hoboken, N.J.: Wiley. p. 43. ISBN 9780470087664.
5. Ballou, David P.; Benore, Marilee; Ninfa, Alexander J. (2008). Fundamental laboratory approaches for biochemistry and biotechnology. (2nd ed.). Hoboken, N.J.: Wiley. p. 235. ISBN 9780470087664.
6. Brakke, Myron K. (April 1951). "Density Gradient Centrifugation: A New Separation Technique". J. Am. Che. Soc. 73 (4): 1847–1848. doi:10.1021/ja01148a508.
7. Oster, Gerald; Yamamoto, Masahide (June 1963). "Density Gradient Techniques". Chem. Rev. 63 (3): 257–268. doi:10.1021/cr60223a003.
8. Ballou, David P.; Benore, Marilee; Ninfa, Alexander J. (2010). Fundamental Laboratory Approaches for Biochemistry and Biotechnology (seconde ed.). University of Michigan: John Wiley & Sons, Inc. p. 213. ISBN 978-0-470-08766-4.
9. Ballou, David P.; Benore, Marilee; Ninfa, Alexander J. (2008). Fundamental laboratory approaches for biochemistry and biotechnology. (2nd ed.). Hoboken, N.J.: Wiley. pp. 238–239. ISBN 9780470087664.
10. Van Holde, K. E. (1998). Analytical ultracentrifugation from 1924 to the present: A remarkable history. Chemtracts – Biochemistry and Molecular Biology. 11:933-943

11. Svedberg, T. (1927). The Ultracentrifuge Nobel Lecture
12. to:[a] [b] Tanford, C., and Reynolds, J. 2001. Nature's robots: A history of proteins. Oxford University Press. pp. 303-305
13. Simoni, D. S., Hill, R. L., and Vaughan, M. (2002). The structure and function of hemoglobin: Gilbery Smithson Adair and the Adair equations. The Journal of Biological Chemistry. 277(31): e1-e2
14. "chromatography". Online Etymology Dictionary.
15. McMurry, John (2011). Organic chemistry: with biological applications (2nd ed.). Belmont, CA: Brooks/Cole. p. 395. ISBN 9780495391470.
16. Hostettmann, K; Marston, A; Hostettmann, M (1998). Preparative Chromatography Techniques Applications in Natural Product Isolation (Second ed.). Berlin, Heidelberg: Springer Berlin Heidelberg. p. 50. ISBN 9783662036310.
17. The first report of Tsvet's discovery was published in 1905:
18. The word "chromatogram" first appeared in print in 1906:
19. Mikhail Tswett (1906) "Physikalisch-Chemische Studien über das Chlorophyll. Die Adsorption."(Physical-chemical studies of chlorophyll. Adsorption.) Berichte der Deutschen botanischen Gesellschaft, vol. 24, pp. 316–326. On page 322, Tsvet coins the term "chromatography":
20. Original : " Wie die Lichtstrahlen im Spektrum, so it is werden in der Calciumkarbonatsäule die verschiedenen Komponenten eines Farbstoffgemisches gesetzmässig auseindergelegt, und lassen sich darin qualitativ und auch quantitativ bestimmen. Ein solches Präparat nenne ich ein Chromatogramm und die entsprechende Methode, die chromatographische Methode."
21. Translation : Like light rays in a spectrum, so the different components of a mixture of pigments are dispersed in the calcium carbonate column following a set pattern, and [they] can be determined in there qualitatively as well as quantitatively. I call such a preparation a "chromatogram" and the corresponding method, the "chromatographic method".
22. Tswett, Mikhail (1906). "Adsorptionanalyse und chromatographische Methode. Anwendung auf die Chemie des Chlorophylls" [Adsorption analysis and chromatographic method. Application to the chemistry of chlorophyll]. Berichte der Deutschen botanischen Gesellschaft. 24: 384–393.
23. Ettre, L. S.; Sakodynskii, K. I. (March 1993). "M. S. Tswett and the discovery of chromatography II: Completion of the development of chromatography (1903–1910)". Chromatographia. 35 (5-6): 329–338. doi:10.1007/BF02277520.
24. «The Nobel Prize in Chemistry 1952". nobelprize.org. Retrieved 25 August 2016.
25. Ettre, L. S. (1993). "Nomenclature for chromatography (IUPAC Recommendations 1993)". Pure and Applied Chemistry. 65 (4). doi:10.1351/pac199365040819.
26. Still, W. C.; Kahn, M.; Mitra, A. (1978). "Rapid chromatographic technique for preparative separations with moderate resolution". J. Org. Chem. 43 (14): 2923–2925. doi:10.1021/jo00408a041.
27. Harwood, Laurence M.; Moody, Christopher J. (1989). Experimental organic

chemistry: Principles and Practice (Illustrated ed.). WileyBlackwell. pp. 180–185. ISBN 978-0-632-02017-1.

28. Anfinsen, Christian B.; Edsall, John Tileston; Richards, Frederic Middlebrook, eds. (1976). Advances in Protein Chemistry. pp. 6–7. ISBN 978-0-12-034230-3.

29. Displacement Chromatography 101 Archived 15 September 2008 at the Wayback Machine.. Sachem, Inc. Austin, TX 78737

30. Bailon, Pascal; Ehrlich, George K.; Fung, Wen-Jian and Berthold, Wolfgang (2000) An Overview of Affinity Chromatography, Humana Press. ISBN 978-0-89603-694-9.

31. Ninfa, Alexander J., David P. Ballou, and Marilee Benore. Fundamental Laboratory Approaches for Biochemistry and Biotechnology. Hoboken, NJ: John Wiley, 2010. Print.

32. Müller, Tobias K.h., and Matthias Franzreb. "Suitability of Commercial Hydrophobic Interaction Sorbents for Temperature-controlled Protein Liquid Chromatography under Low Salt Conditions." Journal of Chromatography A 1260 (2012): 88-96. Web. <http://ac.els-cdn.com/S002196731201268X/1-s2.0-S002196731201268X-main.pdf?_tid=2f7e7186-1c7e-11e6-836f-00000aacb361&acdnat=1463524077_be16e37f7efe44d6b689342fa7669726>.

33. Ren, Jun, Peng Yao, Jingjing Chen, and Lingyun Jia. "Salt-independent Hydrophobic Displacement Chromatography for Antibody Purification Using Cyclodextrin as Supermolecular Displacer." Journal of Chromatography A 1369 (2014): 98-104. Web. <http://ac.els-cdn.com/S0021967314015684/1-s2.0-S0021967314015684-main.pdf?_tid=b4e118f2-1c87-11e6-ba07-00000aab0f27&acdnat=1463528166_e445fcef926f1c22934169b8818c53c6>.

Index